# THE STUDY OF
# THE OCEAN AND
# THE LAND SURFACE
# FROM SATELLITES

# THE STUDY OF THE OCEAN AND THE LAND SURFACE FROM SATELLITES

PROCEEDINGS OF
A ROYAL SOCIETY DISCUSSION MEETING
HELD ON 10 AND 11 NOVEMBER 1982

ORGANIZED AND EDITED BY
J. T. HOUGHTON, F.R.S., ALAN H. COOK, F.R.S.,
AND H. CHARNOCK, F.R.S.

LONDON
THE ROYAL SOCIETY
1983

Printed in Great Britain for the Royal Society
at the
University Press, Cambridge

ISBN 0 85403 211 8

First published in *Philosophical Transactions of the Royal Society of London*,
series A, volume 309 (no. 1508), pages 241–464



Published by the Royal Society
6 Carlton House Terrace, London SW1Y 5AG

# PREFACE

This volume records the contributions to a discussion meeting held at the Royal Society on 10 and 11 November 1982, which addressed a number of applications of satellite observations of the land and ocean surface.

The possibilities of Earth observation from space are very considerable. First of all, the spatial coverage that is achieved is global, and repeated observation in time can also be achieved in an economical manner. Secondly, the whole electromagnetic spectrum is available for observation, from the ultraviolet through the visible and infrared to the microwave and radio region; the potential for the investigation of different properties of the surface through multispectral investigation is large. Thirdly, both passive and active sensors can be employed. Examples of both types of sensors are to be found in this volume; in particular, several papers are included on recent applications of radar observations and radar imaging of the surface.

The first set of papers is concerned with observations of the land and describe applications in geology and mineral exploration, agriculture and land use. A paper is also included on the need for observations of the radiative properties of the land surface as an input to models of the climate system.

Six papers are included that address ocean applications. Exciting new possibilities for ocean measurements from space were pioneered by the U.S. satellite Seasat in 1978, and a lot of thought and effort is currently going into the design and planning of the European Space Agency's first remote sensing satellite ERS-1, which will be devoted to ocean applications and is due for launch in 1989.

The last three papers describe the application of remote sensing to the investigation of sea ice and polar ice sheets. Satellites are an effective means of acquiring the information about the cryosphere that is needed for the World Climate Research Programme.

Finally, an important paper by Dr Kittler on theoretical methods in remote sensing is included in the volume. A great deal of work on techniques for the processing of images and extracting information from them has been carried out in recent years – techniques that have wide application in the remote sensing field.

We are grateful to all the authors and to those who took part in the discussion for their contributions. We appreciate the help of the Royal Society staff, especially Miss C. A. Johnson, who made the arrangements for the meeting, and Dr M. B. Goatly, who prepared the proceedings for publication.

J. T. HOUGHTON
ALAN H. COOK
H. CHARNOCK

*May 1983*

[ v ]

# CONTENTS

[ Nineteen plates ]


PAGE

PREFACE     [v]

E. R. PETERS
The use of multispectral satellite imagery in the exploration for petroleum and
minerals     [1]

P. J. CURRAN
Multispectral remote sensing for the estimation of green leaf area index     [15]
*Discussion:* M. D. STEVEN     [27]

G. C. STOVE
The current use of remote-sensing data in peat, soil, land-cover and crop inventories
in Scotland     [29]

GENERAL DISCUSSION
MONICA M. COLE     [41]

A. HENDERSON-SELLERS AND M. F. WILSON
Albedo observations of the Earth's surface for climate research     [43]

P. H. A. MARTIN-KAYE AND G. M. LAWRENCE
The application of satellite imaging radars over land to the assessment, mapping
and monitoring of resources     [53]

R. K. RANEY
Synthetic aperture radar observations of ocean and land     [73]

J. V. KITTLER
Image processing for remote sensing     [81]

J. D. WOODS
Satellite monitoring of the ocean for global climate research     [95]
*Discussion:* P. K. TAYLOR     [116]

D. E. CARTWRIGHT
Detection of large-scale ocean circulation and tides     [119]

M. J. TUCKER
Observation of ocean waves     [129]
*Discussion:* E. D. R. SHEARMAN, D. E. CARTWRIGHT     [137]

J. E. HARRIES, D. T. LLEWELLYN-JONES, P. J. MINNETT, R. W. SAUNDERS AND
A. M. ZAVODY
Observations of sea-surface temperature for climate research     [139]
*Discussion:* P. WADHAMS, P. K. TAYLOR, J. T. HOUGHTON, F.R.S.     [152]

# CONTENTS

GENERAL DISCUSSION
   E. D. R. SHEARMAN, J. T. HOUGHTON, F.R.S.                                    [155]

T. H. GUYMER
   A review of Seasat scatterometer data                                       [157]

I. S. ROBINSON
   Satellite observations of ocean colour                                      [173]
      *Discussion:* E. G. MITCHELSON                                           [188]

P. E. GUDMANDSEN
   Application of microwave remote sensing to studies of sea ice               [191]

G. DE Q. ROBIN, D. J. DREWRY AND V. A. SQUIRE
   Satellite observations of polar ice fields                                  [205]
      *Discussion:* A. S. LAUGHTON, F.R.S.                                     [219]

ALAN H. COOK, F.R.S.
   Concluding remarks                                                          [221]

*Phil. Trans. R. Soc. Lond.* A **309**, 243–255 (1983)
*Printed in Great Britain*



# The use of multispectral satellite imagery in the exploration for petroleum and minerals

By E. R. Peters

*Hunting Geology and Geophysics Limited, Elstree Way, Borehamwood, Herts WD6 1SB, U.K.*

[Plate 1]

The paper reviews the application of multispectral satellite imagery to mineral and petroleum exploration, from the stage when satellite imagery first became available, with the launch of ERTS-1 (Landsat) just over 10 years ago, to the present day. The operation of Landsat is briefly described, and it is noted that the continuing success of this system for geological application has been in part due to the development of a world-wide network of receiving stations and the application of sophisticated data-processing techniques. Current research into the measurement of infrared spectra for the discrimination of rocks and minerals is discussed. Interpretation techniques are important but their success depends largely upon the experience of the interpreters as geologists. Examples of the use of Landsat imagery in exploration are given, and inter-pretational techniques are reviewed. Combining Landsat interpretation with that of regional geophysical surveys can bring important advantage. Finally the new genera-tion of imaging multispectral satellites is described and the implications with regard to petroleum and mineral exploration are discussed.

## Introduction

Multispectral remote sensing for mineral and petroleum exploration has become incorporated into everyday practice. It is now standard procedure to examine satellite imagery at the area selection stage and to carry out an interpretation at the inception of field studies for integration with the results of geophysical surveys. Real benefits can be demonstrated by explorationists, who at one time were sceptical about the value of satellite remote sensing. The impact on mineral and petroleum exploration is enormous although not yet everywhere appreciated. It represents the greatest single advance in regional and semi-detailed exploration since aerial photographs were first used for geological purposes.

Since the first orbital pictures were obtained during the Gemini programme there has been an increasing recognition of the importance of the view obtained from the satellite-platform. Fred Hoyle predicted that once a photograph of the Earth was taken from the outside, new ideas as powerful as any in history would be let loose. The past 10 years have seen a vast accumulation of papers and presentations on the application of satellite imagery for civilian scientific purposes, including those geological. Therefore, in order to reduce the subject matter to manageable proportions the title of this paper has been deliberately selected to encompass only the applications summarized below.

At present multispectral satellite imagery is virtually synonymous in the minds of most with the Landsat series of satellites, as it is the imagery from these satellites that is most readily available to exploration geologists. Other passive remote-sensing satellites are not discussed and this review has been restricted to the direct application of Landsat imagery to petroleum



20-2

and minerals. Landsat has been employed for many geological purposes, including hydro-geology, geothermal exploration, seismic hazard prediction, and engineering studies. Also excluded from this paper are the indirect use of Landsat in exploration as a mapping base, for terrain evaluation and bathymetric mapping in shallow marine environments.

The Landsat system (originally named ERTS (Earth Resources Technology Satellite) was primarily orientated towards land-use purposes but was also designed to test four geological hypotheses (Fischer 1977):

(1) that some dynamic geologic phenomena could be better viewed in a time-lapse mode, e.g. sedimentation and glaciation;

(2) that colour would prove useful in mapping rock types, geochemical anomalies and alteration products;

(3) that some geological and hydrogeological features are only intermittently visible;

(4) that large structures exist on the Earth's surface which because of their size have gone unrecognized by ground or aerial survey.

For exploration the recognition of the features mentioned in (2) and (4) have proved of greatest importance to the search for petroleum and minerals.

At this stage it is timely to review the practical results achieved so far in this field by the Landsat system, as we shall soon be receiving data from the next generation of civilian land remote-sensing satellites.

## THE LANDSAT SYSTEM

Landsat was originally conceived as the Earth Resources Technology Satellite (ERTS), and ERTS-1 was launched in July 1972. The system was renamed Landsat in 1975 after the launch of Landsat-2. So far, four satellites in the series have been launched. This series of experimental satellites was designed to provide multispectral imagery for the study of renewable and non-renewable resources, and the system is about to become operational.

The satellite weighs 850 kg and is launched into a near-polar orbit at an altitude of approxi-mately 900 km. The orbit is Sun-synchronous, with the satellite passing southwards over the Equator at approximately 9.30 a.m. on each pass. This ensures that there are approximately similar illumination conditions on adjacent tracks, which are one day apart. The image swath is 185 km wide and in this orbit, which takes 103 minutes, the entire surface of the Earth (apart from high polar regions) is covered every 18 days. For convenience each swath is divided up into 185 km segments, designated 'scenes'.

The sensor payloads have varied slightly between each of the first three satellites, and Landsat-4 is an entirely new system. The sensors, the multispectral bands and some comments about their performance are given in table 1. In essence the sensors comprise the multispectral scanner (MSS), recording in four bands, and the Return Beam Vidicon camera (RBV), which is panchromatic. Only the application of the generally available MSS imagery is considered here, as RBV is not generally available because of sensor failure or variable quality of the data.

The Landsat multispectral scanner has an instantaneous field of view of 79 m, and with side lap along each scan line each picture element (pixel) represents an area of 56 m × 79 m on the ground. Each scene is made up of approximately 3240 × 2340 pixels.

Initially, imagery in both real-time and from the on-board recorders was received at the two receiving stations in the United States of America. Since the launch of Landsat-1, receiving stations have been built or are planned in Canada, Sweden, Italy (Earthnet), Brazil, Argentina,

South Africa, India, Thailand, China, Japan and Australia. Each station is responsible for generating and distributing a range of image products, which usually include both photographic and computer-compatible items.

The convention is to compose false-colour composite images from three of the four bands, assigning red to the infrared band. The range of surface coloration and the convention for false-colour imagery are shown in table 1. Colour composite images can be made photographically, or through a colour television monitor with an image display system.

TABLE 1. LANDSAT MSS BANDS

| band | wavelength/μm | type of radiation | filter for colour composite |
|---|---|---|---|
| 4 | 0.5–0.6 | visible green | blue |
| 5 | 0.6–0.7 | visible red | green |
| 6 | 0.7–0.8 | reflected i.r. ⎫ | red |
| 7 | 0.8–1.1 | reflected i.r. ⎭ | |
| 8† | 10.4–12.6 | thermal i.r. | — |

† Landsat-3 only.

Geologists began to use Landsat imagery as soon as it became available in 1972, realizing the importance of the synoptic view, which is of great value in structural geology. The first images used were commonly black and white prints at 1 : 1 000 000 scale of MSS band 5 or band 7. At that stage there were very few laboratories with the facilities for processing computer-compatible tapes (c.c.ts). The importance of working with multispectral bands and colour composite imagery was progressively appreciated. Also at that time there was some resistance to be overcome, because many explorationists could not appreciate the advantages that might be obtained from using Landsat. This in part was due to the over-optimistic claims that were being made at the time, and laid emphasis on the identification of minerals from space, claims that, apart from a few instances, could not be substantiated.

During the late 1970s two important developments took place that contributed to the more widespread use of Landsat imagery. The first of these was the gradual build up of a worldwide network of receiving stations, which means that a large proportion of the land surface can now be imaged in real-time. At first this may not seem of importance for geological applications, in that geologists only need one 'without-repeats' picture, but experience has shown that this is not always true. It is extremely important to obtain the best image for interpretation with regard to season and Sun elevation. More receiving stations mean that the chance of obtaining a clear cloud-free scene is very much improved. Also, it has been shown that images from different dates can contribute to the geological knowledge of a region. This facility will become of increasing importance once geobotanical changes become better understood. Greater choice of imagery has therefore contributed to the increase in geological application.

The second important advance was the development and general availability of image processing and analysis systems coupled with the appreciation of the advantages of working with Landsat c.c.ts. The application of this technology has contributed to many programmes using the current Landsat imagery, and it will become of increasing importance in the application of future satellite imagery. The advantages that could be obtained from the computer processing of Landsat imagery were demonstrated for the discrimination and detection of hydrothermally altered rocks (Rowan *et al.* 1974) and for general regional applications in

Alaska by Albert & Chavez (1975). The early work in this field, principally by the United States Geological Survey, established the image-processing techniques now being generally applied.

## IMAGE PROCESSING

The use of computer-based image processing and analysis systems has gained momentum over the past 5 years, to the situation where nowadays many geologists use computer-compatible data tapes. This groundswell in the use of image-processing systems has resulted from a number of factors that have come together at one time: primarily the development of increasingly powerful mini-computers coupled to image display systems capable of producing results at video rates, and also the development of the software required to produce hardcopy imagery that geologists need, and a desire by exploration geologists to use the best available image to extract the utmost benefit from their interpretations.

The general use of image processing and analysis systems has meant that the full range of data in the Landsat image can be examined to the best effect. The interactive processing of imagery where the results of each stage can be assessed is invaluable, and as a consequence interactive interpretation has been developed as a useful technique. The main functions carried out as part of a programme of geological exploration are image restoration, image enhancement and data extraction.

Apart from replacing lost data, i.e. dropped lines and bad pixels, for purely aesthetic effects, the geologist has been particularly interested in performing geometric corrections. These remove the distortions due to platform movement and correct the data to a cartographic projection by resampling. At the same time the pixel size can be reduced to aid enlargement. Rectification to a cartographic projection is particularly important if the imagery is to be related to the results of geophysical survey techniques where survey stations have been accurately located on the ground. To carry out routine geometric correction of imagery the Canadian Centre for Remote Sensing has built a system dedicated to this task, and the DICS (Digital Image Correction System) was described by Butlin *et al.* (1978). The output from this system is geocoded imagery in the same format as the national topographic maps; these products are required not only for geological use.

In the preparation of enhanced hardcopy the geologists require the best possible display of the data to provide the greatest impact for interpretation. An example is shown in figure 1. This involves the removal of sixth-line banding or striping, which is an inherent defect resulting from drift within the detectors. Also the enhancement will involve a contrast-stretching routine, preferably carried out interactively under the control of the interpreter. Spatial filtering techniques are used to enhance naturally occurring straight-edge features such as fractures. The filters can be omni-directional or if desired can be tuned to a particular direction. The principles for filtering out sixth-line banding and for carrying out edge-sharpening to enhance fracture patterns were described by Chavez *et al.* (1976). These techniques were applied to the interpretation of fracture patterns in southwest Jordan, and several previously unmapped faults were identified.

Information extraction processes are carried out interactively. The processes most favoured by geologists are band ratios, multispectral classification, principal component analysis and colour rotation. These can all be used to enhance specific geological features. Principal component analysis and colour rotation are generally needed to help distinguish lithological

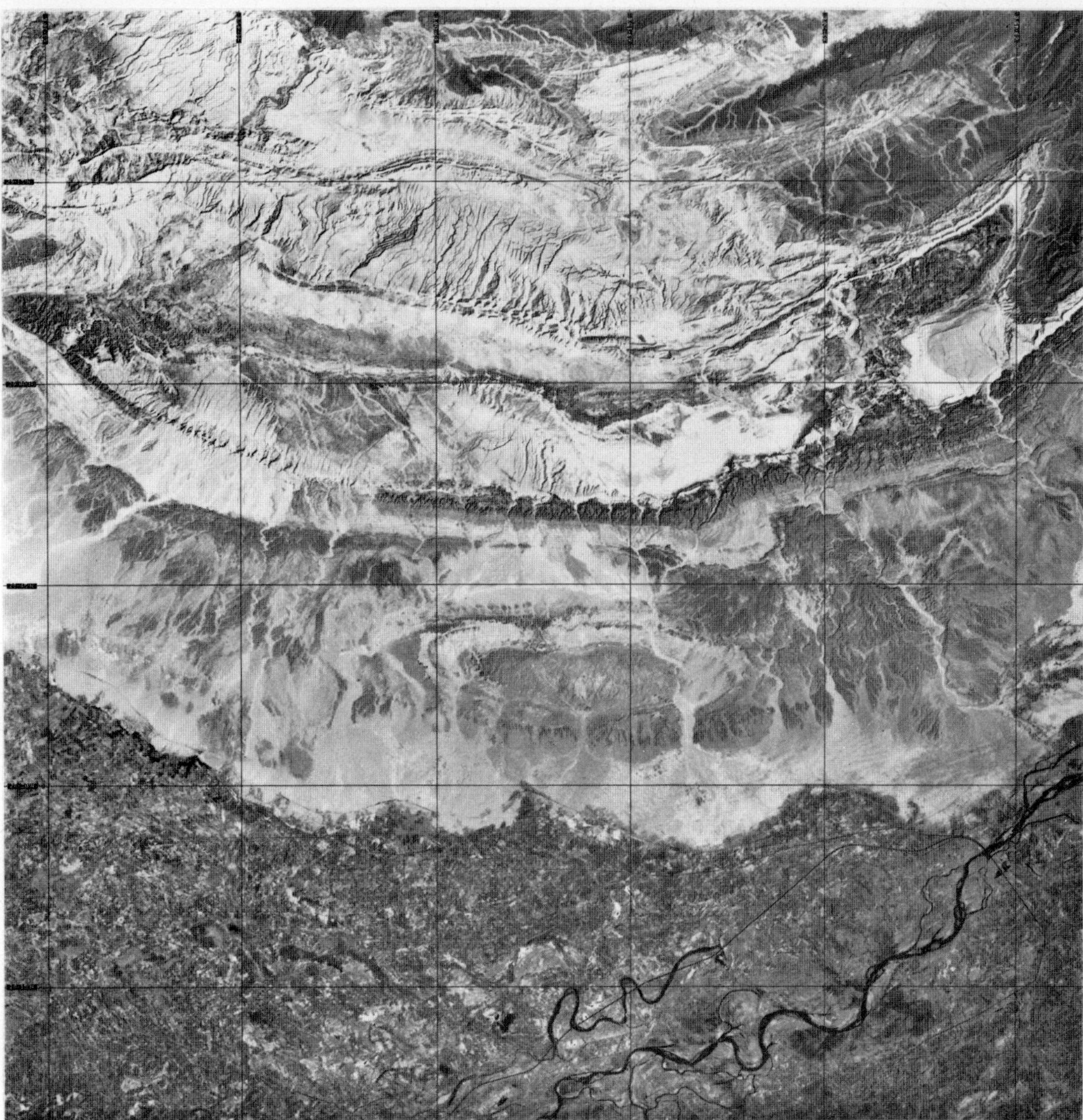

FIGURE 1. An enhanced Landsat image of MSS band 7, Path 163, Row 040, 7 March 1975, showing the fold closure over the Sui gas field in Pakistan (central part of image). The outcrops to the north are folded Mesozoic sedimentary rocks with the irrigated Indus Valley to the south.

differences, whereas ratios are used to suppress illumination differences because regardless of illumination variation a material will exhibit the same ratio value. Ratios are important for the recognition of ferruginous and limonitic cappings, and in addition the drainage pattern tends to be well displayed in band ratios. Supervised and unsupervised MSS classification can prove helpful in exploration, particularly where the target is large or the region is densely vegetated and differences in the vegetation reflect underlying conditions.

The advantages of using a particular method or technique are generally determined by the terrain conditions of the area being explored. These are now better understood by geologists and the programme of image processing is tailored accordingly.

## INTERPRETATION

Geologists are principally concerned with subsurface conditions, whereas satellite multi-spectral imagery provides a measurement of surface reflectance. In the application of multi-spectral imagery to exploration, the geologist is also interested in the indirect indications that can be interpreted from the spatial relations within the imagery, as well as the direct measurement of the spectral reflectance of geological materials. At present we have a situation where routine geological information is obtained by the interpretation of imagery, whereas research is concentrated in the identification of rocks and minerals by reflectance measurements.

Laboratory experiments to determine the characteristics of spectral infrared emissions from rocks and minerals, and whether this technique could be used as an analytical tool, were carried out by Lyon (1965). The conclusion was that composition could be determined with fair precision and this led to the design of a flight-weight spectrometer for remote measurements. Further laboratory tests were conducted by Hunt (1977) with the publication of the visible and near infrared spectra of a large number of rocks and minerals. Experiments carried out with an airborne scanner flown over the Pisgah Crater test site before the launch of ERTS (Landsat-1) enabled Vincent (1972) to reach the conclusion that by ratioing the MSS bands of ERTS images it would be possible to identify iron-rich rocks and minerals. This would be potentially valuable for mineral exploration in the identification of gossans and ferruginous alteration zones in arid and semi-arid regions. The discrimination is due to the iron absorption bands that occur at 1.0 µm for ferric compounds and 0.70 and 0.87 µm for ferrous compounds. After the launch of ERTS (Landsat-1) the application of this technique was demonstrated by Rowan *et al.* (1974) in south-central Nevada and Blodget *et al.* (1978) in Saudi Arabia.

From the results of further test work with the airborne multi-channel scanner the importance of the 2.25 µm part of the spectrum for lithological discrimination was identified. A band in this part of the spectrum has been included in the Thematic Mapper on Landsat-4, and it will be useful for the identification of clay minerals particularly in alteration zones. The N.A.S.A./Geosat test site programme was carried out before the launch of Landsat-4 to simulate the expected results. The application of the Thematic Mapper bands to exploration for porphyry copper deposits has been reported by Abrams *et al.* (1981), to uranium exploration by Conel & Niesen (1981), and to petroleum exploration by Lang & Baird (1981).

Airborne test work to test narrower and narrower bands is continuing, and Siegrist & Schnetzler (1980) identified the optimum bands for rock discrimination, while some test work is also being carried out with narrow-band spectroradiometers as profiling instruments. Goetz & Rowan (1981) describe the aircraft test work carried out before the SMIRR (Shuttle Multispectral

Infrared Radiometer), and Collins *et al.* (1981) describe the results achieved with a 64-channel profiling radiometer. Although further applications of these instruments are likely to be restricted to small areas and be carried on aircraft rather than satellite platforms, the results will inevitably affect the selection of instrumentation to be used in future satellites.

In the practical field there is one basic feature common to all exploration programmes: the need for a geological map. In the use of satellite imagery the explorationist wants to improve his understanding of the geology of the region involved. Only if the geological mapping of the region can be improved in some way will direct benefits be derived from a study of multi-spectral satellite imagery. Despite the increasing application of computer-based analytical techniques, most of the exploration work carried out in the geological field has a substantial interpretational content, where the vital information for application to petroleum and mineral exploration is not detected by a unique spectral signature or by the use of a particular processing option, but by means of visual interpretation to coordinate context, texture, colour, and shape. In addition the experienced field geologist can also bring in previous knowledge of similar regions, which is an important contribution to any interpretation.

At present the practical application of satellite imagery to petroleum and mineral exploration involves an important element of interpretation. This is generally improved by providing the interpreter with the best processed imagery available, and including interactive image analysis, which allows the interpreter to identify specific features with a greater degree of confidence. This was recognized by Raines (1981) when he commented that the interpretative approach has been needed to derive the maximum amount of information. He recognized that the practical application called for systematic discrimination, which comprised the following:

(1) a well defined problem;

(2) the spatial or map approach, which emphasizes characteristics and interrelations, not points;

(3) a knowledge of the application of remote-sensing physics and image-display technology;

(4) a geologist with broad experience.

This succinctly outlines the current practical application of Landsat imagery, which is reinforced by Nash *et al.* (1980) in their paper dealing with photogeology and satellite image interpretation. They detail the steps whereby target areas can be identified and the zone of search is substantially reduced progressively, by using satellite imagery followed by photogeological studies. The paper includes a number of examples from Australia, Africa and Arabia, and the essentials of the approach are a study of existing data, and a Landsat interpretation, to select areas for photogeological study so that targets may be located for detailed ground operations. That paper also emphasizes the need to integrate the interpretation of Landsat imagery with other data, particularly the results of geophysical surveys.

When describing the application of Landsat imagery to petroleum exploration in the western United States, Zall *et al.* (1981) also emphasized the importance of interpretation. They describe the interpretation of Landsat imagery of four regions in the western United States, where they claim that the interpretation made an important contribution to petroleum exploration by reducing the field programmes needed to locate the discoveries. They were able to focus on the target area more quickly by undertaking the following steps:

(1) examination of enhanced imagery;

(2) literature search;

(3) formulation of geological models for entrapment of oil and gas;

(4) integration with geophysical interpretation.

This illustrates the way in which Landsat imagery is being applied to both mineral and petroleum exploration, and the approach in each case is the same. The interpretative approach has until now been the only route possible, but this may be modified once Landsat-4 Thematic Mapper data became available. So far Landsat has proved ideal for evaluating the mineral or petroleum prospects of large regions. This applies equally to unexplored and well known parts of the Earth. Often, by re-examining previously explored territory, structures not previously known can be perceived.

MINERAL EXPLORATION EXAMPLES

A number of different interpretative techniques have been used in mineral and petroleum exploration, and these generally fall into two categories: (1) thematic classification to identify surface features directly related to the target; and (2) structural interpretation, which may give clues to important subsurface features. So far, most of the thematic classification work has applied to mineral exploration, mainly because petroleum deposits are often deeply buried. The more important techniques are described here, with some illustrations from the public domain. Unfortunately, many good examples of the application of Landsat imagery cannot be published for proprietary reasons. The examples given also include areas where a particular technique has been tested over a known area. In some instances this has resulted in additional discoveries being made in the surrounding region.

Landsat imagery has contributed significantly to the advancement of geological mapping in the well known and the remote regions of the world. Although regional geological mapping is a prerequisite for all exploration programmes it is not within the scope of this paper to cover this field. However, it does deserve a brief mention and an example is the Landsat interpretation of the southern Sudan made by Hunting Geology and Geophysics (1976). Here the first geological maps were made of 300000 km² of territory to identify priority areas for mineral exploration. This involved the interpretation of 39 Landsat scenes by using mss bands 5 and 7 at 1:500000 scale; an interpretation map was prepared for each scene at this scale. The interpretation was supported by a rapid reconnaissance of the region to verify the main features identified on the imagery. A lithostratigraphic and a structural synthesis of the region was prepared at 1:2000000 scale and this was used to identify the priority zones. This was followed by more detailed photogeological and field studies in two of the areas chosen.

Ferruginous residual deposits overlying mineralized ground (gossans) were first identified as colour anomalies on enhanced colour ratio composites by Rowan et al. (1974). Most of the gossans identified in this way could not be seen on the standard imagery. This led to the application of this technique to a test site at Wadi Wassat in Saudi Arabia (Blodget et al. 1978). The procedures developed by those authors have now become standard practice for searching for gossan in semi-arid terrain, to the extent that it has been used by the U.S.G.S. in the regional examination of the Walker Lake quadrangle. Here Rowan & Purdy (1981) have used colour ratio composites of enhanced mss imagery, where the composite is made as follows: 4/5 blue, 4/6 yellow, 6/7 magenta. On these images the limonite zones appear green, and the belts of limonitic alteration can be mapped.

The discovery of using mss imagery in this way prompted experimentation with additional bands in the infrared portion of the spectrum, particularly to aid the identification of alteration zones due to mineralization. Airborne tests carried out before the launch of Landsat-4 have proved successful, particularly in the mapping of alteration zones to porphyry copper deposits

[ 7 ]

(Abrams *et al.* 1981), and the delineation of bleached zones in sandstones covering deposits of uranium (Conel & Neisen 1981).

Where mineral deposits or their related alteration zones occur at surface and a known site can be identified on Landsat imagery, and provided that the spectral reflectance of the known site contrasts with its surroundings, it is possible to use multispectral classification techniques to locate sites with a similar spectral response on the image. This technique has proved successful initially at Saindak, Pakistan (Schmidt 1975), where supervised classification based on a 55 km$^2$ training area located 50 similar sites. Of these 50, 5 proved to be due to copper mineralization, which is an acceptable level for the results of field checking.

Similar studies have been carried out on gold-bearing gossans in Chile (Baker 1980), where supervised multispectral classification was used to locate new deposits. This technique has proved of use in the search for calcrete uranium deposits at Yeelirrie in Western Australia and in the Somali Republic.

By far the most common structures identified on Landsat are faults, fractures, linears and circular features. These can often be found in abundance and, following photogeological practice, statistical analysis of fracture traces can be carried out. This, together with some field information, can provide the means of deducing the stress pattern in the area at different periods of geological time. With structural data we are looking for clues to the location of concealed deposits, and not measuring something directly related to the deposit. There are abundant papers dealing with fractures and linears; some of the most notable are mentioned here.

The distribution of mineralization was found to be related to a major lineament in Saudi Arabia identified on Landsat by Moore (1976). The Al Amar fault was traced for 210 km, and a string of metalliferous and industrial mineral deposits form a belt that coincides with the volcano-sedimentary belt controlled by the fault line.

In the study of the Powder River Basin Uranium Deposits, Wyoming, Raines *et al.* (1978) carried out a statistical fracture analysis based on Landsat imagery. They found that the basin axis is marked by a strong linear feature, which is the boundary between two structural zones. There is evidence that this linear feature has influenced sedimentation in the region and later influenced the flow of groundwater, which controlled the distribution of uranium deposits.

Large circular and curvilinear features and their relation to mineralization have been studied by Norman (1980). It is suggested that these features may be related to meteorite impacts that occurred during the Archaean (more than 2600 Ma B.P.). This type of study has been used to investigate impact tectonics, which had previously been advanced to explain the occurrence of the Sudbury nickel deposits in Canada.

A Landsat study including digital analysis was integrated with a field study programme in Nova Scotia (Bruce & Stevens 1981), and nested circular and elliptical patterns were suggestive of convection cells in batholiths. This was verified by the study of the alignment of feldspar crystals on the ground, which followed the curved trend of an elliptical cell. Tin and uranium deposits were located along the cell margins.

In the Powder River Basin in Wyoming, Raines *et al.* (1978) found the local variation in vegetation density to be due to a facies change in the underlying rocks. Here the occurrence of uranium is restricted to one facies, and they were able to identify the differing vegetation density on Landsat imagery. Instances of vegetation stress have been identified on Landsat

[ 8 ]

imagery and have been shown to be related to lithological or mineralogical changes in the substrata. These changes have in some cases been shown to be due to metal toxicity, and this type of study is referred to as geobotany (or remote geobotany); it is an area where more research is required. An example where this technique led to a discovery is described by Cole (1980). Copper deposits were found in Botswana after the investigation of a botanical anomaly identified on Landsat imagery.

### EXAMPLES FROM PETROLEUM EXPLORATION

The use of multispectral satellite imagery does not generally involve looking directly for indications of petroleum on the Earth's surface, although these do occur in the form of natural seepages and tar sands. Because there is little chance of detecting petroleum directly at the Earth's surface most applications for petroleum exploration involve geological mapping and structural analysis, i.e. indirect techniques. Structural analysis forms a major part of all petroleum exploration programmes, as it is necessary to locate a 'closed' structure to trap hydrocarbon gases and fluids in the crust before they escape into the biosphere and atmosphere. In this respect Landsat studies have contributed significantly to many exploration programmes, particularly if undertaken during their early stages, because they provide a means of reducing the ground to be covered so that the prospectors can concentrate their more expensive ground surveys to greater advantage.

Like many other disciplines, petroleum exploration has benefited from the use of Landsat interpretation for terrain analysis in remote regions. This is particularly true throughout many regions of Africa and the Middle East, where the best available maps are geometrically corrected Landsat imagery. In addition the imagery is used in shallow marine areas, where the waters are clear, as a means of producing preliminary bathymetric charts for planning inshore seismic surveys. Landsat imagery is also being used to assist the navigation of airborne surveys and for planning and map-base purposes the imagery has made a very useful contribution to many exploration programmes.

For detailed lithological mapping in petroleum exploration Landsat does not have the resolution needed for plotting the stratigraphy in detail, but it is sufficient for mapping major structural features. Apart from locating folds and salt domes Landsat has been much used in the analysis of fractures and linears. One of the main discoveries from the synoptic view offered by Landsat was the recognition of linear alignments of landscape features that often extend for tens of kilometres. Many studies have been made to assess their geological significance in many parts of the world. It is generally believed that these linears represent the surface expression of fundamental changes in the crust. The dimensions are not apparent to the ground observer and in depth a structural discontinuity is not apparent. Nevertheless, studies have shown (Halbouty 1981) that facies changes occur across these linears throughout geological time, and they are generally aligned parallel to an important regional stress direction. From this evidence it is concluded that linears mark the boundaries of basement blocks, the relative movements of which influence sedimentation and therefore influence the accumulation of hydrocarbons. From the examination of imagery from 15 giant oilfields, Halbouty (1980) concluded that Landsat would have been of invaluable assistance in the search for these fields if it had been available at the time.

Of equal importance is the analysis of small fracture traces that can be observed on satellite imagery. Very often these do not have any associated evidence of movement and can be detected

through a superficial cover of alluvium, soil, etc. The analysis of fracture patterns of this type
can be carried out by visual inspection, or, more often these days, by computer analysis of the
fractures annotated by the interpreter. Several attempts have been made to detect fracture
patterns automatically without a great deal of success. Together with a study of the geological
history of the region and reconnaissance field work to measure joint and fault directions a history
of the stress fields pertaining to different periods of geological time can be worked out for the
region under analysis. This can be used to assist in the selection of structural targets for
petroleum exploration.

Circular and arcuate features of small to medium radius (5–20 km) have also been used for
the location of structural targets for petroleum exploration. Small circular structures are often
due to salt doming, but of equal importance is the possibility that the large structures are due
to basement topography concealed beneath younger sediments. This possibility was recognized
by Zall *et al.* (1981), where circular features and fracture trace analysis assisted in the identification of petroleum targets in four regions in the western United States.

Fracture trace analysis has also been used to detect petroleum reservoirs in fractured rocks
(i.e. where the permeability has been caused by brittle fractures). Such a study is described by
Lang (1982) in the Lost River field of Wyoming. Here Lang used weighted linear density maps
to detect the regions where fractures were most prevalent, and related this to the occurrence of
petroleum reservoirs in the field.

The combination of satellite and airborne synthetic aperture radar was used for the exploration of fractured petroleum reservoirs in the Arkoma Basin in the United States of America
(McDonald *et al.* 1981). In this gas-producing region an attempt was also made to correlate the
production and density of gas wells with fracture density. The researchers concluded that mss
band 7, winter, with uniform contrast stretch was best for linear detection, but band 7 combined
with sar (Synthetic Aperture Radar) is significantly better if the radar look direction is orthogonal to the solar illumination. However, they were unable to demonstrate any relation
between fracture density and gas production that tapped different formations in different
parts of the area.

The importance of combining image interpretation with geophysical potential field techniques has been stressed by many workers. This appraisal of different but related data is
important to petroleum exploration. Hunting Geology and Geophysics (Australia) (1982) have
undertaken a study of the Eromanga Basin, where the Landsat interpretation has been integrated
with gravity and aeromagnetic data, as well as assessing the relation of the movement of
groundwater to petroleum. Many studies where the integration of regional survey data has
played a significant role in petroleum exploration are known (Hunting Geology and Geophysics
Limited, personal communication). In many instances the information obtained from the
satellite imagery is of a relatively simple nature, such as being able to map the contact between
metamorphic Precambrian rocks and the edge of a sedimentary basin that is concealed beneath
superficial deposits and to confirm the existence of basement highs within the basin, or to be
able to pick out fault trends that define the development of horsts and grabens within a
sedimentary basin.

The possibility that natural oil seeps might be detected in the Santa Barbara Channel,
California, from Landsat was investigated by Deutsch & Estes (1980). The fact that slicks
caused by oil spillage could be seen on Landsat imagery had already been established (Estes
*et al.* 1972; Deutsch *et al.* 1976). In this study, oil from the natural seepages was identified on

contrast stretch enhancements of the digital imagery, and the distribution was confirmed by aerial observations made within a few hours of the Landsat pass. The experimenters concluded that this technique might have valuable application for offshore oil exploration.

## New satellites

Landsat-4, the first of the new series of land remote-sensing satellites that will provide multi-spectral imagery, was launched in July 1982. This heralds a new era in the application of satellite remote sensing to resource assessment, as it marks the changeover from an experimental to an operational status. There is also the introduction of a new sensor system with improved spatial resolution, spectral separation, geometric fidelity and radiometric accuracy. This new sensor is designated the Thematic Mapper, and on Landsat-4 it will operate in addition to a multispectral scanner producing imagery at 80 m resolution in the same four bands used on the three previous Landsats.

The Thematic Mapper records in seven spectral bands and these are summarized in table 2. The ground resolution has been improved by reducing the pixel size to 30 m in all but band 6, which has a pixel size of 120 m. As already mentioned, band 7 will be important for geological applications, as it will aid in the discrimination of clays and zones of hydrothermal alteration.

TABLE 2

| band | mss/μm | thematic mapper/μm | SPOT/μm |
|---|---|---|---|
| 1 | 0.5–0.6 | 0.45–0.52 | 0.50–0.59 |
| 2 | 0.6–0.7 | 0.52–0.60 | 0.61–0.68 |
| 3 | 0.7–0.8 | 0.63–0.69 | 0.79–0.89 |
| 4 | 0.8–1.1 | 0.76–0.90 | — |
| 5 | — | 1.55–1.75 | — |
| 6 | — | 10.40–12.5 | — |
| 7 | — | 2.08–2.35 | — |
| pixel size/m | 80 | 30 (B6 = 120) | 20 or 10 panchromatic |

The French Earth observation satellite SPOT (System Probatoire de Observation de la Terre) is due to be launched in 1984, and this system will have two modes of operation: either recording multispectral imagery in three bands with a 20 m pixel size or panchromatic imagery at 10 m pixel size. Furthermore it is designed to obtain stereoscope imagery, as it will be possible to tilt the scan head to repeat coverage of the same area. This will only be possible in areas of stable weather conditions.

## Discussion

In this appraisal of the use of multispectral satellite imagery for mineral and petroleum exploration, the objective has been to demonstrate that the practical use of Landsat imagery has become established. The general application of satellite imagery has come about because it has been possible to demonstrate real benefits to prospectors. This is verified by the number of organizations both in the United Kingdom and throughout the world that are offering services in the geological applications of Landsat remote sensing. These groups are now ready to take advantage of new systems and new developments and to undertake the practical application of current research.

[ 11 ]

In remote sensing the geologist has to make deductions about subsurface conditions on the basis of surface observations, and this means that significant reliance is placed upon interpretative skills. So far with the Landsat system there have been relatively few opportunities to use thematic classification as a means of locating potentially valuable mineral deposits in the Earth's crust. This has been mainly due to the limitations of resolution and recording broad spectral bands. This will be somewhat alleviated once imagery from Landsat-4 is made generally available, as this system will have both greatly improved ground resolution and additional spectral bands specifically requested for geological applications.

In petroleum exploration the emphasis has been on the structural interpretation of imagery, and this will greatly benefit from having stereoscopic data. The proposed launch of the SPOT satellite in 1984 will to a certain degree fulfil this need because it will be possible to acquire stereoscopic images, albeit separated by several days. This will be limited to those regions of the globe where both surface conditions and atmospheric conditions undergo only slow change, as there will be some risk attached to specifying stereoscopic coverage for a specific programme. However, SPOT will also go further towards providing improved resolution, which will be welcomed by the geological remote-sensing community.

Research into the application of narrow band infrared radiometers continues with the SMIRR experiment on the Space Shuttle, and this together with airborne experiments is likely to lead to new discoveries about the geological application of different wavebands. As the application of such imagery becomes understood it is conceivable that the demand will be for specialized satellites; this is not practicable unless the rewards are very great (e.g. the Canadian Radarsat). The need will therefore arise for waveband selection to be under ground control, or by technicians carried on board the Space Shuttle. Other future developments will be the demand for faster access to data, which will be achieved by satellite data links distributing raw data to clients rather than a series of standard products as at present. In the applications field more research is needed in the development of geobotanical techniques, which are equally important in temperate and tropical regions, particularly in understanding the chlorophyll infrared edge shift in chemically stressed vegetation; the importance of the 'red edge' has been emphasized by Horler *et al.* (1981).

One facet of Landsat that should not be overlooked is that within the period of one decade it has contributed greatly to the training of a cadre of applications-oriented remote-sensing scientists. It therefore can be expected that the clever combination of different image data sets will provide considerable application. This will go hand in hand with the increasing use of computers in geology for correlation of exploration data and development of geological models for the occurrence of valuable materials. Work on the digital combination of exploration data with Landsat has been described by Aroidson & Guiness (1981), Missallati *et al.* (1979) and Peters (1980).

I thank the Directors of Hunting Geology and Geophysics Limited for authorizing this paper, my colleagues for critical comment and discussion, and Mrs B. Tubbs and Mrs P. Buckle for production of the paper.

## REFERENCES

Abrams, M., Brown, D., Lepley, L. & Sadowski, R. 1981 In *Digest, International Geoscience and Remote Sensing Symposium, Washington, D.C.*, June 1981, vol. 1, pp. 331–336. Piscataway, N. J.: Institute of Electrical and Electronics Engineers.

Albert, N. R. D. & Chavez, P. S. Jr 1975 *U.S. geol. Surv. prof. Pap.* no. 1015, pp. 193–200.

Arvidson, R. E. & Guiness, E. A. 1981 In *Digest, International Geoscience and Remote Sensing Symposium, Washington, D.C.*, June 1981, vol. 2, pp. 895–896. Piscataway, N.J.: Institute of Electrical and Electronics Engineers.

Baker, M. C. W. 1980 In *Abstracts, Seventh Annual Conference of the Remote Sensing Society.*

Blodget, H. W., Gunther, F. J. & Podwysocki, N. H. 1978 *NASA tech. Pap.* no. 137, p. 34.

Bruce, B. & Stevens, G. R. 1981 *PDA Recorder, Can. Min. J.*, May 1981, pp. 1–6.

Butlin, T. J., Guertin, F. E. & Vishnubhatla, S. S. 1978 Presented at 5th Canadian Symposium on Remote Sensing.

Chavez, P. S. Jr, Berlin, G. L. & Acosta, A. V. 1976 In *Proc. 2nd W.T. Pecora Memorial Symposium*, pp. 235–251.

Cole, M. M. 1980 *Trans. Instn Min. Metall.* B**89**, 73–91.

Collins, W., Chang, S.-H., Kuo, J. T. & Rowan, L. C. 1981 In *Digest, International Geoscience and Remote Sensing Symposium, Washington, D.C.*, June 1981, vol. 1, pp. 337–344. Piscataway, N.J.: Institute of Electrical and Electronics Engineers.

Conel, J. E. & Niesen, P. L. 1981 In *Digest, International Geoscience and Remote Sensing Symposium, Washington, D.C.*, June 1981, vol. 1, pp. 318–324. Piscataway, N.J.: Institute of Electrical and Electronics Engineers.

Deutsch, M., Estes, J. E. & Munchow, C. 1976 In *Program Abs. 19th COSPAR meeting.*

Deutsch, M. & Estes, J. E. 1980 *Photogramm. Engng Remote Sensing* **46**, 1313–1322.

Estes, J. E. & Seuger, L. W. 1972 *Remote Sensing Envir.* **2**, 141–163.

Fischer, W. A., Anguswathana, P., Carter, W. D., Hoshiro, K., Lathram, E. H. & Rich, E. I. 1977 *A.A.P.G. Mem.* **25**, 63–72.

Goetz, A. F. H. & Rowan, L. C. 1981 In *Digest, International Geoscience and Remote Sensing Symposium, Washington, D.C.*, June 1981, vol. 1, pp. 345–346. Piscataway, N.J.: Institute of Electrical and Electronics Engineers.

Halbouty, M. T. 1980 *A.A.P.G. Bull.* **64**, 8–36.

Halbouty, M. T. 1981 In *Digest, International Geoscience and Remote Sensing Symposium, Washington, D.C.*, June 1981, vol. 1, pp. 305–311. Piscataway, N.J.: Institute of Electrical and Electronics Engineers.

Horler, D. N. H., Barber, J. & Barringer, A. R. 1981 In *Remote sensing in geological and terrain studies*, pp. 113–123. London: Remote Sensing Society.

Hunt, G. R. 1977 *Geophysics* **42**, 501–513.

Hunting Geology and Geophysics (Australia) Pty Ltd 1982 *Eromanga Basin study.*

Hunting Geology and Geophysics Ltd 1976 *Mineral exploration in Southern Sudan. Preliminary survey to establish priority areas.*

Land, H. R. & Baird, K. W. 1981 In *Digest, International Geoscience and Remote Sensing Symposium, Washington, D.C.*, June 1981, vol. 1, pp. 589–594. Piscataway, N.J.: Institute of Electrical and Electronics Engineers.

Lang, H. 1982 In *Digest, International Geoscience and Remote Sensing Symposium, Munich*, 1982, vol. 1, pt W4–1. Piscataway, N.J.: Institute of Electrical and Electronics Engineers.

Lyon, R. J. P. 1965 *Econ. Geol.* **60**, 715–737.

McDonald, H., Waite, W., Elachi, C., Borengasser, M. & Tomlinson, D. 1981 In *Digest, International Geoscience and Remote Sensing Symposium, Washington, D.C.*, June 1981, vol. 1, pp. 312–317. Piscataway, N.J.: Institute of Electrical and Electronics Engineers.

McMahon Moore, J. A. 1976 *Mineral Deposits* **11**, 323–328.

Missallati, A., Prelat, A. E. & Lyon, R. J. P. 1979 *Remote Sensing Envir.* **8**, 189–210.

Nash, C. R., Boshier, P. R., Coupard, M. M., Theron, A. C. & Wilson, J. G. 1980 Photogeology and satellite image interpretation in mineral exploration. *Minerals Sci. Engng* **12**, 216–244.

Norman, J. W. 1980 *Trans. Instn Min. Metall.* B**89**, 63–72.

Peters, E. R. 1980 In *Abstracts, Seventh Annual Conference of the Remote Sensing Society.*

Raines, G. L., Offied, T. W. & Santos, E. S. 1978 *Econ. Geol.* **73**, 1706–1724.

Raines, G. L. 1981 In *Digest, International Geoscience and Remote Sensing Symposium, Washington, D.C.*, June 1981, vol. 1, p. 588. Piscataway, N.J.: Institute of Electrical and Electronics Engineers.

Rowan, L. C., Wetlaufer, P. H., Goetz, A. F. H., Billingsley, F. C. & Stewart, J. H. 1974 *U.S. geol. Surv. prof. Pap.* no. 883, p. 35.

Rowan, L. C. & Purdy, T. L. 1981 In *Digest, International Geoscience and Remote Sensing Symposium, Washington, D.C.*, June 1981, vol. 1, pp. 325–330. Piscataway, N.J.: Institute of Electrical and Electronics Engineers.

Schmidt, R. G. 1975 *U.S. geol. Surv. open-file Rep.* (29 pages.)

Siegrist, A. W. & Schnetzler, C. C. 1980 *Photogramm. Engng Remote Sensing* **46**, 1207–1215.

Vincent, R. K. 1972 In *Proceedings, 8th International Symposium Remote Sensing of the Environment*, pp. 1239–1247.

Zall, L., Staskowski, R., Michael, R. & Prucha, S. 1981 In *Technical papers, 48th Annual Meeting of the American Society of Photogrammetry.*

*Phil. Trans. R. Soc. Lond.* A **309**, 257–270 (1983)

*Printed in Great Britain*



# Multispectral remote sensing for the estimation of green leaf area index

By P. J. Curran

*Department of Geography, University of Sheffield,*
*Western Bank, Sheffield S10 2TN, U.K.*

The causal relation between multispectral reflectance and green leaf area index (l.a.i.) has enabled the estimation of green leaf area index by the judicious use of remotely sensed multispectral reflectance measurements.

In this paper three topics are discussed. First, the reflectance properties of a vegetation canopy and the problems of determining the form of the relation between green l.a.i. and red and near-infrared reflectance: these problems include variability in substrate and leaf reflectance and the geometry of the scene and sensor. Second, the methodologies currently employed for estimating green l.a.i.: these methodologies are based on the production of simple, complex or modelled calibration curves. Third, current research at the University of Sheffield: this includes not only studies with multispectral reflectance collected from aircraft-mounted sensors to estimate the green l.a.i. of heathlands and grasslands but also multispectral reflectance collected from satellites to map estimated green l.a.i.

It is concluded that the main applications for this remote-sensing technique are within the fields of agricultural intelligence, agricultural management and ecological research.

## 1. Introduction

Electromagnetic radiation that is reflected or emitted from the Earth's surface can be recorded by a sensor from the ground, from an aircraft or from a satellite and these data can be used to measure and map the Earth's terrestrial, aquatic and atmospheric environments. Most of the data collected by remote sensors have been used in image form to study the spatial disposition of objects. Today an increase in the spatial and radiometric resolution of remote-sensing instrumentation, coupled with an increasing knowledge of the way in which electromagnetic radiation interacts with our environment, have enabled workers to use remotely sensed data to determine for example the amount of soil moisture in a field or the amount of suspended sediment in estuarine waters. One very promising area of remote estimation is the use of multispectral reflectance (which is defined as the spectral radiance standardized by the spectral irradiance) to estimate green leaf area index or green l.a.i., where green l.a.i. is the area of green leaves per unit area of ground. Accurate and timely information on green l.a.i. has application in agriculture for yield estimation and stress evaluation and in ecology for the study of primary production and environmental change.

This paper will review first the reflectance properties of a vegetation canopy; second, the relation between remotely sensed reflectance and green l.a.i.; third, methodologies for the estimation of green l.a.i. from remotely sensed measurements of multispectral reflectance, and fourth, examples of green l.a.i. estimation drawn from work undertaken at the University of Sheffield.

## 2. Reflectance properties of a vegetation canopy

Of the solar irradiance impinging upon a vegetation canopy some is reflected, some is transmitted and some is absorbed. The intensity with which radiation is reflected at any particular wavelength is dependent upon both the spectral properties and also the area of the three main remotely sensed components of a vegetation canopy: leaves, substrate and shadow.

### (a) Spectral properties of leaves

Leaves usually reflect weakly in the blue and red wavelengths owing to absorption by pigments and strongly in the near-infrared wavelengths owing to cellular refraction (figure 1) (Curran 1980c; Kondratyev & Fedchenko 1982).

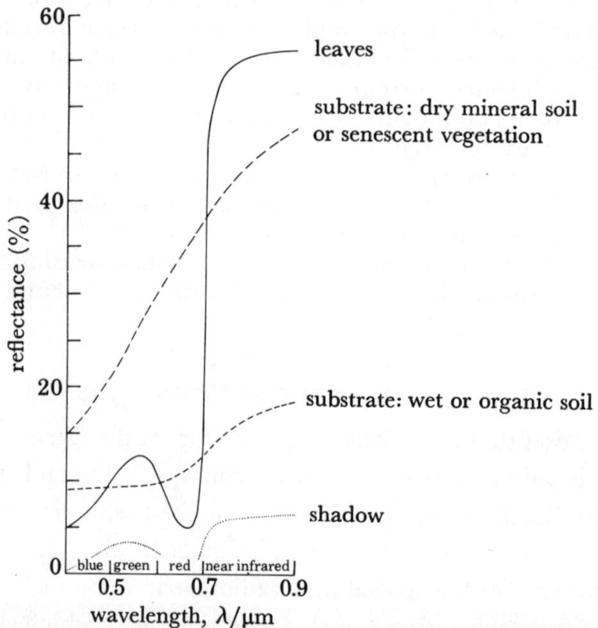


Figure 1. A diagrammatic illustration of the reflectance properties of three components of a vegetation canopy, leaves, shadow and either a light-toned mineral substrate or a dark-toned organic substrate.

### (b) Spectral properties of substrate

The reflectance of vegetation substrates is usually spectrally simple but variable. Three common substrates, senescent vegetation, light-toned mineral soil and dark-toned organic soil, are illustrated in figure 1.

### (c) Spectral properties of shadow

The most common type of shadow in a vegetation canopy results from the transmission of radiation through leaves and its re-radiation from other leaves or substrate. As a result canopy shadow is very dark in visible wavelengths, where most of the visible radiation is absorbed by leaves, and fairly dark in near infrared wavelengths, where little radiation is absorbed by leaves (Colwell 1974).

The relative area of these three spectrally dissimilar canopy components determines the reflectance of the total canopy. The canopy component with most variation in time and space is the area of green leaves and this in turn has most influence on the red and near-infrared reflectance of the vegetation canopy. Workers have therefore sought to use multispectral reflectance, particularly in red and near-infrared wavelengths, as a means of remotely estimating the green l.a.i. of vegetation canopies, especially when the vegetation canopies cover large areas of terrain.

### 3. THE RELATION BETWEEN REMOTELY SENSED RED AND NEAR-INFRARED REFLECTANCE AND GREEN L.A.I.

Green l.a.i. has a negative relation with red reflectance and a positive relation with near-infrared reflectance, as illustrated in figure 2. To express this increasing difference between red and near-infrared reflectance with increasing green l.a.i., a ratio of red to near infrared reflectance is customarily used (Curran 1980$a$). One of the more popular is the vegetation index, $I_v$:

$$I_v = (R_{ir} - R_r)/(R_{ir} + R_r),\qquad(1)$$

where $R_r$ is red reflectance and $R_{ir}$ is near-infrared reflectance.

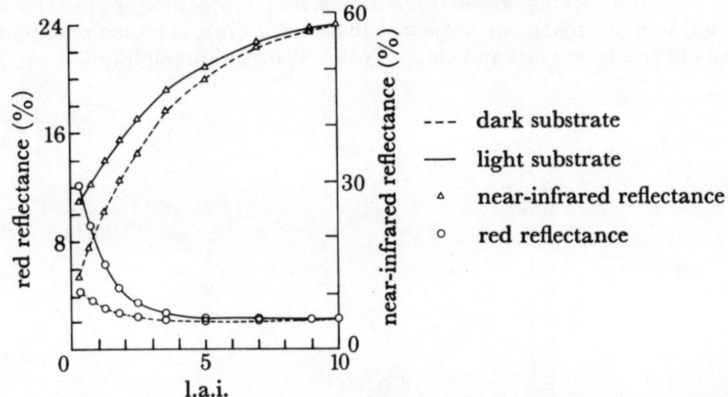


FIGURE 2. The relation between leaf area index (l.a.i.) and red and near-infrared reflectance for a vegetation canopy growing on a light-toned and a dark-toned substrate, derived from modelled data. (Modified from Colwell (1974).)

The form of the relation between the vegetation index and green l.a.i. is generally curvilinear, reaching an asymptote when the substrate is covered by several layers of leaves. The actual form of the relation is dependent upon the species and to a lesser extent the conditions of measurement. For example, in figure 3$a$ the asymptote of the graph, and therefore the point at which the vegetation index ceases to be sensitive to changes in l.a.i., never occurs. For canopies in figure 3$b$ it occurs at l.a.i.s of around 3 and for canopies in figure 3$c$ it occurs at l.a.i.s of over 4. For this reason workers have tended to restrict their attention to cereal canopies with low green l.a.i.s because these canopies are thought to have a near-linear relation between the vegetation index and green l.a.i.

When looking at satellite images of large areas of land it is evident that one of the variations in canopy reflectance is the reflectance of the substrate rather than the leaves. However, it is possible to remove the effect of the substrate from the canopy reflectance by expressing the canopy reflectance relative to the substrate reflectance. This is done by first plotting the substrate line, which is a graph of substrate reflectance in red and near-infrared wavebands (figure 4). This relation is linear because light-toned substrates reflect both red and near-infrared radiation strongly and dark-toned substrates reflect both red and near-infrared radiation weakly (Curran *et al.* 1981).

[ 17 ]

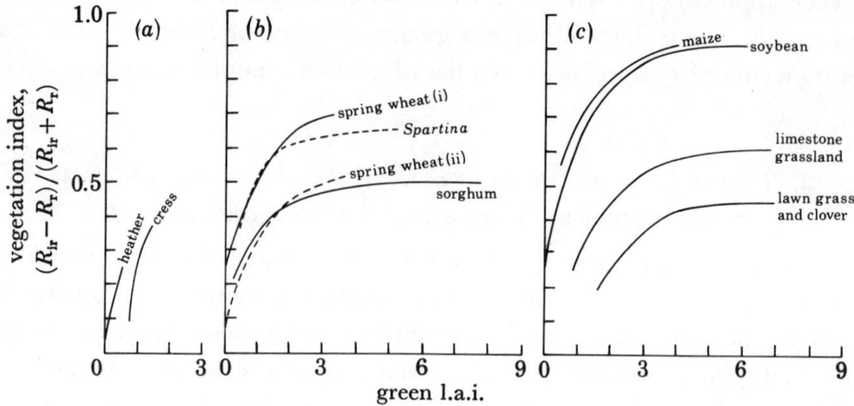



FIGURE 3. The relation between the vegetation index and leaf area index for different vegetation canopies: (a) asymptote not reached; (b) asymptote reached at low l.a.i.; (c) asymptote reached at high l.a.i. Modified from the following sources: for heather, Curran (1981 b); for cress, Curran & Milton (1983); for spring wheat (i), Daughtry *et al.* (1980); for spring wheat (ii), Ahlrichs *et al.* (1979); for *Spartina*, Bartlett & Klemas (1980); for sorghum, Brakke *et al.* (1981); for soybean, Holben *et al.* (1980); for maize, Kimes *et al.* (1981); and for limestone grassland and lawn grass and clover, N. W. Wardley (unpublished).

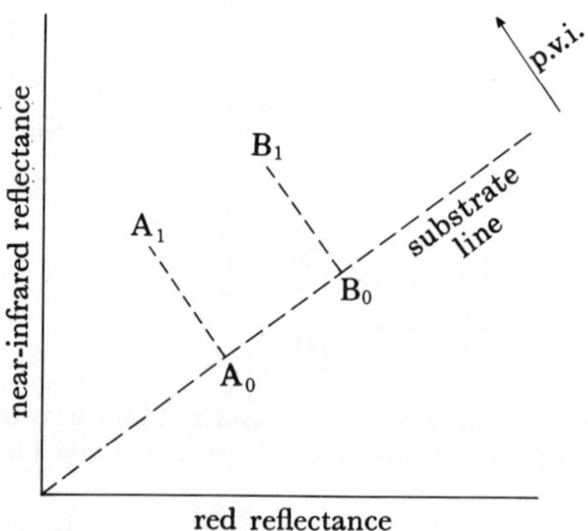



FIGURE 4. The coordinates in a red–near-infrared plot for sites A and B, where $A_0$ and $B_0$ are bare substrate and $A_1$ and $B_1$ are vegetation-covered. The index that can be used to record the spectral change from $A_0$ to $A_1$ and $B_0$ to $B_1$ is the perpendicular vegetation index (p.v.i.) (see text for discussion).

If vegetation were to grow on two sites on this substrate line, for example at sites $A_0$ and $B_0$, then as green leaf area increases the red reflectance will probably decrease and the near-infrared reflectance will probably increase. This will result in the movement of the reflectance coordinates for sites $A_0$ and $B_0$ away from the substrate line to coordinates $A_1$ and $B_1$ (figure 4). The coordinate distance between $A_0$ to $A_1$ and $B_0$ to $B_1$ is directly correlated to the green l.a.i. of the canopy at sites A and B respectively, and can be calculated from remotely sensed reflectance data by using the perpendicular vegetation index (p.v.i) of Richardson & Wiegand (1977):

$$\text{p.v.i.} = \sqrt{\{(R_{s,r}-R_{v,r})^2+(R_{s,ir}-R_{v,ir})^2\}}, \tag{2}$$

where $R_s$ is substrate reflectance and $R_v$ is vegetation reflectance.

These relations between a reflectance ratio and green l.a.i. and between a reflectance transform and green l.a.i. are not entirely dependent upon the area of green leaves but vary with other characteristics of the environment. Some of these characteristics are rarely of great importance (e.g. topography or microclimate), some are predictable and therefore correctable (e.g. phenology and atmospheric effects) (Curran 1982*b*; Slater & Jackson 1982), and some are difficult to predict (e.g. reflectance of the substrate, presence of senescent vegetation and geometry of the scene and sensor). These will be discussed below.

### (a) *Reflectance of the substrate*

The possibility of detecting a change in green l.a.i. by a change in reflectance is dependent upon the reflectance contrast between green leaves and substrate. This is illustrated in figure 2, where it can be seen that (i) reflectance in near-infrared wavelengths is more sensitive to changes in the l.a.i. of vegetation on dark-toned substrates than on light-toned substrates, and (ii) reflectance in red wavelengths is more sensitive to changes in the l.a.i. of vegetation on light-toned substrates than on dark-toned substrates. In some environments the substrate can be so dark as to make red wavelengths insensitve to changing l.a.i., for example on the organic soils of the Somerset levels (Curran 1983) or so light as to make near-infrared wavelengths insensitive to changing l.a.i., for example in arid environments (Curran 1981*b*).

### (b) *The presence of senescent vegetation*

As vegetation senesces, the near-infrared leaf reflectance does not significantly decrease. However, the breakdown of plant pigments causes a rise in red reflectance. Therefore if the amount of senescent vegetation in a canopy increases, the positive relation between near-infrared reflectance and green l.a.i. will probably remain unchanged whereas the relation between red reflectance and green l.a.i. will weaken and probably disappear (Curran 1980*c*). This is a problem in semi-natural vegetation, particularly grasslands, where there is some senescent vegetation in the canopy throughout the year.

### (c) *The geometry of the scene and sensor*

The elevation and azimuth of the Sun and the sensor have a considerable effect on the reflectance of a vegetation canopy.

#### (i) *Solar elevation*

Two interrelated factors contribute to the effect of solar elevation on the reflectance of a vegetation canopy. The first is the degree to which solar radiation can penetrate the canopy, and this is negatively related to solar elevation. The second is the amount of canopy shadow and this is positively related to solar elevation. As a result, most canopies have a negative relation between near-infrared reflectance and solar elevation and a poor relation or none between visible reflectance and solar elevation. This is illustrated for a sedge canopy in figure 5 and is discussed in Curran (1983) and Kondratyev & Fedchenko (1982).

#### (ii) *Sensor elevation*

The elevation of the sensor determines the amount of substrate and shadow seen, for as the elevation moves from the vertical, the area of soil and shadow seen by the sensor decreases, and the area of vegetation seen increases. To move the sensor from the vertical will therefore increase near-infrared and decrease red reflectance, as illustrated in figure 5. The implications of this effect on multispectral reflectance measured at the edge of a Landsat satellite multispectral scanner image or from an oblique SPOT satellite high-resolution visible image are discussed in Curran (1983).

### (iii) *Relative solar and sensor azimuth*

The reflectance of a canopy is usually higher if the sensor is looking onto as opposed to away from the Sun, and away from the Sun as opposed to at right angles to the Sun (Kondratyev & Fedchenko 1982) (figure 5).

For most remote sensing applications where the sensor look-angle is nearly vertical, the effect of the solar azimuth on reflectance increases with a decrease in solar angle and an increase in vegetation canopy roughness, as discussed in Curran (1983).

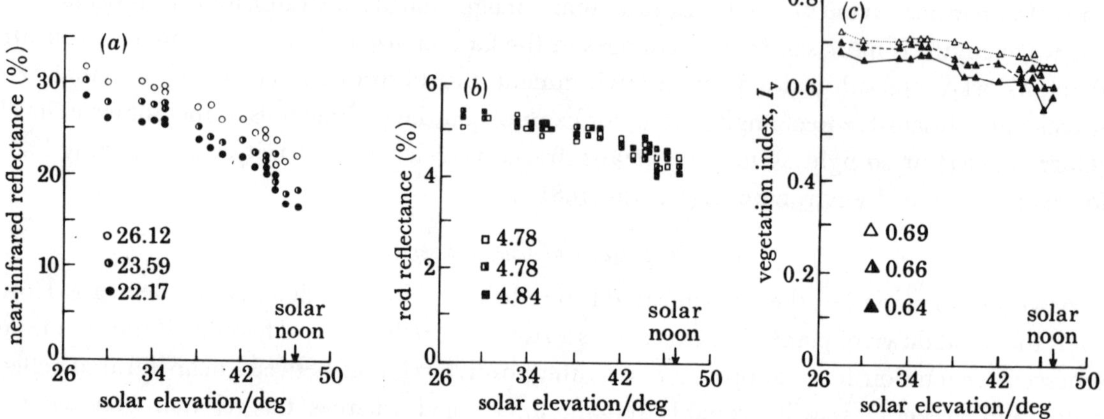

FIGURE 5. The effect of solar and sensor elevation and azimuth on the reflectance properties of a sedge canopy. These data were collected by using a ground radiometer in August 1982 at the Lakkasuo peatland complex near Hyytiälä in Southern Finland (61° 47′ 33″ N, 24° 18′ 40″ E). The vegetation, which has a green l.a.i. of 1.8, is dominated by *Carex lasiocarpa*, *Sphagnum papillosum* and *Sphagnum fallax*, and overlies wet sedge peat. Open symbols, 20° upsun; half filled symbols, 20° downsun; filled symbols, vertical. The numbers in each portion of the figure are mean reflectance values for each of the data sets.

## 4. METHODOLOGIES FOR THE ESTIMATION OF GREEN L.A.I. BY USING MULTISPECTRAL REFLECTANCE

Two stages are involved in the estimation of green l.a.i. from remotely sensed multispectral reflectance: first, the determination of a calibration curve for the relation between multispectral reflectance and green l.a.i.; and second, the use of multispectral reflectance data collected from aircraft or spacecraft as input to this calibration curve. Ultimately the applicability and accuracy of green l.a.i. estimates are dependent upon the quality of either simple, complex or modelled calibration curves (Wiegand *et al.* 1979).

### (a) *Simple calibration curves*

In this method the multispectral reflectance data used for the construction of the calibration curve are collected at the same place and time as the multispectral reflectance used for the prediction of green l.a.i. This can either be a subset of the remotely sensed multispectral reflectance data or can be ground-based measurements of multispectral reflectance. The advantage of the method is that it gives a very high accuracy of green l.a.i. prediction. The disadvantage is that the calibration curve is specific to site and time.

Because the aim of this method is to obtain a very accurate prediction from a specific data set, workers have tended to use any waveband ratio or transform that will give the necessary results;

for example, a study in North Dakota obtained a correlation of 0.93 (significant at the 1 % level) between the observed and predicted l.a.i. of spring wheat (figure 6a). Three wavelengths that had the highest correlation with l.a.i. were selected and applied in a three-variable multiple regression to ground-based radiometric data (Ahlrichs *et al.* 1979). Later studies successfully employed a similar technique with data obtained from the Landsat multispectral scanner (Heilman *et al.* 1977; Pollock & Kanemasu 1979) (figure 6b).

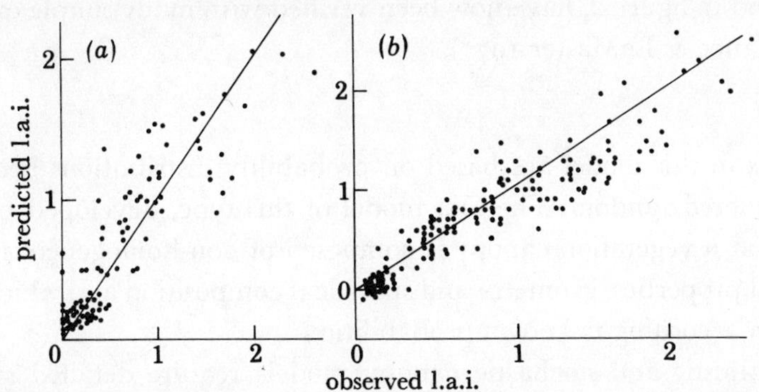


FIGURE 6. A comparison between the observed l.a.i. and the l.a.i. predicted from a simple calibration curve (a) Winter wheat (modified from Pollock & Kanemasu (1979)); (b) spring wheat (modified from Ahlrichs *et al.* (1979)).

Such approaches are most successful when applied to large data sets for simple crops. When working with small data sets and complex crops, workers have found even the simple calibration curve method unsuitable; for example, Chance (1981) used a transform of near-infrared radiance recorded by the Landsat multispectral scanner to predict the l.a.i. of seven sorghum fields in Texas and obtained a non-significant correlation of 0.38 between observed and predicted l.a.i. For grassland canopies in which l.a.i. is linearly related to biomass (Curran 1981 b) it has proved possible to use a simple calibration curve for the estimation of biomass, as has been illustrated by the work of Colwell (1974), Maxwell (1976), Pearson *et al.* (1976), Curran (1980b) and Hielkema (1980).

### (b) *Complex calibration curves*

In this method the reflectance characteristics of a species or species association are measured over space and time to provide calibration curves for the relations between multispectral reflectance, green l.a.i. and other environmental effects. These calibration curves enable multispectral reflectance to be corrected for the environmental effects considered by the operator to be important, before their use in the prediction of green l.a.i. Early work in this field was undertaken in the estimation of grassland biomass where biomass and l.a.i. were linearly related (Deering *et al.* 1975; Deering & Hass 1980; Bartlett & Klemas 1980). A more recent example of green l.a.i. estimation with this method is presented in §5 of this paper.

### (c) *Modelled calibration curves*

The complex calibration curve method, while producing satisfactory results, is very time-consuming because it has to be repeated for each species and species association. To overcome this problem it is possible to model the reflectance properties of different physiological types of vegetation canopy by using a canopy model, which can be deterministic, stochastic or empirical.

### (i) *Deterministic models*

These describe radiation passing through canopies in terms of diffuse flows scattered and attenuated by absorption. These theories are rather abstract and it was the Suits model (Suits 1972) that brought the deterministic model closer to reality. In this model Suits considers a canopy to be composed of mixed components of horizontal and vertical leaves, flowers and stalks, all with their own reflectance properties (Slater 1980). Data from this model, an example of which is reproduced in figure 2, have now been verified with many simple one-species canopies (Colwell 1974; Chance & LeMaster 1977).

### (ii) *Stochastic model*

All of the inputs in this model are based on probability distributions because the processes involved are considered random. The first model of this type, developed by Smith & Oliver (1972), assumes that a vegetation canopy is composed of non-homogeneous layers of material with known optical properties, geometry and statistical composition and which interact with the incoming radiation according to known probabilities.

Both the deterministic and stochastic canopy models require detailed specification of the canopy characteristics. Therefore for semi-natural or forest vegetation the deterministic or stochastic canopy models are unlikely to be a practical option for the construction of calibration curves. However, for agricultural crops where canopy characteristics possess a degree of spatial uniformity it may well be possible to produce calibration curves from canopy models.

### (iii) *Empirical model*

The empirical model is much more useful for the production of calibration curves because it is based on the creation of relatively simple data banks of green l.a.i. and multispectral reflectance data collected for different canopy types under a range of conditions. Such a model is under construction at the University of Sheffield for testing during 1983.

### 5. ESTIMATING GREEN L.A.I. FROM MULTISPECTRAL RADIANCE: EXAMPLES FROM THE UNIVERSITY OF SHEFFIELD

At the University of Sheffield work is currently in progress on a three-phase experiment designed to estimate green l.a.i. remotely. Phase one is a pilot study to determine a methodology for estimating green l.a.i. of a study site by using reflectance measured at aircraft altitudes. Phase two is designed to determine the feasibility of estimating the green l.a.i. of a larger area of terrain by using reflectance also measured at aircraft altitudes. Phase three is the estimation of green l.a.i. of large areas by using radiance measured at satellite altitudes.

Phase one will be discussed in detail and phase two and three will be outlined.

#### (a) *Phase one: pilot study to determine a methodology for estimating the green l.a.i. of a limited area by using reflectance measurements from aircraft altitudes*

### (i) *Study area*

The pilot study was undertaken on Snelsmore Common, a heathland in Berkshire. Four vegetation types were chosen for study: young and mature *Calluna* (heather), *Pteridium* (bracken) and *Pteridium–Calluna* (bracken–heather association). Further details of these are to be found in Curran (1981 c, 1982 a).

[ 22 ]

(ii)  *The collection of ground data for the construction of complex calibration curves*

Snelsmore Common was visited on 12 occasions over a period of 14 months. At each random point, radiometric red and near-infrared reflectance measurements or photographic red and near-infrared reflectance measurements, or both, were taken and on the vegetated areas vegetation samples were collected. These data were expressed as one mean red reflectance value, one mean near-infrared reflectance value and one mean green l.a.i. value per random sample point (Curran 1982 *a*).

(iii)  *Ground data manipulation and the construction of the complex calibration curves*

The diurnal relation between reflectance and solar elevation was determined for each vegetation association and for bare ground. For all of the sites there was no significant deviation in red reflectance from the mean red reflectance (Curran 1982 *a*). However, because the near-infrared reflectance did decrease markedly with solar elevation the reflectance data were corrected to a constant solar elevation.

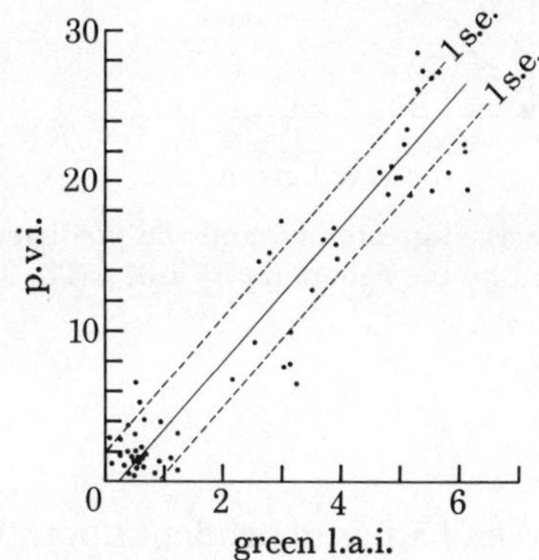


FIGURE 7. The linear relation between the perpendicular vegetation index (p.v.i.) and the green leaf area index (green l.a.i.) for a *Pteridium–Calluna* association. The solid line is described by the equation $x = 0.24 + 0.22\,y$. The s.e. of the estimate is 0.66 and of the forecasts is 0.73 ($y = 20$), 0.72 ($y = 10$) and 0.72 ($y = 2$); $n = 60$; $r = 0.83$; significance level, 1%.

These corrected data were then used to calculate the p.v.i. (equation (2)), which was regressed against the green l.a.i. of the four vegetation associations. In each case p.v.i. was positively related to green l.a.i. over the green l.a.i. range of 0–0.5 for young and mature *Calluna* and over the green l.a.i. range of 0–8 for the *Pteridium* and the *Pteridium–Calluna* (figure 7).

(iv)  *The collection of multispectral aerial photography and observed green l.a.i.*

Near-vertical 35 mm format aerial photography was taken at around 11h00 local time, on 20 dates, from June 1980 to August 1981, from a light aircraft with a camera with the same film–filter combination used for ground multispectral photography (Curran 1981 *a*). The reflectance of the vegetation or soil or both was calculated by using the method proposed by Lillesand & Kiefer (1979). At each random sample point the vegetation was harvested and the green l.a.i. determined by using the methods discussed in Curran (1982 *a*). The results were presented as the mean green l.a.i. within the sample area.

### (v) *Estimating green l.a.i.*

For each flight the aerial photographic red and near-infrared reflectance data were transformed to p.v.i. values. The p.v.i. values were then used to estimate the l.a.i. of each of these areas via the complex calibration curves. The mean of the estimated green l.a.i. range was plotted against the observed green l.a.i. for each site. One of these graphs, that for the *Pteridium–Calluna* association, is reproduced in figure 8. In all cases this expected l.a.i. was linearly related to, and slightly overestimated, the observed l.a.i.

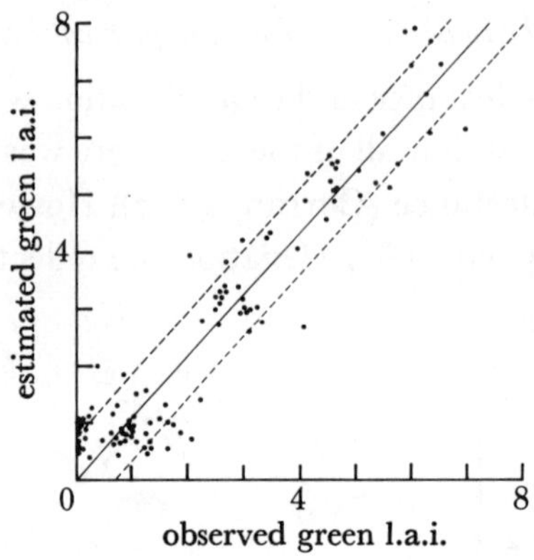


FIGURE 8. A comparison between the observed green l.a.i. and the predicted green l.a.i. for a *Pteridium–Calluna* association. The solid line is described by the equation $y = 1.07x$. The s.e. of the estimate is 0.64; $n = 100$; $r = 0.91$; significance level, 1%.

### (vi) *Discussion*

The accuracy of the estimated green l.a.i. is dependent upon the acceptable error of the estimate, the type of vegetation and the season. If only very low errors are acceptable for each estimate, then accuracies are near to 0%. If very high errors are acceptable for each estimate, then accuracies are near to 100%. Between these two extremes the actual error will be directly related to the initial standard error of the forecast and the accuracy with which the observed l.a.i. can be measured. For example, if the green l.a.i. was observed without error on the *Pteridium–Calluna* site, the accuracy with which this green l.a.i. could have been estimated would have been 68% at an error of 0.72 green l.a.i., as this is the standard error of the forecast given on figure 7.

In practice a user will wish to estimate a wide range of l.a.i. values at a predetermined range of acceptable error. The relation between error range and accuracy is therefore illustrated by means of three possible error ranges, which are comparable between the four vegetation associations (see table 1).

(1) Low error of estimate: this was chosen to represent one eighth of the total green l.a.i. range for each vegetation association and is $\pm 0.05$ green l.a.i. for the young and mature *Calluna* and $\pm 0.5$ green l.a.i. for the *Pteridium* and *Pteridium–Calluna*. (2) Medium error of estimate: this was chosen to represent one quarter of the total green l.a.i. range for each vegetation association and is $\pm 0.1$ green l.a.i. for the young and mature *Calluna* and $\pm 1.0$ green l.a.i. for the *Pteridium* and *Pteridium–Calluna*. (3) High error of estimate: this was chosen to represent one half of the total green l.a.i. range for each vegetation association and is $\pm 0.2$ green l.a.i. for the young and mature *Calluna* and $\pm 2.0$ green l.a.i. for the *Pteridium* and *Pteridium–Calluna*.

An estimate of green l.a.i. can be made with a low error of the estimate to an overall accuracy of 42 %; at a medium error of the estimate the overall accuracy increases to 74% and at a high error of the estimate the overall accuracy increases to 98 % (see table 1).

TABLE 1. THE RELATION BETWEEN THE ERROR OF A GREEN L.A.I. ESTIMATE
AND THE ACCURACY OF THAT ESTIMATE

(For example the green l.a.i. of *Pteridium* can be estimated with an error of ± 1.0, 73 times out of every 100.)

| vegetation association | error of green l.a.i. estimate | accuracy (%) of green l.a.i. estimate at given error level |
|---|---|---|
| young *Calluna* | ± 0.05 (low) | 34 |
| | ± 0.1 (medium) | 62 |
| | ± 0.2 (high) | 92 |
| mature *Calluna* | ± 0.05 (low) | 41 |
| | ± 0.1 (medium) | 73 |
| | ± 0.2 (high) | 100 |
| *Pteridium* | ± 0.5 (low) | 39 |
| | ± 1.0 (medium) | 73 |
| | ± 2.0 (high) | 100 |
| *Pteridium–Calluna* | ± 0.5 (low) | 53 |
| | ± 1.0 (medium) | 89 |
| | ± 2.0 (high) | 98 |

The accuracy of estimation varied with season. The accuracy of the green l.a.i. estimate was highest in the winter and spring, from December to early May, when it is an average of 84%. In the summer and autumn, from late May to November, the accuracy of the green l.a.i. estimate dropped to an average of 52%. The two reasons for this seasonal variation are first the flowering of *Calluna*, which obscures green leaves, and second seasonal canopy changes resulting from the growth and senescence of *Pteridium*

(vii) *Conclusions*

(1) Multispectral aerial photography used in conjunction with complex calibration curves was successfully employed in the estimation of green l.a.i. for four different vegetation canopies.

(2) At an error of estimate of ± 0.1 green l.a.i. for the young and mature *Calluna* and ± 1.0 green l.a.i. for *Pteridium* and *Pteridium–Calluna*, green l.a.i. could be estimated with an overall accuracy of 74%. At higher levels of error the overall accuracy increases.

(3) At times of canopy stability in the winter and early spring and at an error of estimate of ± 0.1 green l.a.i. for the young and mature *Calluna* and ± 1.0 green l.a.i. for the *Pteridium* and *Pteridium–Calluna* the overall accuracy increased to 84%.

(*b*) *Phase two: an experiment to determine the feasibility of estimating the green l.a.i. of a large area by using measurements of multispectral radiance from aircraft altitudes*

During 1981 and 1982 the seasonal change in multispectral reflectance and green l.a.i. were recorded and complex calibration curves constructed for Lathkill Dale, a limestone grassland in Derbyshire. As part of the Natural Environment Research Council's remote-sensing programme MSS-82, multispectral radiance data were collected by a multispectral scanner mounted on an aircraft at the same time that green l.a.i. was measured on the ground. This work took place in

[ 25 ]

September 1982 over an area 8 square miles (2072 ha) in extent centred on Lathkill Dale. Preliminary analysis of these data indicates that accuracies of green l.a.i. prediction may well be superior to those obtained during the pilot study.

### (c) *Phase three: the estimation of green l.a.i. of large areas by using measurements of multispectral radiance from satellite altitude*

During early work at the University of Sheffield, the multispectral radiance data collected by the Landsat multispectral scanner were used to generate simple calibration curves with green l.a.i. The green l.a.i. was then mapped to give estimates of ground cover and green biomass. Initially this work was undertaken for hill and upland areas in the U.K., for example the North York Moors and Peak District (Hastings 1982). More recently this approach has been used abroad and has included monitoring the area of date palm cultivation in Saudi Arabia (Curran & Adawi 1983).

Future work will concentrate on the use of p.v.i. and modelled calibration curves to produce maps of the green l.a.i. of grasslands from satellite measurements of multispectral radiance. The accuracy goal of such maps will be 80 % with an error of $\pm 1$ l.a.i.

The problem that besets this work is the relatively low spatial and radiometric resolution of the multispectral reflectance data collected by the civilian satellites to which we have access (Lillesand & Kiefer 1979). This situation will be improved in 1984 when multispectral reflectance data collected by the Thematic Mapper sensor on the satellite Landsat-4 and the High-Resolution Visible sensor on the satellite SPOT are available.

### 6. COMMENT

Remote estimation of the green l.a.i. of vegetation canopies has been made possible by an increased understanding of how radiation interacts with a vegetation canopy, coupled with improved techniques for measuring that radiation. Today, rapid improvements in the method of green l.a.i. estimation by remote sensing have lead to accuracies that, on a field by field or association by association basis, are acceptable for application in studies where it is difficult or impossible to obtain green l.a.i. measurements by any other method, as would be true of some of the examples discussed in this paper. For this reason the three largest areas of application have been and are likely to continue to be in agricultural intelligence, agricultural management and ecological research. Agricultural intelligence, for the monitoring of your own or a foreign nation's cereal yield, as the U.S.A. monitors their own and the U.S.S.R.'s; agricultural management, for the location of stressed crops in times of drought, or in areas of known disease and ecological research, as an input to primary production studies or as a means of inferring environmental conditions.

I thank the following for financially supporting the work reported in this paper: the Natural Environment Research Council, the Royal Society, the Nature Conservancy Council and the universities of Bristol, Reading and Sheffield. Thanks also to Mr Neil Wardley for access to data on the Lathkill Dale study and to Dr Jim Hansom and Dr Tom Dargie for commenting on an early draft of the manuscript.

# References

Ahlrichs, J. S., Bauer, M. E., Hixson, M. M., Daughtry, C. S. T. & Crecelius, D. W. 1979 In *Proceedings of the International Symposium on Remote Sensing for Observation and Inventory of Earth Resources and the Endangered Environment* (ed. G. Hilderbrandt & H. J. Boehnel), pp. 629–647. Freiburg: International Society for Photogrammetry.

Bartlett, D. S. & Klemas, V. 1980 *Environ. Mgmt* **4**, 337–345.

Brakke, T. W., Kanemasu, E. T., Steiner, J. L., Ulaby, F. T. & Wilson, E. 1981 *Remote Sensing Envir.* **11**, 207–220.

Chance, J. E. 1981 *Int. J. Remote Sensing* **2**, 1–14.

Chance, J. E. & LeMaster, E. W. 1977 *Appl. Optics* **16**, 407–412.

Colwell, J. E. 1974 In *Proceedings of the Ninth International Symposium on Remote Sensing of Environment*, pp. 1061–1085. University of Michigan.

Curran, P. J. 1980*a* *Int. J. Remote Sensing* **1**, 77–83.

Curran, P. J. 1980*b* In *Proceedings of the Fourteenth International Symposium on Remote Sensing of Environment*, pp. 623–637. University of Michigan.

Curran, P. J. 1980*c* *Prog. phys. Geog.* **4**, 315–341.

Curran, P. J. 1981*a* *Small format light aircraft photography*. Geographical Papers, University of Reading, U.K.

Curran, P. J. 1981*b* In *Plants and the daylight spectrum* (ed. H. Smith), pp. 65–99. Academic Press.

Curran, P. J. 1981*c* *Int. J. Remote Sensing* **2**, 369–378.

Curran, P. J. 1982*a* Report to the Natural Environment Research Council on Grant GR3/4076, Swindon, U.K.

Curran, P. J. 1982*b* *Photogramm. Engng Remote Sensing* **48**, 243–250.

Curran, P. J. 1983 In *Ecological mapping from ground, air and space* (ed. R. M. Fuller), pp. 84–100. Institute of Terrestrial Ecology Symposium No. 10, Cambridge.

Curran, P. J. & Milton, E. J. 1983 *Int. J. Remote Sensing* **4**, 246–256.

Curran, P. J. & Adawi, N. 1983 In *First Symposium on Date Palm*, Saudi Arabia: King Fiasal Press. (In the press.)

Curran, P. J., Munday, T. J. & Milton, E. J. 1981 *Int. J. Remote Sensing* **2**, 185–188.

Daughtry, C. S. T., Bauer, M. E., Crecelius, D. W. & Hixson, M. M. 1980 Special Report to the Joint Program for Agriculture and Resource Inventory Surveys Through Aerospace Remote Sensing (AgRISTARS) under code SR-PO-00458, Purdue University, Indiana.

Deering, D. W. & Hass, R. H. 1980 *N.A.S.A. tech. Memo.* no. 87027, Washington, D.C.

Deering, D. W., Rouse, J. W., Hass, R. H. & Schell, J. A. 1975 In *Proceedings of the Tenth International Symposium on Remote Sensing of Environment*, pp. 1169–1178. University of Michigan.

Hastings, S. 1982 *Sheffield Morning Telegraph*, 20 May, pp. 6–7.

Heilman, J. L., Kanemasu, E. T., Bagley, J. O. & Rasmussen, V. P. 1977 *Remote Sensing Envir.* **6**, 315–326.

Hielkema, J. U. 1980 Final Report to the Food and Agriculture Organisation on code CGP/INT/349/USA, Rome.

Holben, B. N., Tucker, C. J. & Fan, C. J. 1980 *Photogramm. Engng Remote Sensing* **46**, 651–656.

Kimes, D. S., Markham, B. L., Tucker, C. J. & McMurtrey, J. E. 1981 *Remote Sensing Envir.* **11**, 401–411.

Kondratyev, K. Ya. & Fedchenko, P. P. 1982 *Spectral reflectivity and recognition of vegetation.* [In Russian.] Leningrad: Gidrometeoizdat.

Lillesand, T. M. & Kiefer, R. W. 1979 *Remote sensing and image interpretation.* New York: Wiley.

Maxwell, E. L. 1976 *J. Range Mgmt* **29**, 66–73.

Pearson, R. L., Tucker, C. J. & Miller, L. D. 1976 *Photogramm. Engng Remote Sensing* **42**, 317–323.

Pollock, R. B. & Kanemasu, E. T. 1979 *Remote Sensing Envir.* **8**, 307–312.

Richardson, A. J. & Wiegand, C. L. 1977 *Photogramm. Engng Remote Sensing* **43**, 1541–1552.

Slater, P. N. 1980 *Remote sensing: optics and optical systems.* Addison-Wesley Publishing Program.

Slater, P. N. & Jackson, R. D. 1982 In *Spectral signatures of objects in remote sensing*, pp. 531–542. Paris: Centre National d'Etudes Spatiales.

Smith, J. A. & Oliver, R. E. 1972 In *Proceedings of the Eighth International Symposium on Remote Sensing of Environment*, pp. 1333–1353. University of Michigan.

Suits, G. H. 1972 *Remote Sensing Envir.* **2**, 117–125.

Wiegand, C. L., Richardson, A. J. & Kanemasu, E. T. 1979 *Agron. J.* **71**, 336–342.

## Discussion

M. D. Steven (*University of Nottingham School of Agriculture, Sutton Bonington, U.K.*). The interaction of light with a crop canopy depends both on leaf area and leaf angle. Relations established between spectral reflectance and leaf area index are species-specific, depending on the leaf angle distribution, and although this may be no great problem in the study of monocultures as in most agricultural crops, it will cause considerable difficulties in the interpretation of reflectance data over mixed canopies or natural vegetation. It may be more appropriate to measure vegetation by an index such as fractional cover that combines both leaf area and angle distributions in an appropriate way.

P. J. CURRAN. There are several species-specific factors that determine the form of the relation between green l.a.i. and the spectral reflectance of a particular species. These factors include the many facets of a plant's chemistry and physiology: for example, stem pigmentation, senescent rates, plant height, leaf thickness and as Dr Steven correctly points out, leaf angle. For the estimation of green l.a.i. there is no need to measure all of these species-specific factors if the spatial sampling used in the determination of green l.a.i. is adequate. The measurements of green l.a.i. must represent the species diversity within the area of canopy that is sensed by the instantaneous field of view (i.f.o.v.) of the sensor. Although this means that one must collect more ground data in complex semi-natural sites, this is not a 'considerable difficulty': it is the cornerstone of spatial sampling. For example, during the Natural Environment Research Council's multispectral scanner flights (reported in my paper), it was found that a canopy sample of 0.25 % of the i.f.o.v. area was adequate to express the green l.a.i. of a homogeneous agricultural pasture. By contrast, a canopy sample of over 2 % of the i.f.o.v. area was required to express the green l.a.i. of some of the heterogeneous semi-natural grasslands.

The use of fractional cover does not dispense with the need for a spatial sampling framework that is linked to canopy diversity. In laboratory studies I have found that fractional cover, when used in conjunction with green l.a.i., can provide a very useful indication of short-term changes in leaf angle, for example during times of drought. However, as a general index of vegetation amount, fractional cover is a most inappropriate measure on all but the sparsest canopies. This is because once the ground is covered by vegetation the index of fractional cover, unlike green l.a.i., is insensitive to further changes in vegetation amount. This is a serious limitation of this index as owing to the near infrared translucence of leaves these further changes in vegetation amount can be sensed remotely, as is reported in my paper.

*Phil. Trans. R. Soc. Lond.* A **309**, 271–281 (1983)
*Printed in Great Britain*

271

# The current use of remote-sensing data in peat, soil, land-cover and crop inventories in Scotland

By G. C. Stove

*Remote Sensing Unit, Department of Peat and Forest Soils,*
*The Macaulay Institute for Soil Research, Craigiebuckler, Aberdeen AB9 2QJ, U.K.*

The remote-sensing methodology developed, at the Macaulay Institute, for natural resource surveys is introduced and some recent mapping and environmental monitoring projects are reviewed. These include peat resource and peatland vegetation mapping in Lewis and North Harris, crop monitoring in Kincardineshire and land-cover mapping in the Buchan Area of Grampian Region, NE Scotland. The current use of remote-sensing data by the Peat Survey Section and the Soil Survey Department is reviewed for peat, soil and vegetation mapping. The principal current projects, which include the D.A.F.S. Bracken Survey of Scotland, the AGRISPINE Experiment for the U.K. National Remote Sensing Centre at R.A.E. Farnborough and the SAR 580 Experiment for the European Space Agency, are discussed.

## 1. Introduction

Successful remote-sensing projects completed in recent years at the Macaulay Institute have invariably resulted through the development of an appropriate remote-sensing methodology for natural resource surveys or environmental monitoring operations. Three phases of development resulted in the formulation of a general remote-sensing methodology for application projects:

(1) equipping and certifying an aircraft with a suitable remote-sensing rig for various survey operations;

(2) developing an automated photogrammetric digitizing and plotting facility for handling 'vector' data in mapping;

(3) developing a digital image-processing system linked to (2) for handling 'raster' data, such as Landsat MSS records.

The net result of the above developments at the Macaulay Institute was to create a novel 'hybrid' automated photogrammetric and image processing system, MAPIPS, for processing ground-acquired, aerial-acquired and satellite-acquired data. This system was specially structured for handling data in vector and raster form and to display information in a variety of ways for interpretation and plotting requirements. The general remote-sensing methodology can be summarized as follows:

(A1) acquisition of imagery, e.g. Landsat data, aerial photography (multiband and multi-level products);

(A2) pre-processing optical and digital data, e.g. enhancement of information, restoration and scaling;

(B1) interpretation of film products and correlation with ground truth data;

(B2) photogrammetric plotting and digitizing thematic information;

(C1) analysis of digital image data – enhancements, measurements, correlations, corrections and classifications;

[ 29 ]

(C2) final feature classifications and storing results;

(D1) data presentation in map or image form, including the compilation and production of reconnaissance maps and detailed maps for reports or publication;

(D2) data presentation in statistical form – final statistics related to feature classifications and areas including significance testing.

## 2. Peat survey mapping in Lewis and North Harris

The production of a medium-scale map illustrating the peat and terrain categories of Lewis and North Harris was the first major application of this general remote-sensing methodology in Scotland. A multilevel sampling strategy was adopted for this survey, which related small-scale, medium-scale and large-scale aerial photography to multispectral Landsat imagery and ground survey data. The accuracy of the peat categories derived from a supervised classification of Landsat data was assessed by comparing the classified areas against the actual ground areas of the features that were photogrammetrically mapped within selected test blocks. Considering all the categories classified over North Lewis, which included intact peatland, eroded peatland, shallow peat and rock complexes, agricultural land and open water, a final average rank area correlation coefficient, $r_s = 0.905$, was computed (Stove & Hulme 1980). This, together with the fact that $r_s$ ranged only from 0.78 to 0.945, suggests that there is a high degree of correlation and consistency in the order of the areal categories from maximum to minimum rank.

The Lewis study demonstrated that a relatively simple image-processing technique (contrast stretching and density slicing the reflectance range of the infrared band, band 7 of Landsat) produced 80 % of the information required to map the peat and terrain categories at the required scale. This map, published in 1981, is primarily a peat resource map that depicts general land-cover types. It also illustrates the present use of peatlands in terms of domestic peat cuttings and land improved for agriculture and forestry. The map is the first of a series of land-cover, peat and other resource maps of Scotland based on Landsat satellite image processing (Stove & Hulme 1980; Macaulay Institute 1981).

## 3. Land-cover mapping in Grampian Region

The aim of this particular work was to test whether simple image-processing techniques and classification methods based on Landsat data would apply to more complex terrain in the Grampian Region, for the mapping of land-cover types in general. A pilot study was conducted in the Laurencekirk–Cairn o'Mount area in 1977 and, following the successful completion of this project, a collaborative study was undertaken for Grampian Regional Council's Department of Physical Planning in 1979 to produce a land-cover map of the Buchan area in NE Scotland, for physical planning purposes.

### (a) The Laurencekirk pilot study

This study was based on the analysis of a single Landsat scene, imaged in early August 1977. The results demonstrated that a single Landsat image was inadequate for identifying all land-cover and crop categories required. It was noted that barley and root crops (such as potatoes and turnips) have unique spectral signatures on certain Landsat bands and that band ratios vary during the growing season, in particular from June to August. By using Landsat data, crops were enhanced successfully during the harvest season in August and September.

### (b) The Buchan Study

The objectives of this study were as follows:

(1) to establish the feasibility of identifying and mapping the following 'land-cover categories', suggested by the planners: (a) farmland, (b) roughland–moorland, (c) hill grazings, (d) wetlands–peat, (e) forestry, (f) vegetated dunes, and (g) bare sand (beaches and exposed dune surfaces);

(2) to establish the optimum time during the growing season for differentiating these categories;

(3) to establish the best enhancement and classification techniques for identifying these categories; and

(4) to establish the most cost effective method of mapping the categories and determining their areal extent.

The study area covered 1500 km² extending from Troup Head on the north coast and following the coastline east and south to the Ythan Estuary. The experimental programme was carried out between September 1979 and October 1980; the Landsat scenes processed were acquired in July and September 1979, respectively. A wide variety of statistical classification techniques were applied to the Buchan imagery, with the use of automated and interactive procedures. Successive crop classifications based on Landsat imagery were tested and compared with known crops at sample locations. The number of individual picture elements (pixels) successfully classified gave an index of accuracy. Results showed that cereal crops could not be clearly outlined on the July imagery. On the September scene, however, a simple density slice of data on the visible red channel (band 5) provided an accuracy greater than 90 % (Stove *et al.* 1980). Analysis of principal components was found to be an efficient interactive enhancement technique for land-cover mapping, particularly where several categories could be highlighted simultaneously.

As a result of this project, a land-cover map of the Buchan area was produced at the same scale as the published soil and land-use capability maps of the area (Stove *et al.* 1981). In addition, by using automated classification techniques, a computer-drawn map illustrating the cereal distribution in September 1979 was produced. In summary, it was demonstrated that all the land-cover categories requested by the planners could be differentiated and mapped entirely from Landsat data. A cost/benefit analysis was made of this work, which showed that the remote-sensing methods employed enabled the work to be done at less than one third the cost of a traditional land use survey by air-photo interpretation and field survey. Furthermore, if repeat surveys were required in subsequent years, this would mean a greater cost saving: with the same techniques, a 1980 map could be produced at one sixth of the cost of the 1979 map.

### (c) Multitemporal studies in the Grampian Region

A joint research programme of work in the Grampian Region has recently been undertaken between the Remote Sensing Units at the Macaulay Institute and R.A.E. Farnborough. The object of this work is to develop an automated land-cover classification procedure for a much larger test block of terrain than Laurencekirk or Buchan. Originally five scenes were selected with a good seasonal range from February to November, covering a 2 year period, but now more scenes are being studied because a number of good images of the Grampian Region were obtained during the AGRISPINE study period in 1982. Change detection is of particular interest

[ 31 ]

in this multitemporal study; in this case scenes are first rectified geometrically to fit the Ordnance Survey grid on regular 50 m × 50 m pixels and then different dates of imagery can be overlaid by the computer. Current investigations include monitoring urban expansion and loss of agricultural land around the city of Aberdeen (Stove 1981).

## 4. Current use of remote-sensing methods in peat survey

The remote-sensing research programme at Macaulay Institute was initiated in 1975 to assist in the routine mapping and assessment of Scottish peatlands, particularly in remote regions where large-area ground surveys proved costly. The primary aim was to establish a data base of peat and terrain information of value to potential peatland development (Stove & Robertson 1979, 1980; Robertson & Stove 1980).

The initial acquisition of remote-sensing imagery related to this programme included aerial photography (panchromatic, true colour and infrared false colour), multispectral Landsat data (optical and digital products) and airborne thermal line-scan products. Adverse weather conditions in Scotland frequently restrict the use and increase the cost of conventional large-format (230 mm × 230 mm frames) air-photo surveys, but with the development of a suitable camera rig for use in light aircraft (Stove 1981), the small-format photograph (with a 55 mm × 55 mm frame) proved ideal for environmental monitoring and peat survey work.

Although 35 mm Nikon cameras are often used in tandem to acquire simultaneous true-colour and infrared false-colour photography of small areas, the standard 70 mm camera employed for block vertical stereo cover of a small area is the Hasselblad 500 EL/M motor-driven camera, with an automatic diaphragm-controlled lens (Zeiss 80 mm planar) system, ideal for variable light conditions. A special interval meter, designed and constructed at the Institute, is used to set the timing of the exposures according to the required forward overlap along the flightpath. Rating tables for a range of flying heights and ground speeds have been computed, the most common scale required for much of the monitoring tasks being 1 : 10 000 scale, as illustrated in table 1. In this example, the flying height is computed to be 800 m above mean ground level, which must be assessed, especially over mountainous terrain. If the ground speed of the aircraft is computed to be 80 knots (148 km h$^{-1}$) and the required forward overlap of frames is 70 %, then the interval meter is set to 4 s.

The distortion characteristics of the 80 mm Zeiss Planar lens have been computed by using a specially constructed goniometer at University College London's Department of Photogrammetry and Surveying, and a 'best-fit' calibrated focal length of 80.23 mm was obtained. This value, known in photogrammetry as the 'principal distance', is used to set up an accurate scaled orientation of stereo pairs in a photogrammetric plotting machine such as the Wild B8S, for precise topographic heighting and contour plotting. Since this value is now known for the camera and lens system used for survey work, precise topographic rectification and plotting of height information is possible. Consequently, this photogrammetric facility is ideal if detailed topographic maps are required over very boggy peat terrain, saturated with small pools (dubh lochans), which negate the possibility of setting up a theodolite and distomat (infrared distance-ranging equipment) on a firm base for accurate tacheometric ground survey. With this air-survey equipment, specialized vertical stereo and oblique air-photos have been acquired for a variety of peat survey tasks. This year, for example, complete vertical stereo cover was obtained over a peatland drainage scheme in Caithness. For this particular application, infrared false-colour photography, taken with a Wratten 12 filter, proved the most appropriate image product for enhancing the required ground detail.

TABLE 1. RATING TABLE FOR THE 80 mm LENS TO GIVE A 1:10000 PHOTO SCALE

(Focal length, 80 mm; image format, 55 mm × 55 mm; scale required, 1:10000; Flying height, 800 m. Note that flying height is set above mean ground level.)

| ground speed | | forward overlap (%) | | | |
| --- | --- | --- | --- | --- | --- |
| | | 60 | 66.6 | 70 | 80 |
| knot | km h$^{-1}$ | time between exposures/s | | | |
| 60 | 111 | 7.2 | 6.0 | 5.4 | 3.6 |
| 65 | 120 | 6.6 | 5.5 | 4.9 | 3.3 |
| 70 | 130 | 6.1 | 5.1 | 4.6 | 3.0 |
| 75 | 139 | 5.7 | 4.8 | 4.3 | 2.8 |
| 80 | 148 | 5.4 | 4.5 | 4.0 | 2.7 |
| 85 | 158 | 5.0 | 4.2 | 3.8 | 2.5 |
| 90 | 167 | 4.8 | 4.0 | 3.6 | 2.4 |
| 95 | 176 | 4.5 | 3.7 | 3.4 | 2.2 |
| 100 | 185 | 4.3 | 3.6 | 3.2 | 2.1 |
| 105 | 195 | 4.1 | 3.4 | 3.0 | 2.0 |
| 110 | 204 | 3.9 | 3.2 | 2.9 | 1.9 |
| 115 | 213 | 3.7 | 3.1 | 2.8 | 1.8 |
| 120 | 222 | 3.6 | 3.0 | 2.7 | 1.8 |

Peatland surveys over large areas such as the Outer Hebrides or northern islands of Scotland are best planned and carried out initially by using large-format, small-scale aerial photography, if available. If multilevel and multitemporal photography is available (e.g. the Lewis survey) then a multistage and multiphase sampling strategy may be selected. Multistage sampling presumes an interest in a single parameter (e.g. peatland), and an estimate of its total distribution. It is most useful at the primary survey stage when problems are encountered in selecting an initial sampling frame. Multistage sampling is a multilevel method progressively apportioning the universe into smaller units in a nested hierarchy, and is most effectively employed when Landsat data is combined with air-survey data and ground information (Stove & Hulme 1980). Multiphase sampling incorporates the use of auxiliary variables in combination with the parameter of interest, to make estimates of population parameters. For example, in the remote-sensing methodology for peatland survey using Landsat data with air-photography, the reflectance values for eroded peatland were classified in a supervised manner through photogrammetrically plotting the width and spacing of erosion channels. In a similar way, P. D. Hulme has shown that detailed statistical analysis of peatland vegetation samples indicates that the categories identified on medium-scale aerial photographs are reasonably distinctive ground units, reflecting different peatland classes. If such ground units can then be interpreted from aerial photographs in the field, this information may again be used to attempt a controlled or supervised classification of Landsat reflectances to separate, for example, eroded peatland from intact peatland, shallow peat and rock complexes and domestic cuttings (Stove & Hulme 1980).

By using the B8S stereoplotter, tonal and textural features on the magnified stereo model created in this instrument can be greatly enhanced and used to delineate topography, hydrology, erosion, vegetation and other factors that must be considered in peatland classification and evaluation (Robertson & Stove 1980). The automated photogrammetric system developed to improve the speed and accuracy of plotting operations (Stove & Ritchie 1982) assisted in the

[ 33 ]

22-2

creation of a data base for peat survey information and paved the way for an interaction of this data in 'vector' form with Landsat multispectral data in 'raster' form (McKay & Stove 1980).

The most significant development in the remote-sensing methodology for peat survey applications came through linking the automated photogrammetric facility with the image-processing system to create a unique hybrid system, MAPIPS (Stove & Ritchie 1982). MAPIPS hardware now consists of two computer systems (the Data General Eclipse and a microcomputer), a Wild B8S stereoplotter fully digitized with a tri-axis locator system, a Ferranti–Cetec System 4 digitizing unit, a Tektronix 4027 colour graphics display and a Wild Aviotab TA flatbed plotting table (Ritchie & Stove 1982). This system can be used not only to display Landsat data and digitized photogrammetric data, but ground survey data can be digitized on the System 4 and, indeed, tacheometric survey data logged from the Wild Distomat system can be manually entered into one of the computers and stored on files. The peat drainage survey area in Caithness was surveyed on the ground with the use of the Wild T1 theodolite with D14 distomat attached; the results were subsequently stored in the peat databank and plotted at the required scale on the Aviotab TA.

It was evident from the close-grid nature of the old peat survey data in map and tabular form that this information would lend itself to geocoding, digitizing and storage in a peat data base. Consequently, a variety of peat bog types throughout Scotland were digitized, the object being to display spatial patterns and relations between topographic and stratigraphic parameters and to correlate the statistical information displayed with Landsat reflectance data.

Ulbster Bog, located on the east coast of Caithness six miles south of Wick, is a typical example of a basin bog that was digitized for this purpose. From the digital topographic data and peat depths, it was possible to compute and display surface and bottom contours, isopachytes, trend surfaces and three-dimensional representations of the surface and basal topography. From the stratigraphic sample records, after peat analysis, surface and bottom moisture contents, ash contents and surface vegetation categories were displayed and cross-correlated. This information was then correlated with Landsat terrain reflectances on all four spectral bands. From the results of this analysis, it was found that the infrared band 6 terrain reflectance categories correlated most closely with the bog vegetation categories, while the infrared band 7 reflectance categories approximated most closely to surface peat moisture variations.

This type of analysis demonstrated the potential of the MAPIPS facility as an interpretation tool to the peat surveyor. It also became clear that access to structured geographical information systems is becoming increasingly important for the Earth scientist and remote-sensing specialist. Furthermore, it was clear that the ability to interpret and fully analyse displays of spatial information (from air, ground and space) statistically should be a prime concern of the remote-sensing data user.

## 5. USE OF REMOTE-SENSING METHODS IN SOIL SURVEY

The main task of the Soil Survey of Scotland is to map and classify systematically the mineral soils of Scotland and to produce land-use capability maps. Through field observations of the soil profile, soil types are identified based on parent materials, pedological drainage and other characteristics. The basic mapping unit is the soil series, defined as a group of soils with similar morphological properties developed on similar geological parent material. Soil series are identifiable with a genetic soil group and drainage class and are grouped into soil associations, which are defined as groups of soil series forming a soil pattern related to parent material and relief. To provide a background to the soil classification, the surveyor has to acquire, from a variety of sources, additional information on the physical environment of the soils. Since the 1950s the main application of remote sensing for soil survey has been through the widespread use of aerial photography for interpretation of terrain and soil characteristics. This additional source of information was initially used in a limited way, but increased availability has led to the adoption of air-photographs as a standard tool for soil mapping. With Landsat satellite imagery now increasingly available on a regular basis for the whole country, current and future image products, in digital form, are being assessed as another additional source of environmental information to the soil surveyor.

Systematic soil survey began in 1947 and until the 1950s work was concentrated in areas of agricultural importance (such as northeast, east, southeast and southwest Scotland). Soil boundaries were drawn in the field on 1 : 25 000 War Edition maps and later edited for publication at 1 : 63 360 scale. In the mid-1950s an attempt was made to evaluate the benefits of aerial photography to soil survey, drawing on experience gained from workers in Holland in particular. Air-photo interpretation methods were subsequently employed in SE Scotland within uncultivated areas. Such techniques were progressively adopted by other soil surveyors who had made extensive use of aerial photography during secondment to Hunting Overseas Surveys. At this time, air-photos were found to be invaluable for the survey of remote hill country, and owing to the inadequacy of the 1 : 25 000 War Edition maps, these photographs were used as base maps as well as interpretative aids. The vertical air-photos were usually borrowed or purchased from the Scottish Development Department's Air Photographs Library. At present, air-photos acquired from Ordnance Survey are used for routine mapping at 1 : 25 000 or 1 : 50 000 scale and for *ad hoc* surveys at 1 : 10 000 scale. Their use to evaluate land with complex terrain and soil patterns in the Western Highlands of Scotland has been assessed (Lawrance *et al.* 1977). All Soil Survey of Scotland work so far has been based on standard panchromatic photography.

Recently the use of Landsat imagery in digital form (in raster-based picture elements or pixels) has been demonstrated to surveyors from all the regional survey teams in Scotland. In addition to MAPIPS, the use of a new high-speed, British-made, interactive image-processing system called GEMS, has been demonstrated for enhancing, interpreting and classifying any remote-sensing data in raster form. The value of the GEMS facility hinges on the GEMSTONE software developed by the U.K. National Remote Sensing Centre at R.A.E. Farnborough for remote-sensing applications. This system, although specially developed for handling Landsat data, can process photography or video data, if raster-digitized, and because of its large number of image stores and overlay planes, is most suitable for superimposing complementary information from ground surveys (as grid data sets), aerial and space imagery. All this information can then be pooled on the GEMS storage planes for correlation and controlled feature classifications. Most of the soil surveyors who examined the potential of Landsat as an additional data source for providing information on the soil environment also had the opportunity to look at the value of Landsat data rectified to the National Grid and resampled at 50 m × 50 m pixels. In this form, ease of location is assured and the zoom and pan facility on GEMS enables rapid training-site identification and assessment.

After a viewing of Landsat scenes acquired at different times of the year on GEMS over the length and breadth of Scotland, some of the conclusions and observations made by the soil surveyors are worth considering. The first general point made was that Landsat imagery at present allows a unique overview of large parts of the country, which can provide added environmental information on surface ground state and crops during the growing season. As a *direct* aid in soil survey, however, Landsat data has limited use because soil survey and classification is concerned with the entire profile, not just the surface horizon. Another observation made was that despite the relatively coarse spatial resolution of Landsat MSS (multispectral scanner) data at present, the superior MSS spectral resolution over broadband panchromatic air-photos means that it can form a valuable complementary source of information, highlighting terrain features that may sometimes not be readily discernible on aerial photographs.

It was also generally stated that Landsat data could provide some positive applications to soil survey as a vegetation indicator to give predictions of soil type. In addition, certain vegetation types could be clearly identified on Landsat data by using GEMS, and the potential of the system as an aid to the subdivision of class 6 land (Land Capability Classification for Agriculture) was suggested. From the results of the Lewis Survey and looking at other scenes of the Western and Central Highlands, it was also accepted that Landsat data could be used successfully to identify peat, shallow peat and rock complexes and rock-dominated land forms. Inter-departmental collaboration is currently being pursued to evaluate the benefit of systems like MAPIPS and GEMS for the production of single-factor maps (Gauld *et al.* 1983). It was concluded that much more emphasis would have to be placed on ground-survey procedures to evaluate fully the applications of Landsat data for soil survey.

## 6. The D.A.F.S. bracken survey

A pilot study on the use of remote sensing to map bracken in Scotland has just been completed for the Department of Agriculture and Fisheries for Scotland. The major objective of this work was to develop a remote-sensing methodology for mapping bracken and to evaluate the cost-effectiveness of the approach in the light of alternative mapping strategies. The test area selected was in central Perthshire, covering an area of 50 km (east–west) by 44 km (north–south). In the SW quarter of this block, centred on the Menteith Hills (25 km × 22 km), the bracken area classified was about 8 % of the total land area, covering some 10 % of the upland grass–moorland area. In the NW quarter of the main block, the bracken area was 10 % of the total land area, accounting for some 13 % of the upland grass–moorland area. Results of this study have shown that bracken does have a unique spectral response at certain times during the growing season and that this signature can best be enhanced by using a ratio of the band 5/b and 7 reflectances. The methodology developed for the pilot study is now being applied to other areas in Scotland; the aim is eventually to complete a bracken inventory of the entire country by remote sensing.

## 7. The agrispine experiment

R.A.E. Farnborough is an active participant in the Space Informatics Network Experiment (SPINE), which is an international project designed to demonstrate the transfer of bulk digital data at high bit rates by using the Orbital Test Satellite (OTS). In 1982, the AGRISPINE project was conceived as a suitable application of the SPINE facility to allow participants to receive Landsat data in less than 24 h for time-dependent studies.

The principal investigating agencies selected for this project include the Forestry Commission the Ministry of Agriculture, the Macaulay Institute, the National College of Agricultural Engineering and R.A.E. Farnborough. The time-dependent applications currently being studied include monitoring crop growth at agricultural and forestry test sites in the Grampian Region and East Anglia, the calving of icebergs from a glacier in Greenland and sea ice characteristics in the Gulf of Finland. All sites are being monitored from 6 March to 27 November 1982, at the time of Landsat-3 overpass every 18 days. To date (September), every scene of the Jacobshavn Glacier and fjord area in West Greenland has proved satisfactory for detailed analysis. The British sites have been less successful, although the Laurencekirk test area in the Grampian Region has been clear on one third of the passes, sometimes only partly cloud-free, but only one good pass has been acquired of the East Anglian site at Thetford.

Test farms in the Grampian site have been monitored since March at the precise time of Landsat overpass. On each pass date, aerial photography and ground measurements, including ground photography of crop and soil conditions, are made. This year has been the first year in which crop stress conditions have been detected on Landsat imagery in the Grampian area, primarily because of the severe frost conditions in winter, which destroyed large areas of winter wheat and winter barley. The severe drought conditions in May and June also caused stress conditions that were detected on the imagery. However, it is unlikely that such stress conditions would have been detected without acquiring every Landsat scene through the SPINE link in less than 24 h and without the execution of air and ground surveys at the time of satellite overpass.

### 8. The SAR 580 (Synthetic Aperture Radar) experiment

The main objective of this experiment is to evaluate the potential of Synthetic Aperture Radar (SAR, an active microwave remote-sensing technique) as an all-weather capability for natural resource surveys and agricultural research applications in Scotland. The Macaulay Experiment (21GB), approved by the European Space Agency, comprised a 245 km transect, approximately 6 km wide, extending from a sea area northeast of Rattray Head (north of Aberdeen) southwards along the coastline to Edinburgh, extending as far inland as the Moorfoot Hills. Other investigators currently studying the SAR data from this site include the Forestry Commission, University College London and the Departments of Geography at Aberdeen, Glasgow and Edinburgh.

On 6 June, 1981 a Canadian Convair 580 aircraft acquired X and C band dual-polarization SAR imagery of the experimental block. Ground data collected before, during and after this overpass involved three separate programmes of work related to agricultural applications, aerial photography and hydrographic survey. A report on the ground data collection programme has been submitted to E.S.A. by the site coordinator and this has been published in the ground data collection programme volume. Optical processed film products in X and C bands have been received for the entire test block, but the digital data have been delayed. Six small digital test areas have been selected from the entire coastal strip; once these data have been received and correlated with the ground information, a final assessment of the SAR imagery will be made.

### 9. Conclusions

The remote-sensing methodology developed for natural-resource surveys has been reviewed through three phases of development. The major applications of this work discussed include peatland surveys in remote areas, land-cover surveys related to agricultural inventories, and environmental monitoring tasks such as bracken mapping. The current use of remote sensing within the Peat Survey is outlined, highlighting in particular the value of MAPIPS (the automated photogrammetric and image-processing system) for various applications. The future potential of remote sensing in the Soil Survey is assessed through the comments of several of the soil surveyors, after conducting Landsat experiments with the high-speed GEMS interactive image-processing system.

After a brief evaluation of the current projects being undertaken by the Remote Sensing Unit, it is interesting to speculate on the future promise of an 'all-weather' monitoring capability with SAR by using the planned European Space Agency's ERS-1 satellite in 1987. Even before then, or indeed the acquisition of higher-resolution multispectral imagery from the latest Landsat-D satellite (with a 30 m × 30 m pixel size) or the planned French SPOT satellite in 1984 (with a 20 m × 20 m multispectral and 10 m × 10 m panchromatic pixel size), a continuation of a timely facility such as the SPINE link to relay Landsat data to the user in less than 24 h is seen as the most important operational development for agricultural remote sensing in Scotland and indeed the U.K. in general. Getting such information on a timely basis and rapidly analysing the data with a GEMS system, for example, would create a very valuable ancillary source of information on land-management practices such as felling, burning, ploughing and harvesting. It is important to realize that the sensor technology is still developing faster than the analytical methods and interpretative skills can be developed to handle the image products (Birnie *et al.* 1982). Much work is still required on the evaluation of radar data compared with MSS data, before an operational 'all-weather' system can be realized and used, in the late 1980s or early 1990s.

I wish to acknowledge the financial support approved by A.R.C. and awarded by D.A.F.S. for the capital equipment required to develop MAPIPS. The financial assistance given by the Department of Industry and the U.K. National Remote Sensing Centre at R.A.E. Farnborough, in the recent demonstrations and evaluation of the GEMS image-processing system for applications in Scotland, is also gratefully acknowledged.

In particular, I wish to acknowledge the encouragement and support given at all times by Mr R. A. Robertson, Head of the Department of Peat and Forest Soils, and Dr T. S. West, Director of the Institute. This support enabled the progressive development of a major remote-sensing research programme and the establishment of a Remote Sensing Unit in 1980.

Apart from other staff in the Department of Peat and Forest Soils, who have provided invaluable support during field survey operations, and the other dedicated members of the Remote Sensing Unit, certain key individuals must be mentioned, who through their personal enthusiasm in remote sensing, have often given freely of their time outside working hours to enable aerial photographic sorties to be completed, particularly at times of Landsat overpass during weekend periods. Such enthusiasts include Dr W. H. Ekin, Robert Gordon's Institute of Technology, who finally developed and gained Civil Aviation Authority certification for the present remote-sensing rig on a Cessna 172 aircraft; Mr J. Mitchell, principal photographer, Macaulay Institute, and Mr J. B. G. Campbell, pilot and amateur photographer, Fordoun Flying Club.

I also wish to thank all the surveyors from the Soil Survey Department, who have commented on the use of GEMS for their particular remote-sensing applications; in particular, Dr J. H. Gauld must be acknowledged for all the additional work he has put into evaluating MAPIPS for the production of single-factor maps.

Finally, thanks are due to Mr R. Grant, Head of Soil Survey, and Dr P. D. Hulme for their comments on the initial draft of this paper.


## REFERENCES

Birnie, R. V., Robertson, R. A. & Stove, G. C. 1928 *Agric. Envir.* **7**, 121–134.

Gauld, J. H. *et al.* 1983 (In preparation.)

Lawrance, C. J., Webster, R., Beckett, P. H. T., Bibby, J. S. & Hudson, G. 1977 *Catena* **4**, 341–357.

Macaulay Institute 1921 *Peat and terrain categories of Lewis and North Harris: map and rectified Landsat false colour composite image at 1:100 000 scale.* Macaulay Institute for Soil Research, and R.A.E. Farnborough.

McKay, D. A. P. & Stove, G. C. 1980 *Comput. Applic. Univ. Nottingham* **7**, 936–965.

Ritchie, P. F. S. & Stove, G. C. 1982 In *Ecological mapping from ground, air and space (Proceedings of the Symposium at Monks Wood Experimental Station, October 1981).* (In the press.)

Robertson, R. A. & Stove, G. C. 1980 In *Proc. Sixth Int. Peat Cong., Duluth, U.S.A.*, pp. 84–87.

Stove, G. C. 1981 In *Remote sensing and strategic planning (Proc. Remote Sensing Society Seminar, Aberdeen, October 1981)* (compiled by R. L. Thomson), pp. 11–40.

Stove, G. C., Birnie, R. V., Cairns, J. C. & Ritchie, P. F. S. 1980 *Land use survey of Buchan based on satellite remote sensing.* Macaulay Institute Report for Grampian Region Department of Physical Planning.

Stove, G. C., Birnie, R. V., Thomson, R. L. & McKay, D. A. P. 1981 In *Geological and terrain analysis studies by remote sensing (Proc. Eighth Annual Conference of the Remote Sensing Society, Plymouth, December 1980)* (ed. J. A. Allan & M. Bradshaw), pp. 91–109.

Stove, G. C. & Hulme, P. D. 1980 *Int. J. Remote Sensing* **1**, 319–344.

Stove, G. C. & Ritchie, P. F. S. 1982 *Photogramm. Record*, **10**, 629–644.

Stove, G. C. & Robertson, R. A. 1979 *ARC Res. Rev.* **5**, 21–27.

Stove, G. C. & Robertson, R. A. 1980 *Telma* **10**, 67–81.

*Phil. Trans. R. Soc. Lond.* A **309**, 283–284 (1983)
*Printed in Great Britain*



# General discussion

MONICA M. COLE (*Bedford College, London, U.K.*). In contributing to a discussion of the use of multispectral satellite imagery in the exploration for petroleum and minerals covered by Mr Peters I wish to emphasize four points, some of which are relevant also to statements made by Dr Curran in his presentation.

The first point is that remotely sensed imagery is a tool and its interpretation a technique to be used as appropriate and integrated with other techniques in mineral exploration. Mr Peters has reviewed the potential of multispectral satellite imagery and emphasized its value in initial reconnaissance studies notably for the identification of geological structures and lithologies. I would emphasize also its value at more advanced stages of exploration when reinterpretation of imagery at large scales and with reference to ground truth data can yield valuable information.

My second point, which follows naturally from the first, is that effective interpretation of remotely sensed imagery requires an appreciation of the geographical environment as well as the geological environment. It is reflectances from the components of the geographical environment that produce the colours and tones seen on the colour composites generated from Landsat imagery. Except in arid areas largely devoid of plant cover, in natural terrain reflectances from vegetation dominate over those from soils and bedrock. Their contribution increases with increasing density of cover. The reflectances from different types of vegetation and from individual plant species, however, vary greatly, depending on the geometry of the canopy, the colour of foliage, the size, shape, angle, etc., of leaves, and the turgidity, water content and nutrient status of leaf cells. It is the differences in vegetation cover producing differing reflectances that permit the discrimination of lithologies and identification of structures on colour composites generated from Landsat imagery. In some areas, however, any or all of relict laterite, superficial cover, former and ephemeral drainage systems, and other physiographic features that are the legacies of geomorphological processes, complicate relations. These need to be understood for effective evaluation of imagery for geological purposes. In this context there is no substitute for field investigations, which are essential for the acquisition of ground truth data needed for effective evaluation of imagery.

The third point that I wish to emphasize is that the relevance of the interpretation of remotely sensed imagery for specific studies varies for different environments. Its value for terrain analysis, geological mapping and mineral exploration in semi-arid and arid lands is acknowledged. Less is known of its potential for studies of humid terrain. I believe, however, that even in mountainous country clothed with tropical and subtropical forests the interpretation of multispectral imagery could yield valuable geological and geographical information provided that sufficient ground truth data are obtained to permit an understanding of the environmental conditions.

Finally, my fourth point is the need to discriminate the significant from the unimportant both in studies of imagery and in the field.

[Professor Cole illustrated her points from four pairs of slides.]

The first pair of photographs shows a large-scale colour composite and an interpretation thereof of a subscene from Landsat imagery of the mountainous humid subtropical Tengchong area of western Yunnan, China, located in a mobile belt characterized by volcanic activity at the collision zone of the Indian and Eurasian plates. Differing reflectances from different types of vegetation cover and different tree species delineate the Mount Ma An volcano and discriminate lavas of differing age and degrees of weathering; and the very bright reflectances (actually occurring in all four MSS bands) that reveal zones of geothermal activity characterized by hot sulphur springs, along fault lines in the area south of the volcano, are noteworthy. The second pair of slides shows an area north of Tengchong. One slide, a colour composite, displays circular structures and complex fault systems that reveal areas where granite has intruded overlying limestone to produce zones of potential multi-metal mineralization. The second, a colour photograph of the terrain, shows a series of geobotanical anomalies over gossans that mark a mineralized belt from which semi-quantitative analyses of soil and rock chip samples indicate the presence of copper, lead, zinc, silver and tin. The imagery revealed the structures; field investigations were needed to identify the vegetation anomalies, gossans and mineralized zones. The third pair of slides shows an enhanced conventional Landsat colour composite and one subjected to rotational procedures of a small area within the Lake Eyre drainage basin of Australia, where there are complex patterns of ephemeral and former drainage channels. These can be identified on the imagery and must be considered in the interpretation of geochemical data for exploration purposes. A greatly enlarged colour rotated composite of part of the area demonstrates how fence lines identifiable on the imagery can assist location of other features both on the imagery and in the field. The final slide carries a warning. An enhanced colour composite of a Landsat subscene covering an area south of Lake Ngami in Botswana, it shows a series of circular features that were initially identified on conventional black and white air photos as prospective kimberlites. Field studies revealed that the vegetation over these features was indicative of acid rocks, not kimberlites. Nevertheless some were drilled. This disclosed that the circular features represented the tops of hills in the Cave Sandstone landscape that existed before the outpouring of the Stormberg lavas that cover the area today. Mr Peters commented on the identification from Landsat imagery of other circular features important in petroleum exploration. Many important structural features can be identified on remotely sensed imagery. The significant must be distinguished from the unimportant, and this requires field investigations.

*Phil. Trans. R. Soc. Lond.* A **309**, 285–294 (1983)
*Printed in Great Britain*



# Albedo observations of the Earth's surface for climate research

By A. Henderson-Sellers and M. F. Wilson

*Department of Geography, University of Liverpool, P.O. Box 147,*
*Liverpool L69 3BX, U.K.*

The primary input of energy to the Earth's climate system occurs at the surface and can be highly sensitive to the surface albedo. Albedo changes have been proposed as one cause of climatic variation, but results from climate models are not yet consistent. It is very difficult to establish an agreed global data set with which to initiate comparative climatic simulations. Albedo observations must be spectrally resolved because reflexion of solar radiation is a strong function of wavelength and incident and reflected beams are modified by the atmosphere. Parametrization of system albedos in energy-balance models draws on satellite data. The use of satellite observations is less easy in general circulation climate models. The removal of atmospheric distortion is particularly difficult. The establishment of a surface albedo data set generally follows one of two approaches: geographical categorization or remote monitoring. Surface albedo specification in current general circulation models is diverse. This paper reviews the ways in which remotely derived albedo measurements are used now and may, in the future, be improved for climate research.

## 1. Introduction

The majority (over 70%) of the solar energy input to the climate system is first absorbed at the surface (figure 1). Changes in the surface albedo, the ratio of reflected to incident radiation, provide a fundamental source of variability within the climate system. Climate model simulations are sensitive to the input values and parametrization of surface albedo (see, for example, Charney *et al.* 1977; Hummel & Reck 1979; Hansen *et al.* 1981). Globally the absorbed solar radiation, a function of the albedo, $\alpha$, is balanced by the emitted thermal infrared radiation plus the fluxes of sensible and latent heat, $\Phi_s$ and $\Phi_l$ (figure 1), so that

$$(1-\alpha)\,R_S^{\downarrow} = e(\sigma T_S^4 - R_L^{\downarrow}) + \Phi_s + \Phi_l, \tag{1}$$

where $e$ is the infrared emissivity and $R_L^{\downarrow}$ and $R_S^{\downarrow}$ downward longwave and shortwave radiation fluxes. The global balance disguises local and regional inhomogeneities.

The reflexion of incident solar radiation from many surfaces is a strong function of wavelength: the albedo of plants is *ca.* 0.10 in the wavelength region below 0.7 μm, rising to *ca.* 0.5 at near-infrared wavelengths, whereas snow is observed to have high albedos *ca.* 0.8 in the visible region, decreasing to *ca.* 0.3 at longer wavelengths (figure 2). Surface albedos are found to depend upon both the incident angle and spectrum of the received radiation and upon the state of the surface (Henderson-Sellers & Hughes 1982). Snow albedo is a function of age, depth, compaction and purity; ice albedos depend upon structure and surface puddling of the ice (Robock 1980; Warren & Wiscombe 1980); the albedo of foliage alters as it matures (Monteith 1973); ocean albedos depend on latitude and wind speed (Cogley 1979) and the albedo of bare soil surfaces is dependent upon the soil moisture (Kondratyev 1969; Ahmad & Lockwood 1979). The considerable range in observed surface albedos makes parametrization of surface albedo into climate models very difficult. Observational work could be planned better if there were an understanding of the requirements of the modelling community.

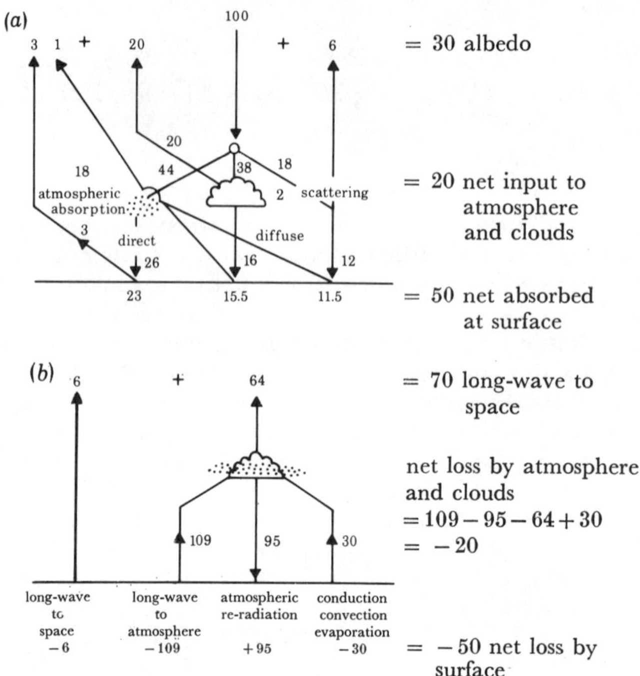


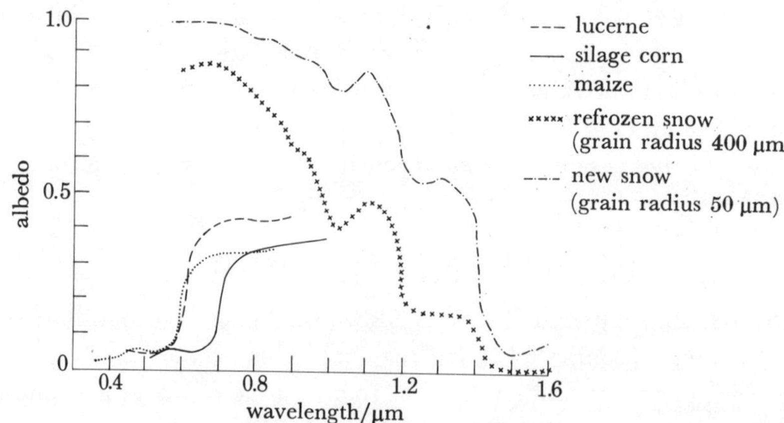


FIGURE 1. Schematic diagram of the globally and annually averaged components of the shortwave (a) and thermal infrared (b) radiation streams.



FIGURE 2. Spectral albedos of plant and soil surfaces (data from Kondratyev 1969) and snow albedos (redrawn from Wiscombe & Warren 1980).

## 2. ALBEDO FOR CLIMATE MODELS

The term albedo implies integration of reflectivity over the full solar spectrum (*ca.* 0.15–2.00 µm). As instruments are frequently restricted to narrow wavelength regions, estimation or calculation (see, for example, Warren & Wiscombe 1980) often replaces integration. The albedo measured by satellites (the system albedo) encompasses surface and atmospheric phenomena (Winston *et al.* 1979). The global albedo is dominated by cloud reflectivity (figure 1).

The method of albedo representation employed varies amongst climate models. The simplest climate models (EBMS) consider an energy balance in latitudinally averaged zones (see, for example, Budyko 1969; Sellers 1969; Cahalan & North 1979; North *et al.* 1981). The albedo parametrization is generally of the form

$$A = \begin{cases} A_{\text{ice}} - f(T(\theta)) & \text{for} \quad T > T_{\text{critical}} \\ A_{\text{ice}} & \text{for} \quad T \leqslant T_{\text{critical}}. \end{cases} \tag{2}$$

There is no physical basis for parametrization of the system albedo as a function of the surface temperature alone. All other model types include surface and atmospheric reflexion separately. Surface temperature calculations for the averaged globe in one-dimensional radiative convective (RC) models, where the dimension is height, are sensitive to the surface albedo (see, for example, Hummel & Reck 1979; Hansen *et al.* 1981). Statistical dynamical models, formulated in terms of latitude and height, parametrize the energy balance of land and ocean separately within each zone. They also respond to surface albedo changes (see, for example, Potter *et al.* 1975, 1981).

The full three-dimensionality of the climate is parametrized in general circulation models (GCMS). The effects of clouds and atmospheric gases and particulates are treated so that the surface incident radiation is well specified. The surface parametrization includes turbulent fluxes of sensible and latent heat and the radiation term dependent upon the surface albedo. Carson (1981) reviews the land surface albedos used in a wide selection of GCMS. He groups the treatment of snow-free surface albedos into three categories (see table 1): (i) a single fixed value for 'bare' land; (ii) land albedo specified as a function of latitude only; (iii) specified geographical distribution of albedo. The general trend in all GCMS is towards category (iii).

Representation of snow-covered and ice-covered surfaces is diverse, partly because observational data are sparse. The albedo of snow-covered surfaces depends on the type, density and roughness of vegetation and the depth of snow cover (Kukla & Robinson 1980). Snow albedo can increase with cloudiness as a result of near-infrared absorption (Grenfell & Maykut 1977). Carson (1981) identifies three types of snow-covered or ice-covered surfaces (table 1) used in GCMS: (i) surfaces with an instantaneously variable depth of snow either predicted or implied; (ii) permanent or seasonally prescribed snow-covered and ice-covered land surfaces; (iii) permanent or seasonally prescribed areas of sea ice. There is generally a simple dependence of albedo on snow depth (Holloway & Manabe 1971). Since one goal of most modelling groups is to simulate cryosphere–climate feedback, the diversity among the methods of assigning surface albedos and of predicting areas of ice and snow is disturbing. The effect of altering the 'frozen surface' albedo in EBMS (equation (2)) is dramatic (Warren & Schneider 1979; Cahalan & North 1979), whereas GCM simulations of even the present-day cryospheric extent are extremely poor (see §4).

### 3. GLOBAL SURFACE ALBEDO DATA

The urgent need for an agreed global surface albedo inventory for climate modelling has been identified (see, for example, GARP 1975). There have been a limited number of attempts to produce such a data set but uncertainty over surface albedos persists. The difficulties encountered in trying to construct an acceptable surface albedo data set are considerable. The choice of a 'typical' albedo for a land class from observational data is difficult. Spectral

TABLE 1. SURFACE ALBEDOS USED IN ATMOSPHERIC GENERAL CIRCULATION CLIMATE MODELS
(Updated from Carson (1981)).

| example centre (reference) | model | snow-free and ice-free surfaces | snow-covered and ice-covered surfaces |
|---|---|---|---|
| Atmospheric Environment Service (Canada) (Boer & McFarlane 1979) | AES | no specific details but implied geographical distribution based on Posey & Clapp (1964) | follows Holloway & Manabe (1971), equation (3): see GFDL |
| Australian Numerical Meteorology Research Centre (McAvaney *et al.* 1978) | ANMRC | latitudinal variation based on Posey & Clapp (1964) | snow albedo prescribed as latitudinal variation of $\alpha$. $\alpha$ of sea ice = 0.07 |
| Computing Centre, Siberian Academy of Sciences (Marchuk *et al.* 1979) | CCSAS | $\alpha = 0.2$, bare ground<br>$\alpha = 0.1$, ocean | snow: $\alpha = 0.2 + 0.4\, d_{sw}$<br>$\quad \alpha \leqslant 0.6$<br>(same as NCAR)<br>ice: $\alpha = 0.6$ |
| Geophysical Fluid Dynamics Laboratory (Holloway & Manabe 1971) | GFDL | geographical distribution based on Posey & Clapp (1964) | snow: equation (3)<br>$\alpha = \alpha_1 + (0.6 - \alpha_1)\, d_{sw}^{\frac{1}{2}} \quad d_{sw} < 1 \text{ cm}$<br>$\alpha = 0.6 \qquad\qquad\quad d_{sw} \geqslant 1 \text{ cm}$ $\Big\}$ (3)<br>poleward of 75° lat., albedo for land and pack-ice 0.75 |
| (Manabe & Stouffer 1980) | GFDL | geographical distribution based on Posey & Clapp (1964) | sea ice: 0.5 for lat. < 55°,<br>0.7 for lat. > 66.5°,<br>0.45 if top melting |
| Goddard Laboratory for Atmospheric Sciences (Halem *et al.* 1979) | GLAS | Feb.: geographical distribution based on Posey & Clapp (1964)<br>Aug.: Charney *et al.* (1977) vegetated land 0.14, desert 0.35, ocean 0.07 | snow and ice 0.70<br><br>Holloway & Manabe (1971) equation (3): see GFDL |
| Goddard Institute for Space Studies (Hansen *et al.* 1983) | GISS | land: 8 vegetation types have seasonally varying albedos for < 0.7 μm and ≥ 0.7 μm | snow-free ice: 0.45 ocean,<br>0.5 land<br>snow albedo: $\alpha_s = 0.5 + e^{-a/5}$<br>ground partly snow-covered albedo:<br>$\alpha_1 + (\alpha_s - \alpha_1)\,(1 - e^{-d_{sw}/d_{mask}})$<br>two spectral regions |
| Meteorological Office, U.K. (Corby *et al.* 1977) | UKMO | 5-level model: snow-free land values vary with latitude range 0.150–0.223 | snow:<br>$\alpha = \alpha_1 + 0.38\, d_{sw}^{\frac{1}{2}}$<br>$\alpha \leqslant 0.6$<br>similar to Holloway & Manabe (1971), equation (3): see GFDL<br>sea ice and permanent snow cover:<br>$\alpha = 0.8; \ T_0 < 271.2 \text{ K}$<br>$\alpha = 0.5; \ T_0 \geqslant 271.2 \text{ K}$ |
| (Saker 1975) | UKMO | 11-level model: land 0.2, sea (where effective) 0.06 | transient snow cover 0.5, permanent snow cover, land ice and sea ice 0.8 |
| National Center for Atmospheric Research (Washington & Williamson 1977) | NCAR | originally:<br>geographical distribution based on Posey & Clapp (1964) | originally:<br>snow or ice:<br>$\alpha = 0.2 + 0.4\, d_{sw}$ and<br>$\alpha \leqslant 0.6$ |
| (Dickinson *et al.* 1981, and unpublished) | | third-generation model: two spectral regions < 0.7 μm and ≥ 0.7 μm; dependence on vegetation type and extent | third-generation model: snow albedo function of age and depth of snow: two spectral regions |
| Oregon State University (Schlesinger & Gates 1979) | OSU | geographical distribution based on Posey & Clapp (1964) and model's 9 surface types | fixed value |
| Rand Corporation (Gates & Schlesinger 1977) | RAND | geographical distribution based on Posey & Clapp (1964) | — |
| University of California at Los Angeles (Arakawa 1972) | UCLA | bare soil 0.14, ocean 0.07 | snow-covered 0.7, ice-covered soil or sea water 0.4 |

Symbols: (i) $d_{sw}$ is the water equivalent depth of snow (centimetres); (ii) $a$ is the snow age (days); (iii) $d_{mask}$ is the vegetation snow masking depth equivalent thickness of water (centimetres); (iv) $\alpha_1$ is the snow-free land albedo.

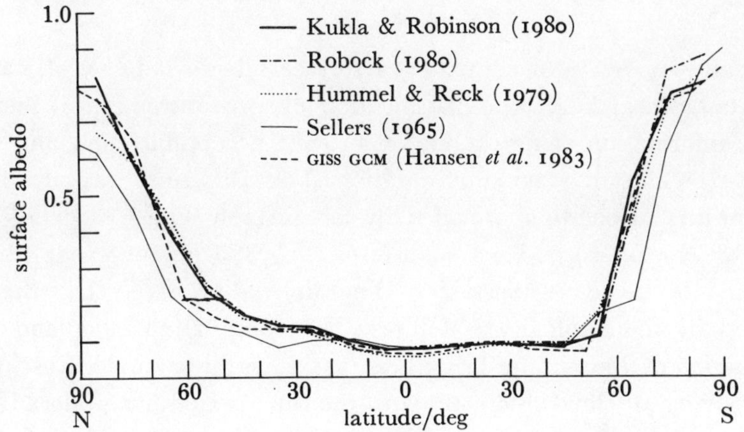


FIGURE 3. Latitudinally averaged cross sections of annual averaged albedos from a variety of sources.

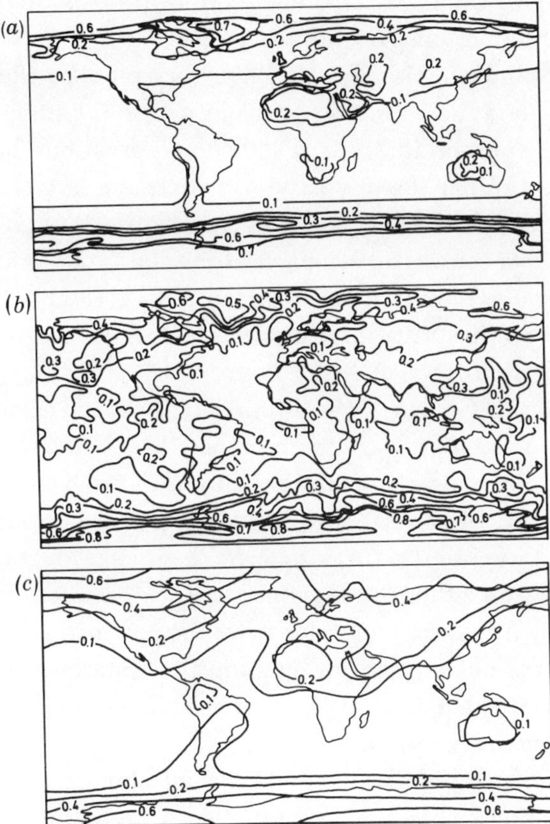

FIGURE 4. (a) GFDL surface albedos (redrawn from Preuss & Geleyn 1980); (b) surface albedos derived from satellite-observed minimum albedos by inversion (after Preuss & Geleyn 1980); (c) surface albedo map drawn from annual average surface albedo values of Hummel & Reck (1979) (from Henderson-Sellers & Hughes 1982).

variation (figure 2), the effects of shading (see, for example, Dozier & Frew 1981; Otterman 1981) and the addition of snow to vegetated environments modify surface albedos. These effects alter as the solar elevation and azimuth vary (Nkemdirim 1973; Ahmad & Lockwood 1979). Atmospheric scatter tends to increase measured clear-sky system albedos. It is therefore preferable to make observations close to the surface. However, satellites may provide the only data in inhospitable areas.

[ 47 ]

### (a) Surveys

There are basically two ways of compiling a surface albedo data set: (i) a satellite data composite; (ii) geographical land type classification plus measurement at selected 'typical' locations. Satellite information alone cannot yet provide acceptable snow and surface albedo data and will never be able to provide information about diffuse surface albedos under total cloud cover. At the high resolution provided by for example Landsat, detailed case studies abound (Moscher & Norton 1977; Rockwood & Cox 1978; Robinove 1982), but a complete global compilation is a massive undertaking. The alternative is to select the lower spatial resolution provided by meteorological satellites. The compilation of a land-surface survey requires the production of a minimum-brightness map (see, for example, Raschke *et al.* 1973). The chance of removing all clouds increases with the time period but so does the 'noise' produced by the surface variability (snow, moisture, vegetation) and the effect of real radiometer noise. Rossow (personal communication 1982) has found that for N.O.A.A. sr data the 3 month absolute minimum value observed has a value approximately 1.6 standard deviations below the true clear-sky minimum. A continually updated minimum albedo will continue to decrease because of instrument noise. In the 'inventory plus specific observation' method of compiling global albedo data there are two problem areas: (*a*) composing the global classification and (*b*) obtaining appropriate 'typical' albedo observations for each selected category.

The resulting albedo sets differ (figures 3 and 4). There is a large region of the globe ($40°$ N to $40°$ S) for which the latitudinally averaged data sets (figure 3) are in agreement. Middle and high latitude values are less consistent. Kukla & Robinson (1980) emphasize the effect of the seasonal cryosphere whereas Hummel & Reck (1979) give greater weight to the seasonality in vegetation albedos. Figure 4 illustrates three global albedo fields. Figure 4*a* is the global model based on ground and aircraft observations of Posey & Clapp (1964); figure 4*b* shows surface albedo derived by Preuss & Geleyn (1980) from satellite data calculated by Raschke *et al.* (1973); and figure 4*c* is the annual albedo map proposed by Hummel & Reck (1979) based on land-survey and ground observations updating the work of for example Posey & Clapp (1964) and Schutz & Gates (1972). In general, low-latitude and mid-latitude values are lower in the satellite-derived data set (figure 4*b*) over deserts. A greater degree of spatial variation is observed in figure 4*b* than in either the Hummel & Reck data set (figure 4*c*) or the Posey & Clapp model (figure 4*a*). At extreme high latitudes in the Northern Hemisphere the satellite data are lower than surface observations, whereas in the Antarctic satellite derived values are higher than the two ground based data sets. Contamination of the satellite data appears to have been caused by cloud cover in a number of areas.

A major advantage in using satellite-derived data for climate models is that this albedo is ready-integrated information for the grid box sampled. Careful interpretation of satellite data could solve the problem of heterogeneous land classes within a specified gcm grid. However, there are clearly significant discrepancies between surface and satellite-derived data sets.

### (b) Limitations of data sets

There are three categories of effects that lead to discrepancies between different observations of surface albedos: (i) instrumental, (ii) viewing geometry, and (iii) environmental, especially atmospheric, influences. Instrumental limitations in satellite surveys result from deficiencies in the optical system, satellite instability, sensor design and degradation, and the orbital configuration. Radiometers not subject to in-flight recalibration drift from their original settings (see, for instance, Winston *et al.* 1979). The influence of the waveband of the sensor upon the observations can be critical, and comparison of measurements made in different wavelength regions is difficult (see, for example, Duggin *et al.* (1982) and the discussion of figure 5*b*). It is generally found that minimum clear-sky albedo occurs at midday (Nkemdirim 1973). Kowalik *et al.* (1982) give an analysis of the relation between Landsat observed radiances and the solar

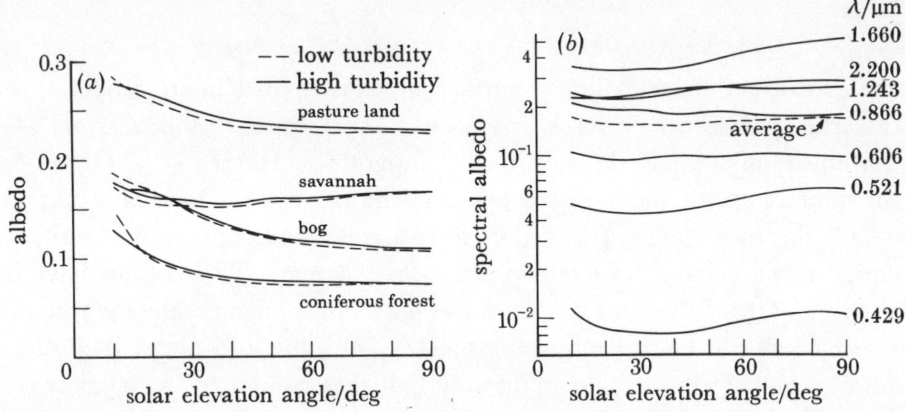


FIGURE 5. (*a*) Four natural surface albedos, wavelength-integrated from observations for low and high turbidity atmospheres as a function of solar elevation angle (after Kriebel 1979); (*b*) spectral albedos and wavelength-integrated albedo for savannah as observed on 10 August 1971 (after Kriebel 1979), with permission of Elsevier Science Publishing Co.

zenith angle at the time of observation for ten sites in Nevada. Although significant, the zenith angle effect poses less of a problem for Landsat data than for those data from meteorological satellites.

The non-Lambertian nature of land surfaces (Stowe *et al.* 1980; Kimes & Kirchner 1982) and the influence of the scene geometry upon observed albedos can be considerable (Monteith 1959; Idso *et al.* 1975). Dirmhirn & Eaton (1976) suggest that the reflectance observed through a small aperture, as by satellites, can be inverted from the observed system albedo only if the degree of non-isotropy is known. However, Ueno (1981) suggests that over a complete canopy and mixed surface types, the anisotropic reflective properties of individual elements may be masked and the assumption of isotropy reasonable. Dozier & Frew (1981) and Otterman (1981) derive relations between planar and non-planar configurations. Kriebel (1976) shows that, for low solar zenith angles, shadowing effects are small and that only slight azimuthal anisotropy occurs.

The problem of removal of atmospheric contamination (figure 5*a*) (Turner 1978; Otterman *et al.* 1980) is important for all except very low-altitude flight measurements (Kung *et al.* 1964). Bauer & Dutton (1962) demonstrated that albedo measurements made up to heights of approximately 300–400 m are within 0.05 of measurements made at the surface. This difference is as large as the contrast between some land type classes. Recent theoretical calculations by Briegleb & Ramanathan (1982) illustrate some features of clear-sky albedos and their relation to surface albedos. Duggin *et al.* (1982) show that removal of cloud contamination is difficult. Kriebel (1976, 1979) considers the effect of aerosol loading on satellite-sensed 'clear-sky' albedos. His results suggest a greater dependence upon solar zenith angle than upon atmospheric turbidity (figure 5*a*). Figure 5*b* shows that there is a dependence of spectral albedo upon zenith angle. It seems to be fortuitous that the dependences cancel when the wavelengths are integrated.

[ 49 ]

The purpose and skill of the originator of the albedo file can be important: cloud contamination, ignored in the original because of detailed local knowledge, may be archived. Moscher & Norton (1977) publish an estimate of the error on their calculated values of surface albedo (*ca.* 11 % of the retrieved surface albedo). It is apparent that all current surface albedo data sets should be treated cautiously.

### 4. CONCLUSIONS AND RECOMMENDATIONS

The effect of incorrect specification of surface albedos upon climate model sensitivity is difficult to ascertain. Hummel & Reck (1979) state that surface albedo errors of $\pm 0.025$ produce uncertainties of $\mp 2.5$ K in the surface temperature. Hansen *et al.* (1981) calculate, from their one-dimensional RC model, a temperature change of 1.3 K for a land-surface albedo alteration of 0.05. Sagan *et al.* (1979) calculate a man-generated temperature change of 0.2 K resulting from an albedo change of 0.001 over the last 25 years. Their results have been disputed by Potter *et al.* (1981). Revised calculations suggest that Sagan *et al.* (1979) tend to overestimate the surface albedo by neglecting the recovery of some arid areas. Inclusion of these revisions leads to a global albedo increase of only half that computed by Sagan *et al.* (1979), with a resultant temperature decrease of *ca.* 0.1 K.

The sensitivity of current GCMs to surface albedo changes is a function of the type of change, its geographical location and the nature of the GCM itself. The parametrization of surface albedo in GCMs differs (§2 and table 1). Preuss & Geleyn (1980) make a comparison between very short (10 day) atmospheric integrations by using the surface albedo data sets in figure 4*a*, *b*. They observe that, although computed surface temperature differences were small, at the cessation of their experiment the atmospheric column temperature showed no tendency to equilibriate.

There are two regions for which GCM climate sensitivity to surface albedo changes has been established: deserts and the cryosphere. In 1975 Charney proposed a biogeophysical feedback mechanism as the cause of extending drought regions in the Sahel. Increased albedo, resulting in a net radiative loss, produced general subsidence, which decreased cloud formation and rainfall (Charney 1975; Charney *et al.* 1977; Sud & Fennessey 1982). It is unclear how surface modification will modify the local climate régime in humid regions. Preliminary results from a simulation of the effects of tropical deforestation in the Amazon region with the GISS GCM show no significant alteration in local or regional surface temperature, although local annual precipitation decreases by 200 mm. The ice–albedo feedback mechanism is believed to be significant for climatic perturbation on a large number of timescales (see, for example, GARP 1975). It is the fundamental forcing effect on EBMs (equation (2)) and is included in all other climate models (see, for example, Wang & Stone 1980; Manabe & Stouffer 1980), although the lack of success of GCMs in simulating the current sea-ice extent (Manabe & Stouffer 1980) could affect computed temperature sensitivities.

It is important that interchange between the remote-sensing and climate-modelling communities be encouraged. Surface albedo variations, caused by changes in the cryosphere extent and vegetation pattern, have been invoked as important causes of climatic change, but modelling results are not yet consistent and no agreed global data set is available with which comparative simulations can be made. Currently the albedo differentiation and level of accuracy derivable solely from satellite measurements are much lower than those being examined in climate-sensitivity experiments. The development of understanding of our global environment requires constructive and coordinated development of climatic modelling and satellite and surface observational programmes.

M.F.W. holds an N.E.R.C. studentship; part of this work was undertaken while A.H.-S. was an N.R.C. (U.S.A.) Visiting Research Associate at the Goddard Institute for Space Studies, New York. K. P. Shine, N. A. Hughes and J. Jones are thanked for their help.

## References

Ahmad, S. B. & Lockwood, J. G. 1979 *Prog. phys. Geog.* **3**, 510–543.

Arakawa, A. 1972 *Numerical Simulation of Weather and Climate*, Tech. Rep. no. 7. Met. Dept., U.C.L.A.

Bauer, K. B. & Dutton, J. A. 1962 *J. geophys. Res.* **67**, 2367–2376.

Boer, G. J. & MacFarlane, N. A. 1979 In *Report of the JOC Study Conference on Climate Models* (GARP no. 22), vol. 1, pp. 409–460.

Briegleb, B. & Ramanathan, V. 1982 *J. appl. Met.* **21**, 1160–1171.

Budyko, M. I. 1969 *Tellus* **21**, 611–619.

Cahalan, R. F. & North, G. R. 1979 *J. atmos. Sci.* **36**, 1178–1188.

Carson, D. 1981 In *Report of J.S.C. Study Conference on Land Surface Processes in Atmospheric General Circulation Models, Greenbelt, U.S.A., 5–10 January 1981*, pp. 67–108. World Climate Research Programme, I.S.C.U./W.M.O.

Charney, J. G. 1975 *Q. Jl. R. met. Soc.* **101**, 193–202.

Charney, J. G., Quirk, W. J., Chew, S. M. & Kornfield, J. 1977 *J. atmos. Sci.* **34**, 1366–1388.

Cogley, J. G. 1979 *Mon. Weath. Rev.* **107**, 775–781.

Corby, G. A., Gilchrist, A. & Rowntree, P. R. 1977 In *Methods in computational physics*, vol. 17, pp. 67–110. New York: Academic Press.

Dickinson, R. E., Jaeger, J., Washington, W. M. & Wolski, R. 1981 *Boundary subroutine for the NCAR global climate model* (NCAR Tech. Note no. 0301/78-01). Boulder, Colorado.

Dirmhirn, I. & Eaton, F. D. 1976 In *Proc. Symp. Radiation in the Atmosphere* (ed. H. J. Bolle), pp. 435–437. Garmisch Partenkirchen Scientific Press.

Dozier, J. & Frew, J. 1981 *Remote Sensing Envir.* **11**, 191–205.

Duggin, M. J., Schoch, L. & Gray, T. I. 1982 In *Proc. SPIE, Symp. East, Washington, D.C., 6–7 May 1982*.

Garp 1975 *The physical basis of climate and climate modelling* (GARP Publication Series no. 16). (265 pages.) Geneva: World Meteorological Organization.

Gates, W. L. & Schlesinger, M. E. 1977 *J. atmos. Sci.* **34**, 36–76.

Grenfell, T. C. & Maykut, G. A. 1977 *J. Glaciol.* **18**, 445–463.

Halem, M., Shukla, J., Mintz, Y., Wu, M. L., Godbole, R., Herman, G. & Sud, Y. 1979 In *Report of the JOC Study Conf. on Climate Models* (GARP no. 22), vol. 1, pp. 207–253.

Hansen, J. E., Johnson, D., Lacis, A. A., Lebedeff, S., Lee, D., Rind, D. & Russell, G. 1981 *Science, Wash.* **213**, 957–966.

Hansen, J. E., Russell, G., Rind, D., Stone, P., Lacis, A. A., Lebedeff, S., Ruedy, R. & Travis, L. 1983 *Mon. Weath. Rev.* (In the press.)

Henderson-Sellers, A. & Hughes, N. A. 1982 *Prog. phys. Geog.* **6**, 1–44.

Holloway, J. L. Jr & Manabe, S. 1971 *Mon. Weath. Rev.* **99**, 335–370.

Hummel, J. R. & Reck, R. A. 1979 *J. appl. Met.* **18**, 239–253.

Idso, S. B., Jackson, R. D., Reginato, R. J., Kimball, B. A. & Nakayama, F. S. 1975 *J. appl. Met.* **11**, 109–113.

Kimes, D. S. & Kirchner, J. A. 1982 *Remote Sensing Envir.* **12**, 141–149.

Kondratyev, K. Ya. 1969 *Radiation in the atmosphere.* (912 pages.) Academic Press.

Kowalik, W. S., Marsh, S. E. & Lyon, R. J. P. 1982 *Remote Sensing Envir.* **12**, 39–55.

Kriebel, K. T. 1976 In *Proc. Symp. Radiation in the Atmosphere* (ed. H. J. Bolle), pp. 445–450. Garmisch Partenkirchen Scientific Press.

Kriebel, K. T. 1979 *Remote Sensing Envir.* **8**, 283–290.

Kukla, G. J. & Robinson, D. 1980 *Mon. Weath. Rev.* **108**, 56–67.

Kung, E. C., Bryson, R. A. & Lenschow, D. H. 1964 *Mon. Weath. Rev.* **92**, 543–564.

McAveney, B. J., Bourke, W. & Puri, K. 1978 *J. atmos. Sci.* **35**, 1557–1583.

Manabe, S. & Stouffer, R. J. 1980 *J. geophys. Res.* **85**, 5529–5554.

Marchuk, G. I., Dymnikov, V. P., Lykosov, V. N., Galin, V. Ya., Bobyleva, I. M. & Perov, V. L. 1979 In *Report of the JOC Study Conference on Climate Models* (GARP no. 22), vol. 1, pp. 318–370.

Monteith, J. L. 1959 *Q. Jl R. met. Soc.* **85**, 386–392.

Monteith, J. L. 1973 *Principles of environmental physics.* (241 pages.) London: Edward Arnold.

Moscher, F. R. & Norton, C. C. 1977 In *3rd NASA Weath. and Climate Rev.* (NASA Pub. no. 2029), pp. 171–175.
Nkemdirim, L. C. 1973 *Agric. Met.* **11**, 229–242.
North, G. R., Cahalan, R. F. & Coakley, J. A. Jr 1981 *Rev. Geophys. Space Phys.* **19**, 91–122.
Otterman, J. 1981 *Adv. Space Res.* **1**, 115–119.
Otterman, J., Ungar, S., Kaufman, Y. & Podolak, M. 1980 *Remote Sensing Envir.* **9**, 115–129.
Posey, J. W. & Clapp, P. F. 1964 *Geofisica int.* **4**, 33–48.
Potter, G. L., Elsasser, H. W., MacCracken, M. C. & Ellis, J. S. 1981 *Nature, Lond.* **291**, 47–50.
Potter, G. L., Elsasser, H. W., MacCracken, M. C. & Luther, F. M. 1975 *Nature, Lond.* **258**, 697–698.
Preuss, H. & Geleyn, J. F. 1980 *Arch. Met. Geophys. Bioklim.* **29**, 345–356.
Raschke, E., Vonder Haar, T. H., Bandeen, R. W. & Pasternak, M. 1973 *J. atmos. Sci.* **30**, 341–364.
Robinove, C. J. 1982 COSPAR Sess. paper no. 10.2.4., Ottawa, Canada, 31 May–2 June.
Robock, A. 1980 *Mon. Weath. Rev.* **108**, 267–285.
Rockwood, A. A. & Cox, S. K. 1978 *J. atmos. Sci.* **35**, 513–522.
Sagan, C., Toon, O. B. & Pollack, J. B. 1979 *Science, Wash.* **206**, 1363–1368.
Saker, N. J. 1975 Unpublished Met. Off. Met. 020 Tech. Note no. II/30.
Schlesinger, M. E. & Gates, W. L. 1979 In *GARP/JOC Study Conf. on Climate Models* (GARP no. 22), vol. 1, pp. 139–206.
Schutz, C. & Gates, W. L. 1972 *Global climatic data for surface, 800* mb, *400* mb. The Rand Corporation.
Sellers, W. D. 1965 *Physical climatology.* (272 pages.) Chicago: University of Chicago Press.
Sellers, W. D. 1969 *J. appl. Met.* **8**, 392–400.
Stowe, L., Jacobowitz, H. & Taylor, V. R. 1980 In *Proc. Int. Rad. Symp., Fort Collins, Colorado, 11–16 August.*
Sud, Y. C. & Fennessey, M. 1982 *J. Climatol.* **2**, 105–125.
Turner, R. E. 1978 In *Proc. of the Twelfth Int. Symp. on Remote Sensing of the Environment*, pp. 783–793.
Ueno, S. 1981 In *Processes in marine remote sensing*, vol. 12 (ed. P. J. Vernberg & F. B. Diemer), pp. 452–509. University of South Carolina Press.
Wang, W.-C. & Stone, P. H. 1980 *J. atmos. Sci.* **37**, 545–552.
Warren, S. G. & Schneider, S. H. 1979 *J. atmos. Sci.* **36**, 1377–1391.
Warren, S. G. & Wiscombe, W. J. 1980 *J. atmos. Sci.* **37**, 2734–2745.
Washington, W. M. & Williamson, D. L. 1977 In *Methods in computational physics*, vol. 17, pp. 111–172. New York: Academic Press.
Winston, J. S., Gruber, A., Gray, T. I., Varnadove, M. S., Earnest, C. L. & Mannello, L. P. 1979 *Earth atmosphere radiation budget analyses derived from NOAA satellite data June 1974–February 1978*, vols 1 and 2. Washington, D.C.: Meteor. Sat. Lab. NOAA-NESS.
Wiscombe, W. J. & Warren, S. G. 1980 *J. atmos. Sci.* **37**, 2712–2733.

*Phil. Trans. R. Soc. Lond.* A **309**, 295–314 (1983)
*Printed in Great Britain*



# The application of satellite imaging radars over land to the assessment, mapping and monitoring of resources

By P. H. A. Martin-Kaye and G. M. Lawrence

*Hunting Geology and Geophysics Limited, Elstree Way, Borehamwood, Herts. WD6 1SB, U.K.*

[Plates 1–12]

Airborne imaging radars have been put to substantial civilian use over the past 15 years, accumulating aggregate areal ground coverage of imagery of about $18 \times 10^6$ km². Much of this coverage has been from cloud-ridden equatorial regions that pose difficulties for systematic air photography, capitalizing upon radar's capacity to acquire data virtually without regard to daylight and cloud conditions. The work was required mainly for regional development, land-use studies, petroleum and mineral exploration or geothermal energy studies, and was carried out by a few established groups.

The Seasat sar, operating in 1978, had a coverage essentially restricted to North America and Europe. Although not particularly suited to overland purposes it introduced many additional users to regional radar imagery. The Shuttle imaging radar sir-a, which flew in November 1981, achieved a coverage of about $10^7$ km² in the aggregated 8 h of operation, looking at a greater variety of terrain, particularly among arid lands, and geologic type than had been accumulated previously. Examples presented here show the even illumination of sir-a, and a depiction of geographic and geological features superior to Seasat and in some cases superior to Landsat. Interpretability for regional geological purposes is good and is adequate for general land use and other purposes. In arid regions new applications for studies on desertification and hydrogeology are highlighted. An example of subsurface backscatter is presented. Once wide-covering multifrequency and multitemporal radar become available from N.A.S.A.'s sir-b and samex, planned for 1984 and 1987 respectively, and from other major radar satellites to be launched by the European Space Agency, the Canadian Centre of Remote Sensing and the Japanese Space Agency, and from the German Spacelab sar, spaceborne radar imagery will join Landsat as a primary tool in thematic mapping of large areas of the Earth's surface.

## 1. Introduction

Imaging radars are unaffected by conditions of daylight, haze and most cloud, any of which can defeat scanners and cameras. The images from survey programmes are interpretable to various levels: for general cartography, terrain morphology, land use, natural vegetation and geology; for ice, and sea states; and for other similar purposes. The combination of operational reliability and practical application led to some $18 \times 10^6$ km² of coverage by commercial airborne mapping radars (slar) between 1969 and the present. In 1978 N.A.S.A.'s ocean-monitoring satellite, Seasat, in little over 100 days, inspected about $10^8$ km² by radar, both over land and sea. In November 1981, the Space Shuttle Columbia obtained radar images of about $10^7$ km², about $6 \times 10^6$ km² of which were over land, with the sir-a (Shuttle Imaging Radar) instrument.

The slar surveys were flown widely about the world but emphasized the cloud-ridden tropics. Seasat's recovered data were restricted to the European and North American areas whereas the Shuttle's radar looked at a great diversity of terrain in the tropics and the near-tropics including, very importantly as it transpired, arid regions. Both the side-looking airborne radar (SLAR) programmes and the imaging from space have shown that radars have valuable resource mapping and monitoring roles to fulfil.

## 2. RESOURCE SURVEYS BY SIDE-LOOKING AIRBORNE RADAR (SLAR)

The principal commercial side-looking radar systems and their surveys for resource evaluations are listed in table 1. Commercial activities were triggered off by the two-stage Project RAMP (Radar Mapping of Panama) in 1967 and 1969, which publicized mapping radar's ability to recover data from clouded areas. Later, the RADAM project covered the Amazon basin and then, in Projecto RADAM BRASIL, the whole of Brazil. Similar operations spread in South America and extended to southeast Asia and to Africa. A comparatively recent review of the developing use of SLAR has been given by McDonald (1980).

The large SLAR programmes of the 1970s were contracted by Governments for thematic interpretative mapwork in the guidance of development planning. Surveys were also commissioned by mining and especially oil companies needing quick initial reconnaissance of remote areas. Nuclear and geothermal power plant siting studies brought about some imaging in North America, Japan and the Philippines. The requirements of oil exploration and the monitoring of ice and the movement of ice flows resulted in substantial use of radar off northern Canada and Alaska.

The characteristics and economics of the commercial SLAR systems operating before the launch of Seasat (1978) dictated that the principal applications were to large-area surveys for 1:200000–1:250000 scale image and cartographic products of poorly mapped, logistically and photographically difficult regions. For semi-detailed inventory and monitoring purposes the systems worked well. The reliability of data acquisition allowed fixed-price, fixed-schedule programmes. The images were able to provide an up-to-date pictorial mosaic map which, for numbers of operational purposes, was often much better than had previously existed. The shadowing effects given to landscape by the oblique illumination enabled comparatively ready interpretation of geological structures that are often reflected in morphology. Although the resolution was low (mostly 10–40 m), for the detail required in land-use and natural vegetation mapwork it was sufficient for the mapping scales that are adopted for regional surveys.

A substantial literature on airborne radar imaging and the interpretation of imagery exists. Bibliographies include Bryan (1973, 1979), Moore *et al.* (1974) and Walter (1968). The major SLAR projects are, however, not all well documented publicly, the data of several being unpublished or in summaries. Wing (1971), Wing & Dellwig (1970) and Wing & MacDonald (1973) give accounts of RAMP. The imagery of the Brazilian projects is available with a series of 1:500000 scale interpretative maps in geology, land use and natural vegetation, together with accompanying reports; general accounts have been given by Azevedo (1971), Moreira (1973) and Correa (1980). A five-volume report (Diazgrandados 1979) has been issued for the PRORADAM project of Columbia. Martin-Kaye & Williams (1973) presented a summary of the radar-geological mapwork for Nicaragua, and Parry & Trevett (1979) and Hunting Technical Services Limited (1978) discussed the land-use and natural vegetation work of the NIRAD project in Nigeria. From a more recent programme in Africa, Dellwig (1980) describes the updating of maps of Togo. Of other regional programmes, e.g. Peru (see, for example,

## TABLE 1. COMMERCIAL SLAR SYSTEMS

(A full list of operating SLAR systems is given in Schlude (1981).)

| operator | equipment | type | radar wavelength | signal polariza-tion | image swath km | nominal resolution/m along track (range) | across track (azim.) | platform | areas covered |
|---|---|---|---|---|---|---|---|---|---|
| Westinghouse | Westinghouse AN/APQ-97 | real aperture | Ka; 0.86 cm | HH, HV | 21 | 25 | 8 | DC6 B | *ca.* $1.7 \times 10^6$ km² over North America, Panama (RAMP), Colombia, Ecuador, Nicaragua, Australia, Papua–New Guinea, Indonesia, British Solomon Is. Ceased commercial operation in 1973. |
| Aeroservice Corp. | Goodyear AN/APQ-102 (G.E.M.S.) | synthetic aperture | X; 3 cm | HH | 37 | 15 | 15 | Caravelle | *ca.* $12 \times 10^6$ km² over North America, all Brazil (RADAM), southern Venezuela. (Amazonas), Colombia (PRORADAM), southern Peru, U.K., several areas of southeast Asia incl. Japan, Taiwan, Philippines; Gabon. |
| MARS Inc. | AN/APS-94 D Dual-look | synthetic aperture | X; 2.5 cm | HH | 100 | 116 | 30 | Grumman Gulf Stream | *ca.* $3.5 \times 10^6$ km² over North America, Central America, parts of southeast Asia, U.K., several countries of tropical Africa incl. Togo, Equatorial Guinea, Nigeria (NIRAD). Aircraft and ownership recently changed. |
| Intertech Ltd. (for C.C.R.S.) | Willow Run ERIM; now SAR-580 | synthetic aperture | simultaneous X (3 cm) & L (23 cm); experimental C (10 cm) | HH normally HV & H & V | 5 5 18 | 1.5 (X) 3 (L) 10 | 2 (X) 5 (L) 10 | Convair 580 | *ca.* 750,000 km² over North America, especially Canada and Arctic, Europe. |

Martin-Kaye *et al.* 1980) or southeast Asia (see, for example, Froidevaux 1980), relatively little has been published.

Seasat, SIR-A and the European SAR 580 campaign (figure 2*a*) reflect and have encouraged recent interest in radar, which manifests itself in the increasing number and variety of radar papers in remote sensing journals and symposia. The 1981 and 1982 International Geoscience and Remote Sensing Symposia of the I.E.E.E. (Keydel *et al.* 1982), the Snowmass Workshop (Harrison 1980) and E.S.A./EARSeL Workshop proceedings (Guyenne & Lévy 1981) produced particularly significant contributions.

The application of imaging radar specifically to regional resource surveys has been discussed in a general way by Martin-Kaye (1973, 1980).

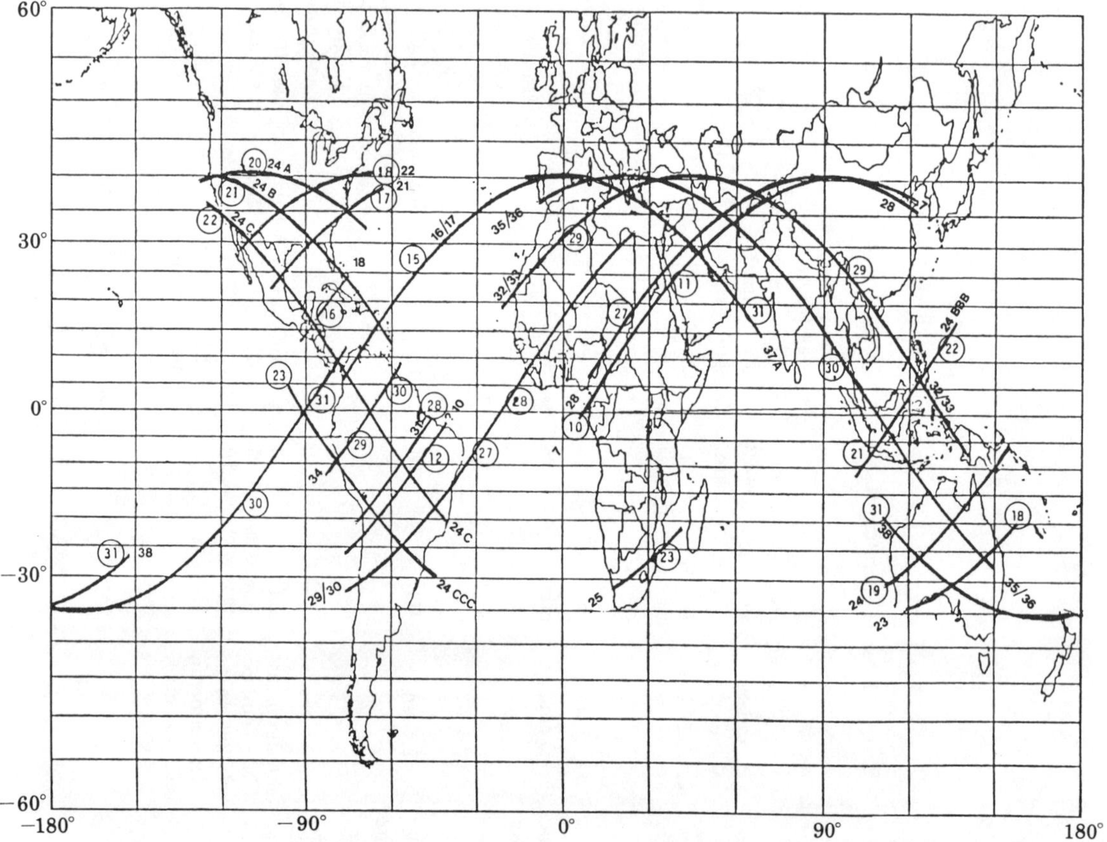

FIGURE 1. SIR-A coverage: circled numbers are reference numbers of orbits; uncircled numbers are reference numbers of SIR-A data-takes. Map prepared by N.A.S.A./J.P.L.

## 3. PLANNING FOR IMAGING RADAR SATELLITES

In 1964, well before the large commercial regional SLAR surveys had been flown, N.A.S.A. commissioned a study of satellite applications for active microwave systems. It was almost immediately concluded that radar had important potential. By the mid-1970s, further examination (Nagler & McCandless 1975; McCandless & Miller 1975; Matthews 1975, 1978; Rouse 1977), decided upon the launch of an oceanographic radar satellite. This became Seasat-A, put into orbit in 1978.

The need for improved ocean environmental monitoring was, and remains, quite clear, for such purposes as weather forecasting, hazard warning, ice surveillance and detection of pollution.

The potential of satellite radar also interested the European Space Research Organisation (E.S.R.O.), now the European Space Agency (E.S.A.), which commissioned feasibility studies for a European Sarsat in the early 1970s (Noel & Pelleau 1973).

The E.S.R.O. studies established that there were good arguments for the satellite radar monitoring of dynamic events, particularly for Europe, owing to frequent and widespread cloud cover. The interest has been continued by the E.S.A., which plans an oceanographic radar satellite ERS-1 for the late 1980s (Honvault 1981).

## 4. Seasat

Seasat was launched from Vandenberg Air Force Base, California, on 26 June 1978. After 1503 revolutions, on 10 October (in U.K.), the system failed and could not be reactivated.

Details of Seasat's operation and application are found in McDonough & Deane (1979), Ford *et al.* (1980), E.S.A. (1981), Sabins *et al.* (1982) and McDonough *et al.* (1983).

Reviews of spaceborne imaging radars including Seasat are given by Elachi (1980) and the National Academy of Sciences (1977).

The principal sensor of Seasat was an L-band, (23 cm) SAR, the data from which were telemetered in real time to the ground. There was no provision for on-board recording, and data recovery was controlled by the location of receiving stations. This limited the coverage to the North American and European regions (table 2).

The total SAR data collected (approximately $20 \times 10^6$ Mbits) corresponded to about 2500 min of operation, equivalent to approximately $10^8$ km$^2$ of ground coverage of the radar imagery. An engineering performance evaluation (Held *et al.* 1982; Curlander 1982) concluded that resolution to 6 m × 21 m, position location to 200 m and relative calibration of image data within ± 2 dB are achievable with the Seasat data set.

The data were either optically or digitally correlated to images, the former being a fairly rapid and elegant process, the latter gaining access to all the data but presenting a large computing requirement.

Data from 53 passes over the European area were recovered by the Royal Aircraft Establishment's receiving station at Oakhanger. The 272 min of recording made by the R.A.E. represents a ground coverage of about $11 \times 10^6$ km$^2$. The image swaths 100 km wide range southward from Greenland to northern Africa; Iceland and the United Kingdom received almost complete coverage but elsewhere there are gaps between the image strips.

Despite the premature failure the Seasat mission had major successes. The SAR was the first such instrument to provide synoptic high-resolution radar images of the Earth's surface. Although the mission was not directed at land purposes overland, imagery was awaited with keen interest. Ford *et al.* (1980) give a review of results in which images are allowed to speak for themselves; some are placed for comparison alongside Landsat MSS, and, in one case, an RBV scene. European, particularly United Kingdom, overland results have been documented for the Royal Aircraft Establishment (Hunting Geology and Geophysics Ltd 1981), in a detailed but limited-edition report that may be inspected at the National Centre for Remote Sensing at Farnborough. An example of Seasat mosaicing and interpretation is given in figure 3, plate 2, and figure 4.

TABLE 2. SPECIFICATION OF MAIN ELEMENTS OF N.A.S.A.s SEASAT SYSTEM

| | |
|---|---|
| active period | launched 26 June 1978; failure of power system on 10 October 1978 (expected lifetime was one year) |
| average orbital altitude | 790.17 km ± 50 m; period 100.75 min; 14.3 orbits per day; orbit repeat 152 days; nearly circular, non-Sun-synchronous |
| orbit inclination | nominally 108°; range 104°–108° |
| total data acquisition time | 60 min/day; direct readout to receiving station total 2500 min |
| coverage | approximately equivalent to $10^8$ km², between 72° N and 72° S: North and Central America, Iceland, Europe and north Africa; the United Kingdom's receiving station operated by the Royal Aircraft Establishment at Oakhanger Hampshire, acquired *ca.* $11 \times 10^6$ km² over Europe between 1 August 1978 and 10 October 1978 (53 passes, 272 mins total recording time) |
| alignment | sensor aligned to within 0.07°; pointing angle to within 0.035°; unobstructed view of the Earth |
| image swath on ground | 100 km |
| image resolution on ground | about 25 m × 25 m (about 40 m for rapid optical correlation, about 25 m × 25 m for precision optical correlation and 6 m × 21 m for digitally correlated imagery) |
| number of looks | depended on processor; four in most cases |
| antenna size | 10.74 m × 2.16 m fixed |
| frequency of transmitted signal | 1.275 GHz |
| wavelength of transmitted signal | L-band; 23.5 m |
| bandwidth of transmitted signal | 19 MHz |
| polarization of transmitted signal | HH |
| depression angle of transmitted signal | 70° |
| incidence angle at surface of ground | 23° ± 3° across swath, from vertical (normal to surface) |
| peak power transmitted | 1000 W |
| transmitted pulse length | 33.4 μs |
| pulse repetition rate | 1463–1640 s$^{-1}$ |
| time–bandwidth product | 634 |
| data recorder bit rate (on ground) | 110 Mbit s$^{-1}$ (5 bits per word) |
| range scale factor | nominally 1:690000 for optical correlated imagery |
| scale range | ± 3 % |
| attributable costs | $400M; launched by Atlas-F rocket |
| accompanying experiments | radar altimeter, radar scatterometer, visible/infra-red radiometer scanning multi-frequency microwave radiometer (SMMR) similar to that on Nimbus G. Two tape recorders for non-SAR data sufficient to store data from two orbits at 25 kbit s$^{-1}$ |
| ground stations (all real-time reception for SAR) | Fairbanks, Alaska; Goldstone, California; Merrit I., Florida (all N.A.S.A.); Shoe Cove, Newfoundland (C.C.R.S.); Oakhanger, United Kingdom (R.A.E.); Canary I. (both E.S.A.) |

DESCRIPTION OF PLATE 1

FIGURE 2. (*a*) Comparison of SLAR bands X (3 cm) and L (23 cm), Buxton, U.K., SAR 580. Optical correlated imagery. Scale approximately 1:100000. Resolution is 1.5 m × 2.1 m for X-band and 3 m × 5 m for L-band. The mesa-like hill is formed from mid-Carboniferous gritstone more or less horizontally bedded and covered by a thin blanket of hill peat and moorland low scrub. Buxton, dry-stone walls, roads, trees, quarries are well represented on both X and L images, although X-band better delineates individual radar reflectors. The generalization of features on L-band can help rapid recognition of features, e.g. stands of trees, the outline of Buxton.

    (*b*) Seasat image of mountains and glaciers in Greenland. The direction of look is from the top of the image; 'chevrons' and conspicuous layover (see particularly the layover of the mountain peaks adjacent to the glaciers) are produced by the steep depression angle of Seasat.

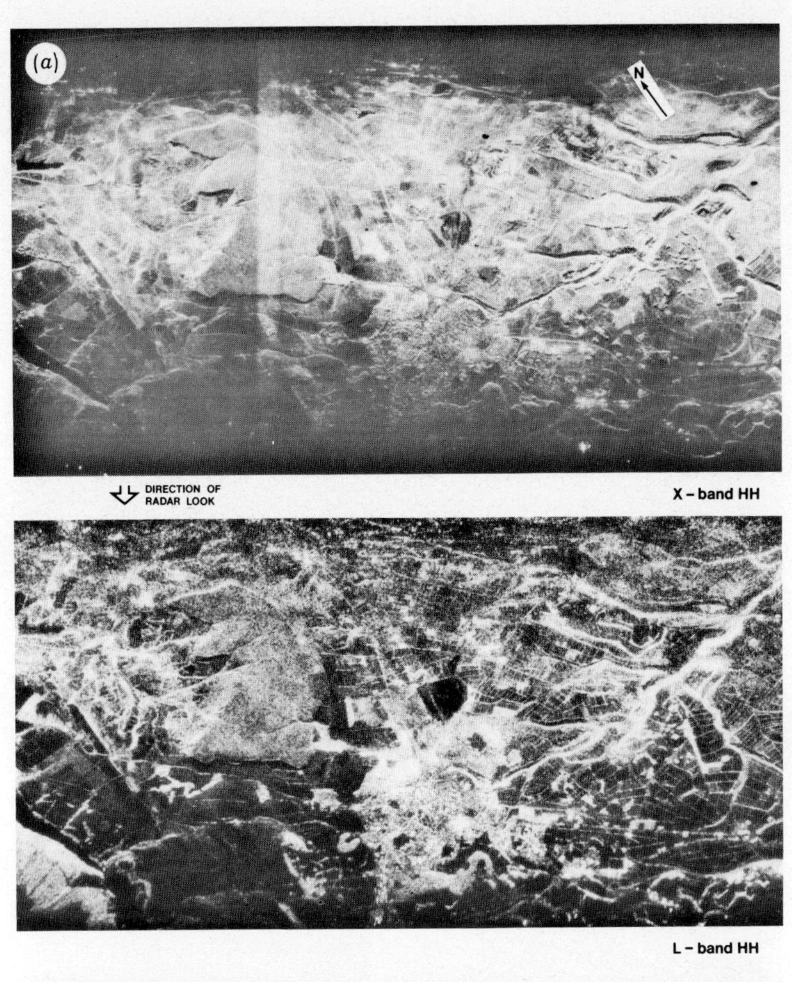


FIGURE 2. For description see opposite.

FIGURE 3. Seasat radar mosaic of the United Kingdom and adjacent Ireland, constructed by Hunting Surveys Limited for the R.A.E./E.S.A.-Earthnet and N.E.R.C. (N.A.S.A.'s Seasat 1 SAR data acquired at R.A.E. Oakhanger and optically correlated by E.R.I.M., U.S.A.).

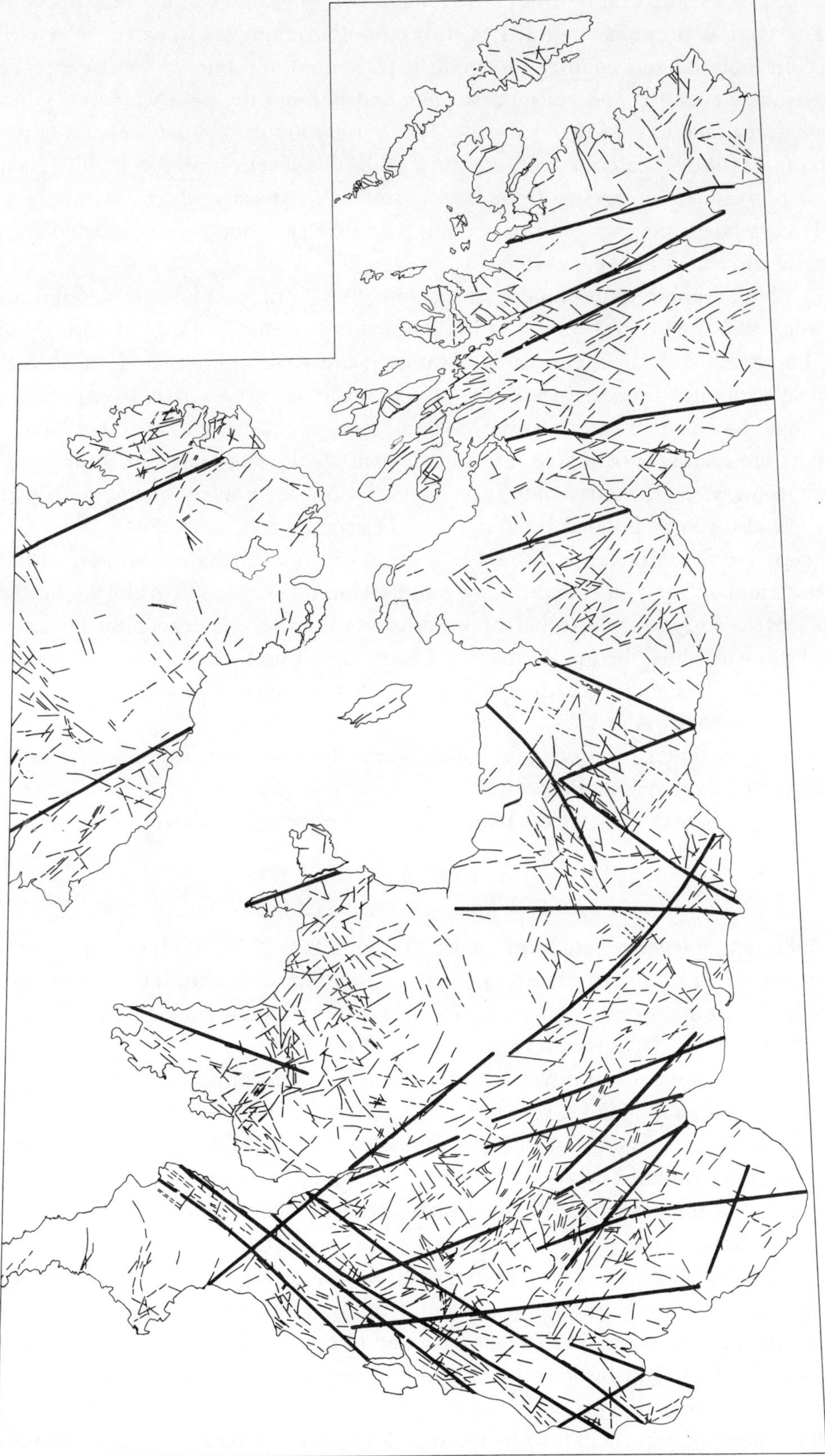

FIGURE 4. Rapid preliminary interpretation of geological lineaments of Seasat mosaic of United Kingdom.

The Seasat SAR configuration was for oceanographic work and was not fully appropriate over land. The large depression angle led to severe foreshortening and layover effects (figure 2$b$, plate 1). In mountainous country these effects prevented any but the most generalized interpretations. However, for areas of modest relief and flatland the data were amenable to interpretation at regional to semi-detailed scales (i.e. 1:100000 to 1:500000 scale for geological and 1:50000 to 1:100000 scale for land-use purposes). The level of interpretability relates to the mode of processing, improving from 'survey mode' material, which is rapidly processed, optically correlated imagery (bulk processing), to precision optically correlated imagery and to digitally correlated imagery (table 2).

Major physiographic units (land and sea, mountain ranges, plateaux, coastal and inland plains, etc.) and the general nature of the landscape (e.g. flat, rolling, irregular, hummocky, mountainous; forested, arid, cultivated, swampy), are well differentiated on Seasat imagery. Geological structures (major lineaments, faults and folds) are sometimes expressed very distinctly (figures 3 and 4). Lines of trees, woodland areas over 5 ha, and heathland and moorland over 50 ha are readily identifiable. Man-made constructions of 100 m or so and more and in sufficient contrast to the surrounds (e.g. airport runways, embankments, long piers, major bridges, canals or large isolated buildings) can be seen.

Less clearly differentiable, or variably so, are lakes, intermediate drainage, lowland coastlines, towns and villages, pasture, meadow and parkland. Where solid geology is masked by any form of cover (e.g. soil, vegetation or urbanization), lithologies can only be inferred from morphology, which may be uninformative. Lesser roads usually cannot be seen; even railways often cannot be distinguished although some are conspicuous. (See also figure 2$a$, comparing L-band with X-band radar.)

Interactive computer processing of digital Seasat data brought little improvement of interpretability. The data were monospectral, and the number of processing options is hence limited. There was more benefit for land-use mapwork than for geology, at least in temperate European circumstances.

## 5. Landsat and Seasat

The information level in data from the MSS of Landsats 1, 2 and 3 is mostly greater than from Seasat SAR (Pala *et al.* 1980), despite the better resolution of Seasat. This is certainly true for areas of high mountain relief because of distortions due to radar ranging, but is also generally so elsewhere. Sometimes, as shown in some of the illustrations of Ford *et al.* (1980), there is more detail obtainable from Seasat. Over ice, lava, drumlin fields, braided drainages, meander flats and variable deserts (hamadas or sand seas), differentiation may be better, and the same applies in particular circumstances for many other features. Owing to radar's sensitivity to slope, geological structures expressed only subtly by relief may be seen better on radar imagery.

On account of the distortions in side-looking radar imagery, complete registration between Landsat and Seasat images can only be approached for flatland conditions. Numbers of such digital composites have now been reported upon (see particularly the co-registrations of Elachi *et al.* (1982$a$), Ford (1982) and Blom & Daily (1982)). The procedure involves the resampling of Seasat digital data to compatible pixel size and to Landsat swath orientation. The two data sets have individual contributions that are mostly best considered alongside each other rather than in combination.

Digital superimposition of data from different Seasat SAR overpasses to compare multitemporal information is readily achieved. Locations and areas of change may be speedily discriminated by colour-coding the two outputs appropriately, as for example in Martin-Kaye *et al.* (1983).

The evaluation of Seasat overland led to several clear conclusions. In the European context, the distribution of the majority of topographic, cultural and geological features of the continent is well known. There is no need to resort to radar satellites to establish their existence or whereabouts, although interesting new perspectives can be obtained from space and the synoptic views. Seasat did yield new geological information in the depiction of unsuspected linear features in the relative flatlands of southern England and doubtless could do the same elsewhere (figure 4). So far as can be seen, it is not necessary to inspect these with systematic regularity, although images from different seasons might provide some complementary information. Purposes of this nature do not make a convincing case for an unmanned radar satellite over European land areas. The value would mainly lie in crop inventory and the monitoring of crop state, leading to improved crop yield forecasting, a purpose that SAR has some capacity to satisfy.

## 6. Shuttle imaging radar: SIR-A

### (a) The system

SIR-A, the Shuttle imaging radar, was one of the seven experiments of the first scientific payload of the Space Shuttle Columbia, carried on the second flight launched on 12 November 1981 (table 3). The payload concerned remote sensing of land resources, environmental quality, ocean conditions and meteorological phenomena (figures 5 and 6). The flight was curtailed, causing abandonment of the original SIR-A coverage plan. Nevertheless, the full 8 h capacity of the on-board optical film recorder was used and about $10^7$ km² was imaged. The on-board recording freed the enterprise from the tether of ground receiving stations, which limited Seasat. SIR-A swaths spanned continents and oceans between about 40° N and 40° S latitudes, greatly extending the variety of geographic coverage and of terrain types inspected by mapping radars (figure 1). The experiment was a notable success in geological application, to which it was primarily directed, particularly in the demonstration of the extent of radar's capacity for lithology-related discriminations in bare rock areas, its sensitivity to subtle, possibly buried, features in dry desert sands, and its sensitivity to moisture in bare soils.

The radar was of Seasat type, also operating in the L-band (23 cm). There were important differences in configuration and the philosophy of the experiment. An operational monitoring role suits the repetitional global-ranging performance of non-geostationary satellites. The single-mission capacity and flexibility of the Shuttle lend it to experiment and once-only or occasional inventories. Columbia was launched with the SIR-A antenna fully deployed in the cargo bay (figure 5), thus ensuring an integrity not easily equalled in unfolding structures. For the geological objectives the critical difference in the system between Seasat and SIR-A was the smaller depression angle of SIR-A (table 3).

### (b) Preliminary results and general comment

Only a few of the first results of SIR-A studies have yet been published. The first preliminary accounts of SIR-A have been given by Elachi *et al.* (1982*b*), Elachi (1982), Covault (1982), Taranik (1982), Estes (1982), Kobrick (1982) and Hubbard (1982). The present comments are based upon those sources and studies in progress in the U.K. Most of the non-U.S. coverage by SIR-A is held in the United Kingdom by official observers to the experiment.

SIR-A imagery is impressive. Detailed examination shows that, geologically, it is interpretationally satisfactory at the delivery scale (nominally 1:500000) and for some areas carries significant new information. Detail can be excellent (figure 7, plate 3).

TABLE 3. SPECIFICATIONS OF MAIN ELEMENTS OF SHUTTLE IMAGING RADAR (SIR-A) SYSTEM

(Optical recording only.)

| | |
|---|---|
| active period of Shuttle | launched 12 November 1981; returned 14 November 1981: 54 h |
| average orbital altitude of Shuttle | circular; 257–266 km (av. 137 naut. miles) |
| orbit inclination | 38° |
| total data acquisition time | 8 h |
| coverage | see figure 1; parts of 60 countries between 40° N and 40° S |
| area covered | *ca.* $10^7$ km²; *ca.* $6 \times 10^6$ km² over land |
| image swath on ground | 50–55 km |
| image resolution on ground | 40 m × 40 m |
| number of looks | 4–7 (4 or 5 in practice) |
| antenna size | fixed, 2.1 m × 9.35 m |
| frequency of transmitted signal | 1.3 GHz |
| wavelength of transmitted signal | synthetic aperture; 23.5 cm (L-band) |
| bandwidth of transmitted signal | 6 MHz (optical recorder) |
| polarization of transmitted signal | HH |
| depression angle of transmitted signal | 43° at centre of swath |
| variation of depression angle across swath | $\pm 3°$ from centre |
| incidence angle at surface of ground | $50° \pm 3°$ from vertical |
| peak power transmitted | 1000 W |
| range scale factor (slant range imagery) | nominally 1:500000 |
| scale range | 1:485000 to 1:515000 |
| signal film type | 3650 ft Kodak RAR 3493, 70 mm wide |
| film speed | nominally 37 mm s$^{-1}$ |
| range focal length | 300 mm |
| azimuth focal length | 2 m |
| range spatial frequency range | 3–34 mm$^{-1}$ |
| azimuth spatial frequency maximum | 44 mm$^{-1}$ |
| designed surface backscatter cross section response | $-8$ to $-28$ dB |
| attributable cost, OSTA-1: $11.6M plus $4.4M integration | |
| attributable cost, SIR-A: $9M | |
| accompanying experiments: | ocean colour experiment scanner (OCE), scanning microwave infrared radiometer (SMIRR), FILE, MAPS (figure 3) |

Speckle is quite pronounced in the bulk-processed images currently available. At the delivery scale it is not obtrusive but enlargement beyond 1:200000 scale renders it conspicuous (figure 12, plate 8).

High mountain relief produces the usual distortions resultant from radar ranging, but the effects are notably less severe than for Seasat, where the depression angle was larger (figure 16, plate 12).

### (c) Results over tropical regions

Some tropical forested areas were imaged from space for the first time by SIR-A. The cloud cover of the Pakaraima Mountains, Guyana (figure 14a, b, plate 10), for example, has resisted the endeavours of Landsat since 1972. In such regions geological interpretation is essentially based upon morphology. However, because of the good depiction of morphology by radar shadowing, some areas are remarkable for the geological data content in spite of the forest cover. The Rio Uaiauka area of southern Venezuela (figure 13, plate 9) is a particularly good example. More subtly expressed information from tropical rainforest is shown in figure 14e.

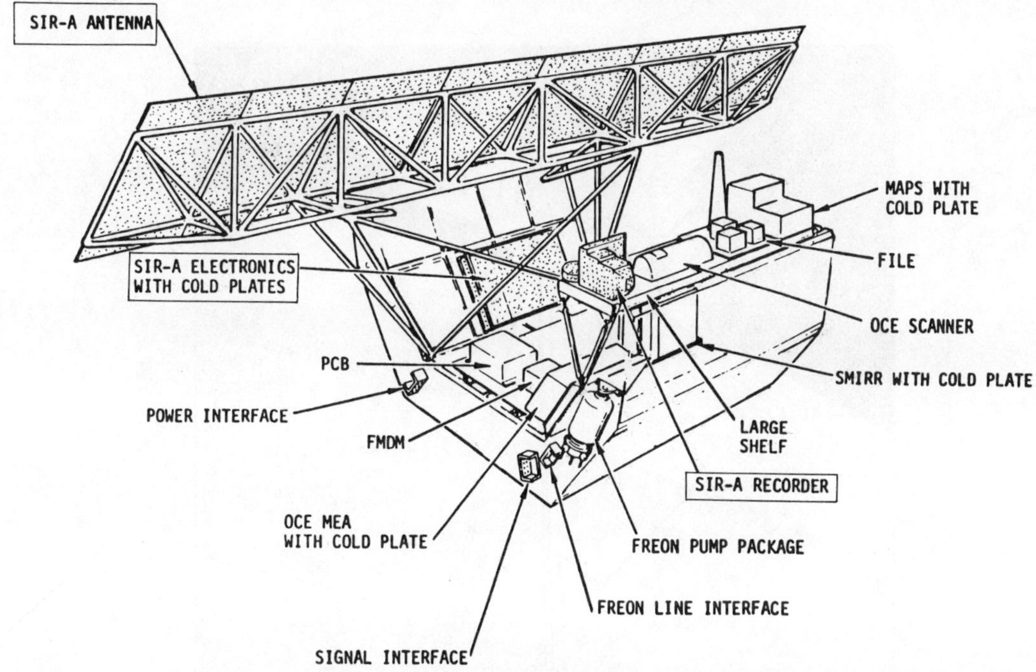


FIGURE 5. OSTA-1 pallet on board the second Shuttle flight, November 1981, showing SIR-A equipment. (After Taranik (1982).)

New information is carried in the detail of these illustrated examples of tropical rainforested regions. Systematic global coverage at this level of quality would be an enormously important baseline data set. An example of an instant application is given in figure 14c, d, which shows part of the Gran Sabana of Venezuela, near to the Guyana border. Strong lineaments converge on circular features that seem hitherto to have been unrecorded. They may mark intrusive bodies that have escaped notice among outcrops of the large basic sills interspersed in the quartzites and conglomerates of the Roraima Formation. An attractive speculation is that they are diamond-bearing pipes, the source of the diamonds distributed about the Roraima plateaux; this source is at present unknown. Structural information of immediate use in hydrocarbon prospecting can be seen in the example from West Irian (figure 14f) (see for comparison Froidevaux (1980)).

[ 63 ]

### (d) Results over arid regions

SIR-A produced a surprise in the unexpected diversity and detail of geological information from arid areas.

Dry sands, even some dunes, show little radar feature if without vegetation (Blom *et al.* 1982); they appear as dark image tones representing low or no backscatter (figure 10, plate 6; figure 15*c, d*, plate 11). Drainage courses are well depicted where fine dry sand fills the channels (figure 8, plate 4; figure 9*b*, plate 5). This precise delineation of drainage and clear depiction of sand sheets suggests that spaceborne imaging radar can have an important role to play in desertification and hydrogeological studies.

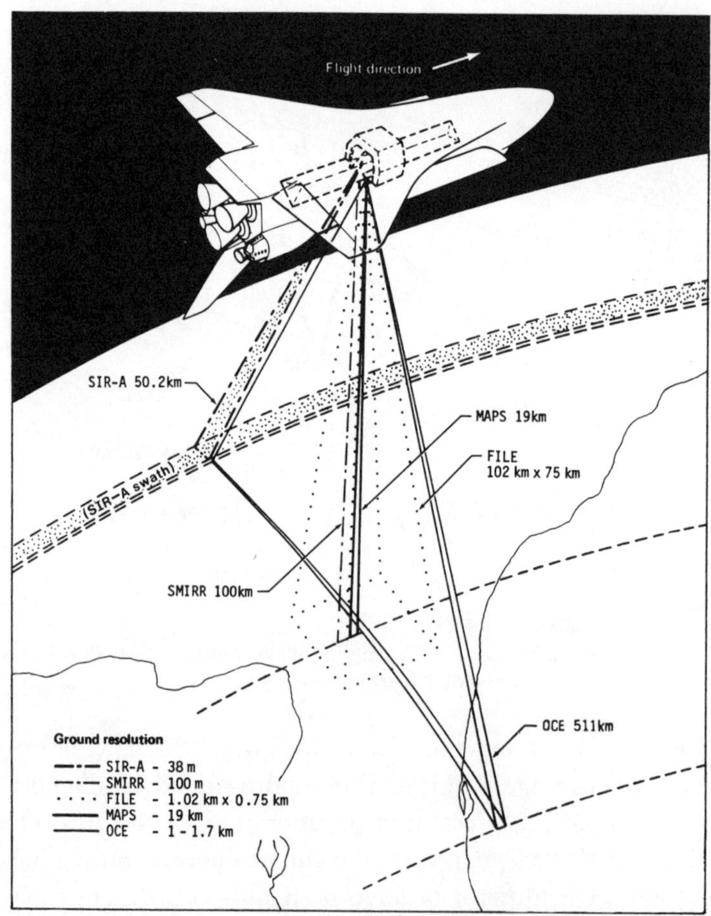


FIGURE 6. OSTA-1 experiments on board the second Shuttle flight, November 1981, showing SIR-A swath. SMIRR, Shuttle multispectral infrared radiometer (a non-imaging experiment to determine the most suitable i.r. bands for lithology recognition); FILE, feature identification and location experiment (uses ratio of visual red and near-infrared reflectance for data management purposes); MAPS, measurement of air pollution from satellite (determines CO in the troposphere); OCE, ocean colour experiment (measures blue-green colour variation via scanner). (After Taranik (1982).)

The absence of vegetation can lead to images in which geological structure is spectacularly displayed (figures 10, 11 and 15*a–c*). The presence of evaporites may complicate the interpretation as in the Grand Kavir salt desert of Iran (figure 15*a*). Salt plugs provoke a variable but distinct speckly bright to grey tone of distinctive context (see also Carver & Bush 1979).

[ 64 ]

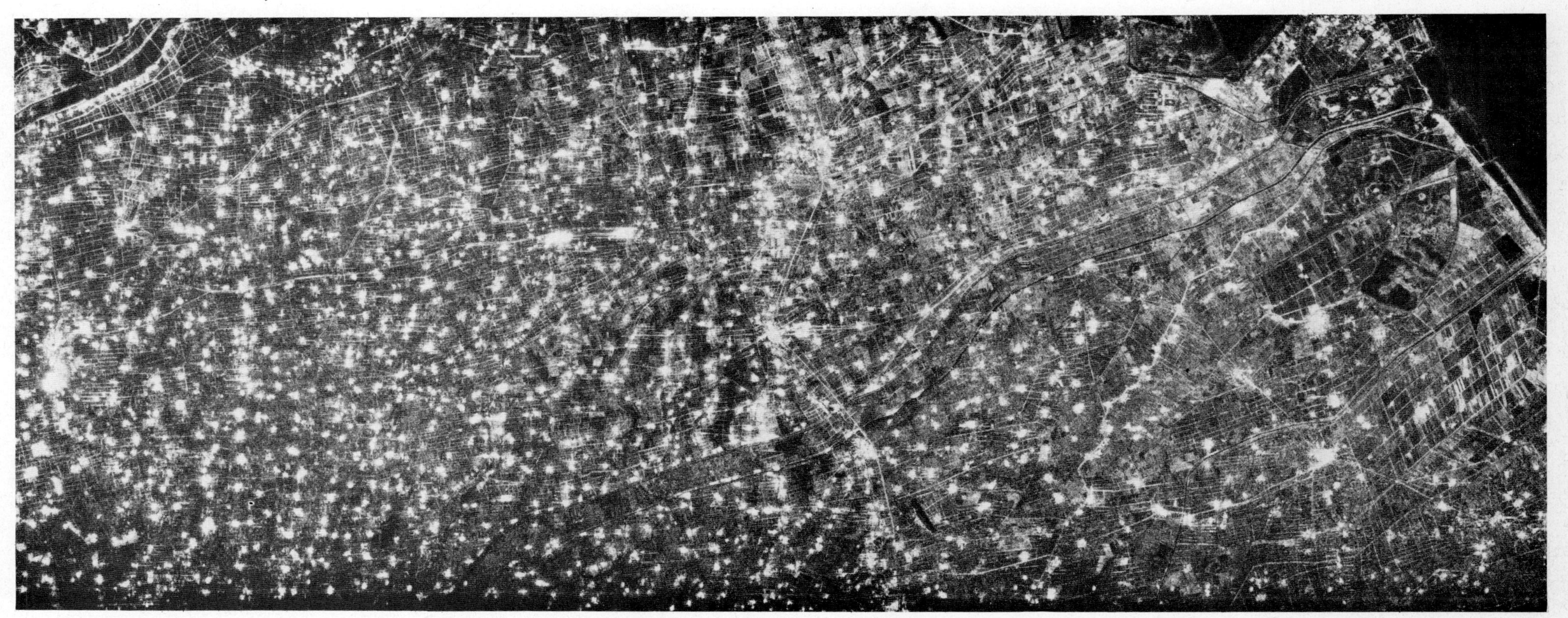

FIGURE 7. SIR-A image (data-take 7) over densely populated and cultivated alluvial/loess plain of north China, near Peking. Nominal scale 1:500000. The bright reflectors are towns and villages. Roads and tracks show clearly from the black tones of flat waterlogged fields. North is to bottom of image.

FIGURE 8. SIR-A swath 29/30 over Adrar des Iforas, northeast Mali. Scale nominally 1:500 000. Bright areas are rough-surfaced outcropping rock; black tonal areas are thick sand. 'Younger Granite' ring-intrusions and associated dyke-swarms and the Iforas granulite unit are particularly distinct. The radar illumination is from top of image, north is to bottom left corner. The area of the subscene in figure 9b is shown. The margins are not rectangular because the subscene has been 'warped', on an image processor, to co-register with Landsat.

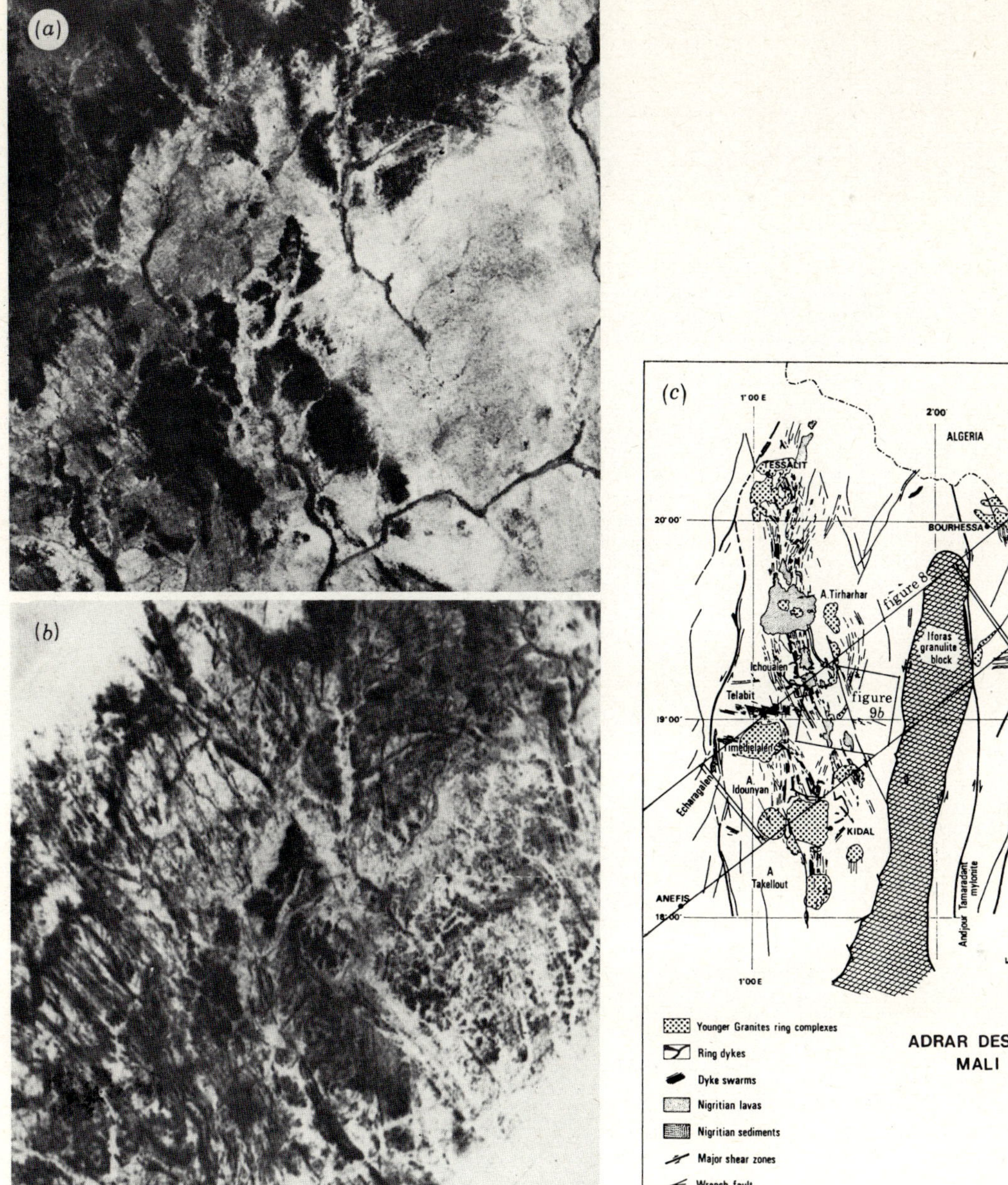



FIGURE 9. (*a*) Adrar des Iforas; Mali Landsat MSS subscene enhanced and warped to co-register with SIR A image on Hunting's image processor (reproduced in monochrome).

(*b*) Equivalent SIR-A image reproduced as negative (bright radar tones reversed to appear black). The bright-reflecting low-relief sand/gravel areas of the Landsat compare with minor drainages and outcrop and perhaps sub-outcrop on the SIR-A (especially the bottom right corner). The dyke swarm shows much more clearly on the SIR-A and a possible small circular feature is identified within a larger circular (top left). The SIR-A imagery responds to small outcrop and boulders better, resulting in more detail, than Landsat because of the former's better resolution (40 m × 40 m compared with 50 m × 80 m for Landsat) and its response to surface roughness. In this very dry area there is also possible shallow penetration and backscatter return of the SIR-A radar beam from subsurface bedrock.

(*c*) Geological sketch of Adrar des Iforas, northeast Mali, showing SIR-A swath (data-take 29/30) and subscenes of figures 8 and 9*b*. (After Black *et al.* (1979).)

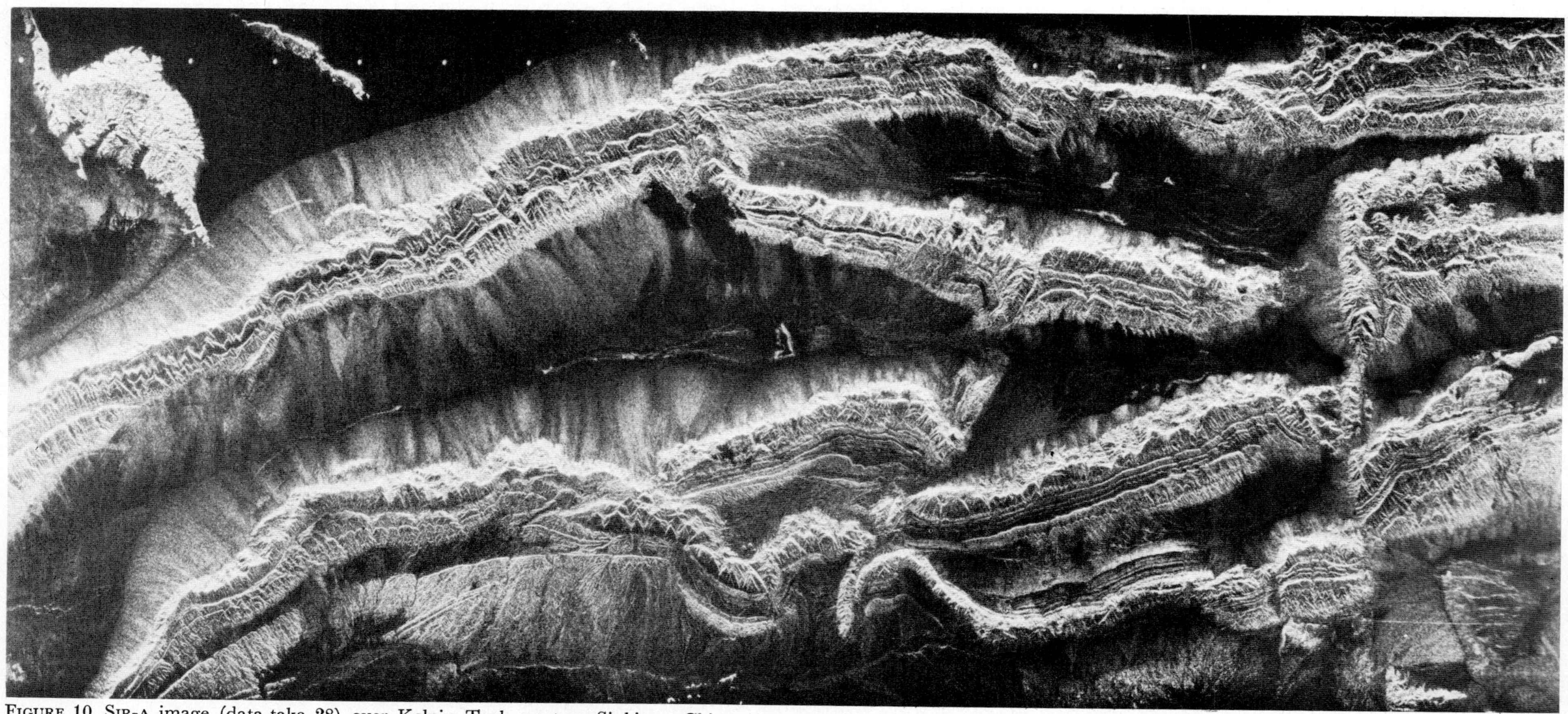

FIGURE 10. SIR-A image (data-take 28) over Kelpin Tagh, western Sinkiang, China, between Tien Shan ranges and Takla Makan desert. Scale approximately 1:500 000. In this arid area folded and thrusted sedimentary rocks are cut by prominent transcurrent faults. Thrust planes dip northward (towards bottom of image). Large cuestas are surrounded by alluvial fans whose rough cobble and boulder surfaces provoke a bright radar response. The boundary between fans and desert sands, which have almost no radar backscatter, is distinct (top left of image). Radar illumination from top of image at approximately 50° incidence angle; north is approximately to bottom left of image. The white dots are fiducial marks representing each second of the Shuttle's orbit.

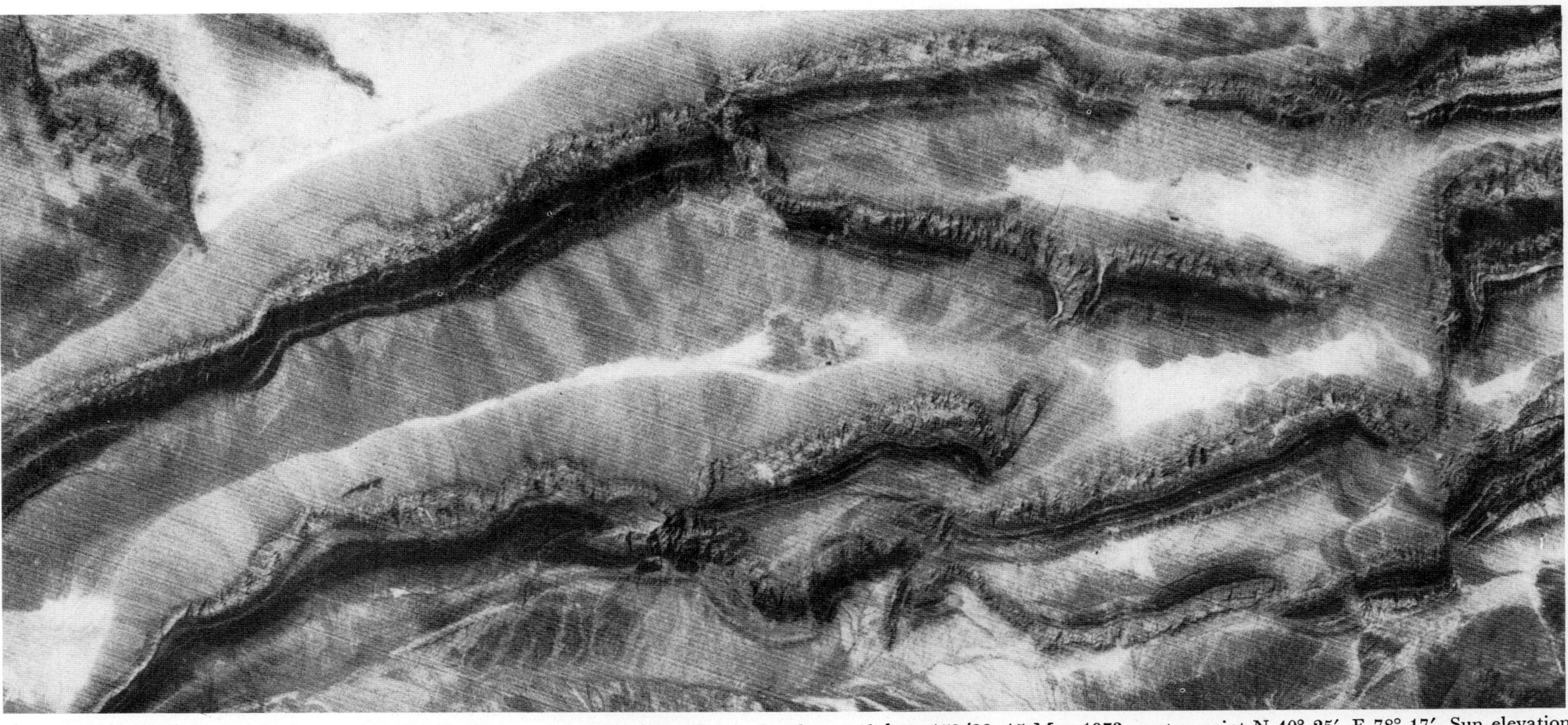

FIGURE 11. Landsat MSS band 7 image over Kelpin Tagh, western Sinkiang, China. Landsat path/row 159/32, 15 May 1973, centre point N 40° 25′, E 78° 17′. Sun elevation 59° from southeast (top left-hand corner). The better resolution and depiction of surface roughness of the SIR-A results in better definition compared with Landsat.

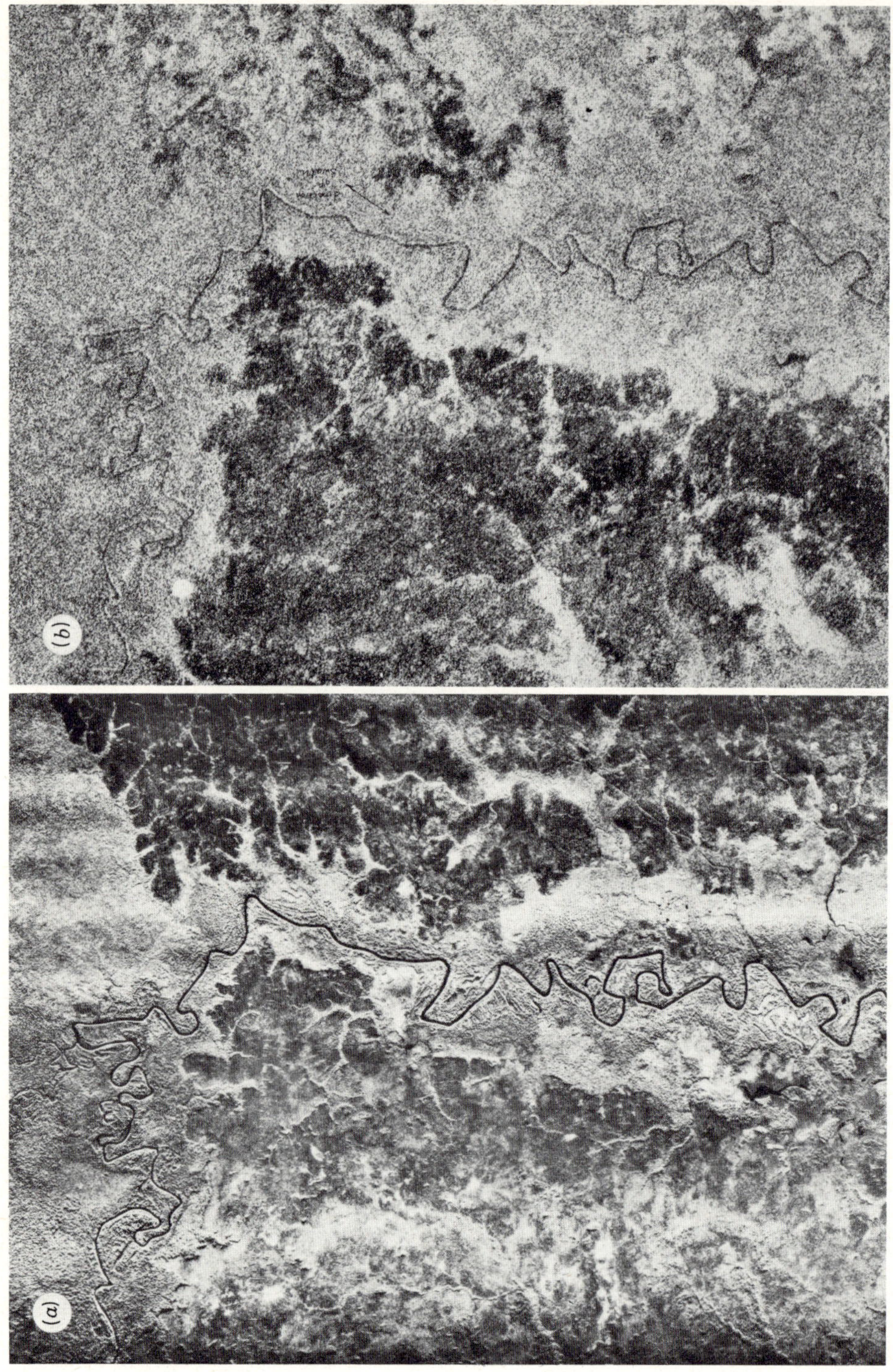

FIGURE 12. For description see opposite.

FIGURE 12. Comparison of (*a*) Westinghouse Ka (0.86 cm) SLAR and (*b*) L-band (23 cm), SIR-A (data-take 24C) over low-relief area of northeast Nicaragua. Scale nominally 1:200000. Area of forest and cleared forest savanna underlain by clays and gravels. Although the SIR-A has distinctly inferior resolution (40 m compared with 10 m) laterite-surfaced roads can show well. Differences of tone on the imagery are thought to be due to temporal changes (the two images are 10 years apart, but both of wet season) rather than radar wavelength differences. The dark interfluve areas are probably waterlogged savannah. The radar illumination is from the right-hand side of the image for the SLAR, at an incidence angle of *ca.* 60°; that for the SIR-A is from the top left-hand corner of the image at an incidence angle of 50°. Note that the marked difference of radar illumination and depression angle do not materially change the delineation of radar geological units.

*Note added in proof* 16 (*March* 1983). Radar depression angle is, of course, substantially variable across the SLAR swath (*a*), being larger in near range (right-hand side) and smaller in far range. The depression angle at centre of swath is equivalent to that of SIR-A but from approximately the opposite direction. The cleared-forest grasslands appear darker at the larger depression angle of the near range of SLAR. The SIR-A (*b*) may be showing a similar but smaller effect in the reverse direction; the depression angle variation across swath of SIR-A is only 5–6° compared with the 30–40° of SLAR.



FIGURE 13 (*overleaf*). (*a*) SIR-A image (data-take 34) over part of southern Amazonas, Rio Uaiauka, Venezuela. Nominal scale 1:500000. Inaccessible low-relief area of tropical rain forest underlain by well fractured rocks (left) mapped as acid extrusive rocks set in Lower–Mid-Proterozoic granites and migmatites of the Guiana Shield. The pronounced circular feature is probably a granite ring-intrusion (not shown on maps). The distinct mesa-like form comprises probable Lower Roraima Formation (or an older unit) arenites of ?Mid-Proterozoic age, draped, and possibly gently warped, over the older rocks. Remnants of the tabular arenites on the well fractured Basement rocks account for the preservation of these older rocks. A faulted fold of metasediment or volcanic horizon emerges prominently from featureless Basement rocks (bottom right). Note that the image over this forested area contains no direct radar-signature of lithologies. Context, morphology and inferred geological structure impart, however, substantial indirect information on lithology. The scarp of the Lower Roraima is emphasized by its being perpendicular to the radar signal. The white dots are fiducials representing each second of the Shuttle's flight. Radar illumination is from the top of the image and north is to the bottom left corner.

(*b*) Preliminary geological interpretation of (*a*) at 1:000000 scale. X, Rio Maraca shear zone in Basement. Full key given in the legend to figure 14.

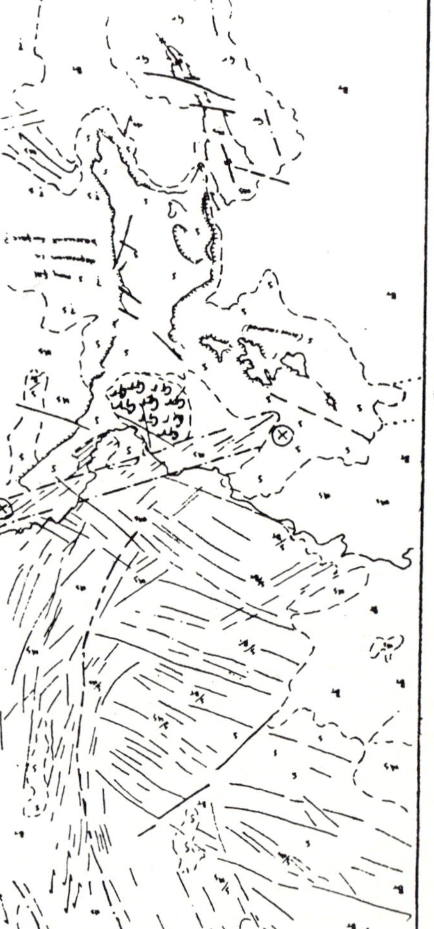

FIGURE 13. For description see previous page.

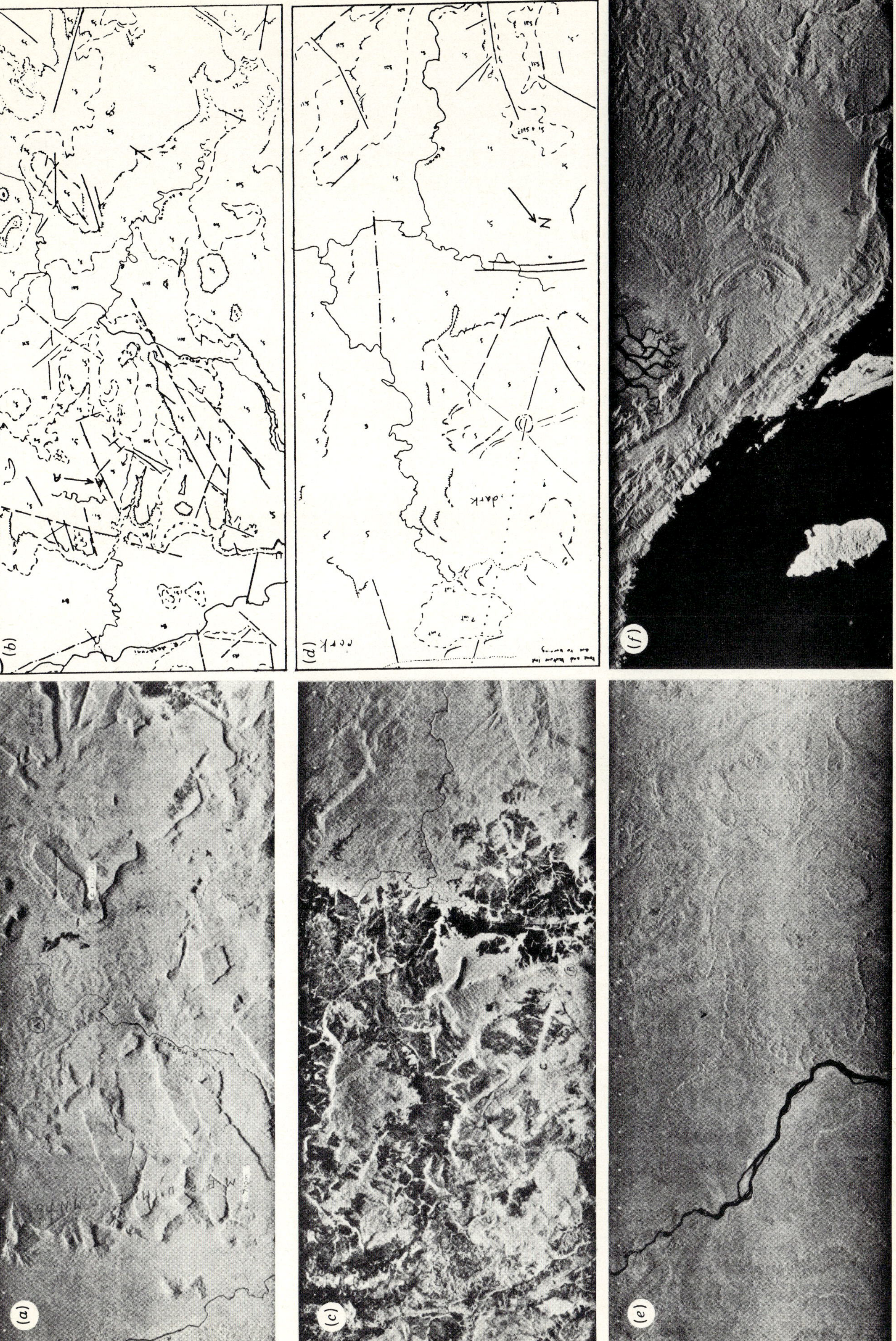

FIGURE 14. For description see overleaf.

DESCRIPTION OF PLATE 10

FIGURE 14. SIR-A over tropical regions at nominal 1:1000000 scale; radar illumination is from approximately southward, i.e. from top of each image.

(a, b) Western Guyana and eastern Venezuela (data-take 34). Forested Pakaraima Mountains and River Mazaruni Basin. Proterozoic clastic arenites of Roraima Formation form massif slabs with steep scarps; basaltic sills at A and partly waterlogged savannah (dark low relief areas) of La Gran Sabana at B Low ground on left underlain by gneisses of Guiana Shield.

(c, d) Eastern Venezuela (southeast Bolivar), and La Gran Sabana and Rio Caroni (data-take 34). Dark areas are partly waterlogged upland plateau savannah at 1000–1500 m. Note strong linear fracture A-B-A' and circular features at C. Alluvial diamonds in the region have not been traced to source and one can speculate that these circular features are traces of kimberlite intrusion.

(e) Mid-Amazon Basin, Brazil (data-take 24C). Dense rain forest and subtle traces of ?Carboniferous sediments.

(f) Lengguru fold belt, Vogelkop Peninsula, west Irian (data-take 32/33). Folded Tertiary sediments in dense tropical rain forested mountains.

Key to figures 13b and 14b, d. S, sedimentary formations, subscript indicates superposition, $S_1$ being oldest. Roraima arenaceous formations. Ms, Precambrian metasediments; d, dykes; sill, basaltic sills of Roraima Formation; int, igneous intrusions; Gr, Precambrian granitic plutons; Gns, Precambrian gneiss; BT, deeply weathered 'Basement' migmatites and granites, etc., masked by lateritized soils and vegetation.

DESCRIPTION OF PLATE 11

FIGURE 15. SIR-A over arid regions at nominal 1:1000000 scale; radar illumination is from approximately southward, i.e. from top of each image.

(a) Grand Kavir salt desert, north central Iran (data-take 28). Prominent but low-relief faulted anticlinal fold of mid-Tertiary continental sandstones and shales (producing alternate rough and smooth surfaces) apparently pierced by a salt-plug (not shown on geological maps). Surrounding areas are flat-lying Quaternary playas, silt and sand. Conspicuous salt plugs (Neogene) are especially well delineated by the L-band radar (right of image).

(b) Western Sahara (data-take 32/33). Remarkable sequence of regularly bedded Devonian arenaceous sediments showing minor faulting and drag-folding.

(c) Chaîne d'Ougarta, western Algeria (data-take 32/33). Cores of Upper Proterozoic sediments and volcanics surrounded by Lower Palaeozoic sediments. A dune field shows as a distinct black tone with dune features picked out by moderate backscatter from coarser materials and interdune areas.

(d) East of Ndjamena, Chad (data-take 28). Barchan-dune field derived from Lake Chad ancient deposits. Older long linear vegetated dunes can be seen at top right.

(e) West Queensland, Australia (data-take 35/36). Braided drainage of the Rivers Diamantina (left) and Hamilton (drains to Lake Eyre), crossing a mostly low-relief arid savannah underlain by Mesozoic platform-cover sediments outcropping at A. The bright tones of the drainage and outcrop results from radar backscatter from rough surfaces, in excess of a few centimetres trough-to-peak 'relief' according to Rayleigh formula. The black tones result from lack of backscatter from flat-lying waterlogged, and sand- and clay-covered, surfaces.

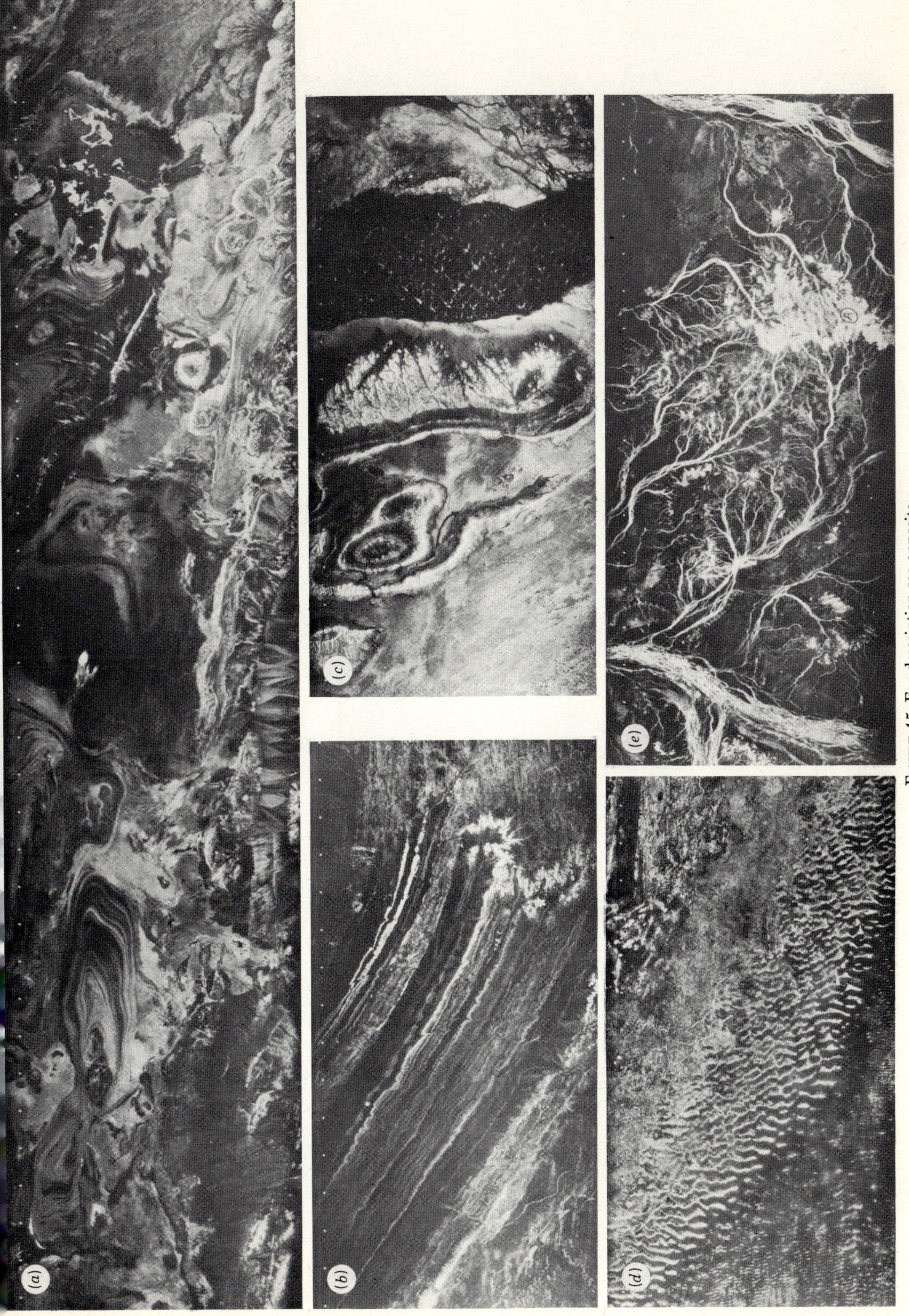

Figure 15. For description see opposite.

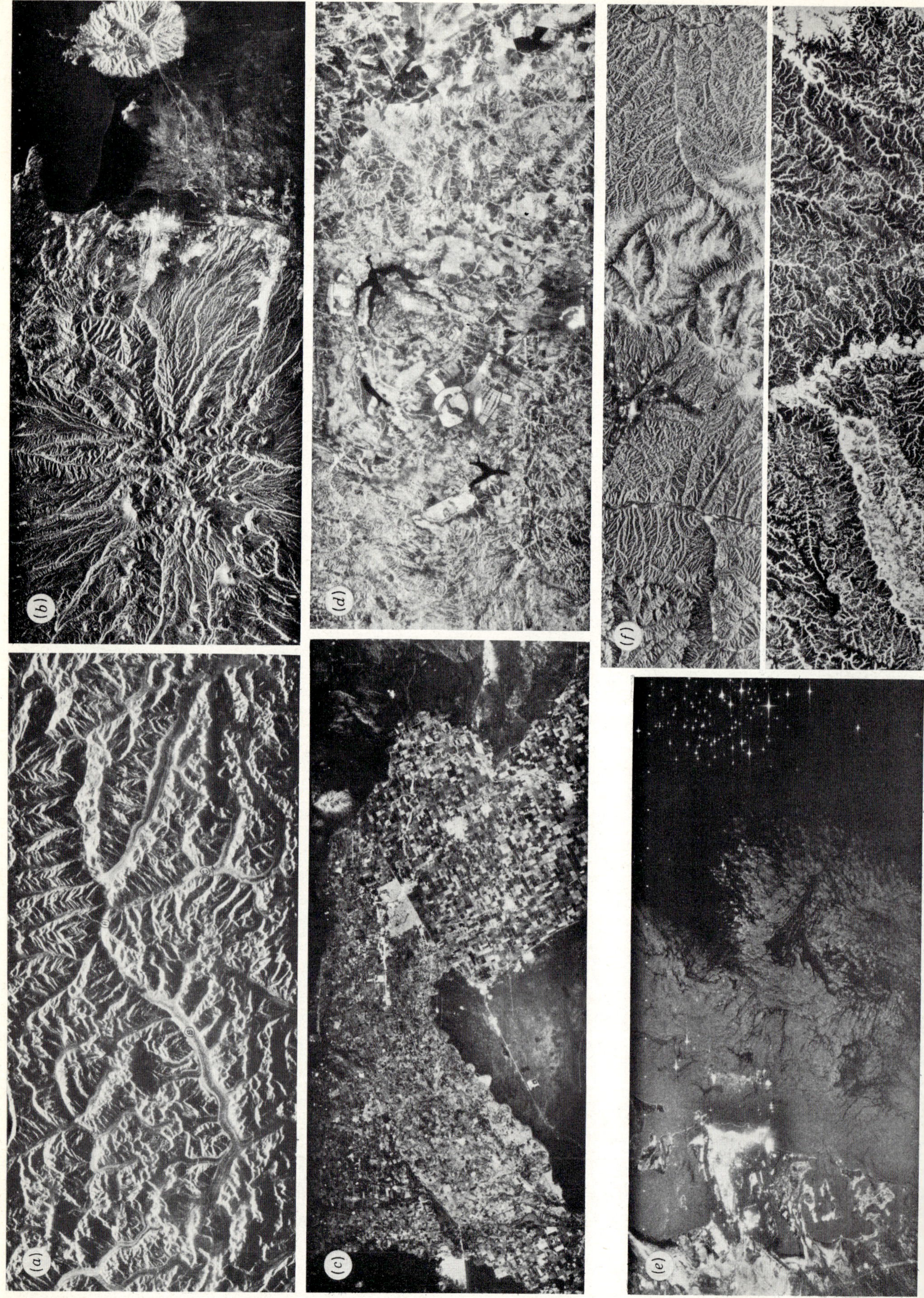

In low-relief arid regions the roughness of the surface is the predominant control of radar backscatter, as has been previously recognized from smaller areas inspected by SLAR and Seasat (Schaber *et al.* 1976; Daily *et al.* 1978; MacDonald & Waite 1973), and seen over wide areas on SIR-A images (e.g. figures 10 and 15*e*).

In the very arid Adrar des Iforas, northeast Mali, surface roughness is particularly well recorded on SIR-A imagery (figures 8 and 9*a–c*). The Adrar des Iforas forms a southwest extension of the central Saharan Hoggar Shield, and comprises Upper Proterozoic and reworked older basement sliced by major north-trending shears and comprehensively intruded by granitoid intrusives and volcanics. The central shear-bounded Iforas granulite unit (Eburnean age) is well shown on both Landsat and SIR-A (figures 8 and 9). To the west of the granulite unit, late Proterozoic granitic batholiths are intruded by prominent ring-intrusions of alkali granites, rhyolite and associated dyke swarms, which are especially well depicted on the SIR-A image (figure 8).

The Iforas SIR-A image presents substantially more information than the equivalent Landsat image, in part owing to the better resolution of SIR-A (40 m × 40 m compared with 50 m × 80 m of Landsat). There is also a possibility that the SIR-A L-band is penetrating dry surface sands and fine gravel, and backscattering from shallow-buried sub-outcrop (see for example the case of Sudan/Egypt (Eberhart 1982; Elachi *et al.* (1982*b*)). Work in progress in the U.K., however, suggests that this phenomenon may be limited to small areas and that it is the response to surface roughness and the better resolution that result in the greater data content of SIR-A (Lawrence & Martin-Kaye 1983). Examples similar to southern Egypt have been noted for central and northwest Saudi Arabia, central China and elsewhere.

### (*e*) *Studies for agriculture, land use and other purposes*

The SIR-A experiment was designed for geological purposes and no systematic land-use studies have yet been reported. The resolution of the bulk-processed material, 40 m × 40 m, is too coarse for any but general inventory studies at much the same level as that achievable from

---

### DESCRIPTION OF PLATE 12

FIGURE 16. Topography, land use and coastal features on SIR-A imagery at nominal 1:1000000 scale; radar illumination is from southward, i.e. from top of each image.

(*a*) Karakoram Range, northwest Pakistan/southwest China (data-take 32/33). Peaks of typically 8000 m in the highest part of the Himalayas. Although lay-over and foreslope brightening are widespread and occasionally severe, there is little interfering radar shadow. The high altitude of the sensor results in an even illumination across the swath even in mountainous terrain; morphological detail, e.g. glaciers, is good although the content of bedrock geological data is poor. The thick snow of the whole area is apparently completely penetrated by the L-band radar.

(*b*) Kuh-e-Sahand/Boz Dagh (3700 m) volcano, Tabriz, northwest Iran (data-take 35/36). A massive Quaternary andesite volcano (over 5000 km²) with several cones, mostly of dacite. Tuffs and ashes occupy the lower slopes. The Kuh-e-Choruglu Neogene basalt volcano (on right) is surrounded by the lake of Daryacheh Yerezaiyeh (top right corner).

(*c*) Mexicali, northern Mexico (data-take 24C). Dense cultivation in semi-arid, low relief area showing fields, roads and towns, and marked difference of field patterns across the U.S. frontier to the north.

(*d*) Brasilia, eastern Brazil (data-take 24C). Roads, buildings, reservoirs, minor drainage channels, forest and cultivated areas show clearly in this example from tropical savannah.

(*e*) Arabian Gulf and Abu Dhabi (data-take 37A). Features offshore Abu Dhabi (large bright aggregate reflexions) include petroleum drilling rigs (bright stellate reflectors) and sea surface disturbance probably influenced by bottom topography.

(*f*) Examples of two dense drainage networks from Rio Meticá, Colombia (data-take 24C), and Shansi Graben, China, upper Huang-ho River (data-take 28).

### TABLE 4. PROPOSED SPACEBORNE IMAGING RADAR PROGRAMMES BEGINNING BEFORE 1990

| | N.A.S.A.<br>SIR-B | N.A.S.A.<br>SAMEX | E.S.A.<br>ERS-1 | C.C.R.S.<br>Radarsat | Japan S.A.<br>ERS-1 | E.S.A./<br>D.F.V.L.R.<br>SAR facility | Dornier<br>M.R.S.E. |
|---|---|---|---|---|---|---|---|
| expected launch | 1984 | 1987 | 1988 | 1990 | ?1990 | ?1986 | ?1983–1984 |
| orbit altitude/km | 225 | 225 | 675 | 1000 | 570 | 250 | — |
| orbit inclination | 57° | 57° and polar | quasi-polar, Sun-synchronous; high angle; 3 day cycle | inclined polar, Sun-synchronous, 99.5° | ?polar, Sun-synchronous | — | — |
| frequency/GHz | 1.3 | ?1.3 and 9.6 | 5.3 | — | 1.2–1.3 | 5.3 and 9.6 | 9.62 |
| wavelength | 23 cm (L-band) | L and C synthesized X-band | X (and L?) | 5.7 cm (C-band) and possible 23 cm (L-band) | 23 cm (L-band) | X, C or simultaneous X-C | C-band |
| transmitted bandwidth/MHz | 12 (×2) | — | — | — | 13 | 14 | — |
| polarization | HH | HH, HV, VV | HH | — | — | ?VV; poss. dual polarization | HH/VV |
| depression angle | variable | variable | — | — | 33° | 46° (survey mode); 24° ('pointing' mode) | — |
| incidence angle (from vertical) | variable, between 15° and 60°, controlled from ground | 15–60° | *ca.* 35° | 30–45° | — | — | — |
| image swath width/km | 30–60 | 30–60 | 75–80 | 150–250 | 75 | 75 or 30+30 for simultaneous X and C | 8.5 |
| antenna | folding 8 panels; 4 m stowed; 10.5 m × 10 cm; tiltable | ?as SIR-B | *ca.* 10 m × 1 m | 16.5 m × 1.5 m | — | folding; 10 m × 1.6 m extended | 2 m long |
| image resolution/m | 17–55 | ?17–55 | 100 (1 dB) or 30 (2.5 dB) | 25–30 | 25 | 22–35 | 25 |
| number of looks | 4 | — | — | 3–4 | 4 | 5 | — |
| peak power transmitted/W | 1000 | — | (mean < 400) | — | 1500 | — | — |
| data acquisition time | 25 h digital plus simultaneous 8 h optical | — | — | continuous | — | — | — |
| area covered | 31 × 10⁶ km² in total | eventually global in several missions | ground track spacing of 900 km at equator; coverage could be increased by changing 3 day repeat cycle ×2 if use both ascending and descending passes; global coverage possible | — | — | — | small |

| | N.A.S.A. SIR-B | N.A.S.A. SAMEX | E.S.A. ERS-1 | C.C.R.S. Radarsat | Japan S.A. ERS-1 | E.S.A./ D.F.V.L.R. SAR facility | Dornier M.R.S.E. |
|---|---|---|---|---|---|---|---|
| recording | digital and optical; digital relayed by TDRSS at 46 Mbit s⁻¹ | digital and optical; possible real-time option | digital, at *ca.* 100 Mbit s⁻¹, and optical | direct transmission to 2 receiving stations with SAR processors, in Canada to serve as Arctic Pilot Ice mapping station; onboard recorders have limited space; use of data will need receiving station | optical and digital | digital; data stored on board at 120 Mbit s⁻¹ and relayed at 32 Mbit s⁻¹ | digital and optical |
| processing | possible real-time processor by 1985 (all data in 7 days at 15 m resolution); quick-look at 100 m available immediately; full resolution optical processed data need 2½ years to process | — | — | near real-time processing to produce images 4 h after over-flight (MDA developing technology) | — | — | — |
| other equipment | large-format vertical camera with 15 m resolution | — | SAR wave mode; C-band scatterometer, radar altimeter; ocean colour mapper (OCM) | wind–wave scatterometer; optical imaging system/ thematic mapper; scanning microwave radiometer; altimeter | — | — | — |
| platform | Shuttle | Shuttle; about 4 flights every 6–9 months, missions of 5–7 days each; Spacelab at later date | satellite | satellite with 3–5 year life; 3–4 needed for operations 1990–2000 | satellite | Shuttle | Spacelab; now developed into SAR Facility programme; may be incorporated on Shuttle |
| reference | Carver (1982) | Carver (1982)† | Honvault (1981) | Raney (1982) | Matsumoto *et al.* (1982) | Braun (1982) | Theile (1982) |

† N.B. FIREX (free-flying imaging radar experiment) is proposed to follow SAMEX on satellite or Spacelab.

Seasat. Nevertheless there are striking examples of how a radar of SIR-A type could monitor or assist studies in urbanization, forestry and forest-destruction, agricultural development, terrain and geomorphological analysis, development planning, irrigation and others.

Figure 7, showing of the North China Plain near Peking, illustrates the detail achievable by SIR-A as well as giving insight into the dense population of China. The image is a part of two swaths of similar density of towns and villages, over 75 000 km². Figure 16*c, d*, depicting croplands, contrasts strongly with the remote terrains of figure 16*a*, *b*, *f*. Figure 16*e*, of Abu Dhabi, depicts interesting urban and coastal features.

### 7. Future spaceborne imaging programmes

The significance of the results from SIR-A is that they illustrate the Shuttle's aptitude for experiment, and probably also for operations, with imaging radars. This was important because there are many questions yet to be resolved as to the optimum spaceborne radar system(s) for monitoring the Earth's land surfaces. The Shuttle appears to be the ideal vehicle with which the necessary experiment can be made. It combines global performance with the opportunity for varying the instrumental parameters.

SIR-B, again with L-band radar, but with variable-depression antenna and digital data transmission, is planned by N.A.S.A. for launch in 1984. This will permit the evaluation of the effect of depression angle in geological, natural vegetation and land-use objectives. The current proposal in N.A.S.A. is that SIR-B should be followed in the second half of the 1980s by a multi-polarization, multifrequency radar. This experiment, SAMEX (Shuttle Active Microwave Experiment) promises to be as important for agricultural and land-use purposes as for geological studies. It may not be until SAMEX flies that an overland operational radar satellite can be justified.

The E.S.A., the Canadian Centre of Remote Sensing (C.C.R.S.) and the Japanese Space Agency have major satellite imaging radar programmes planned during the 1980s. Summary details are given in table 4.

Multifrequency and multipolarization radar imaging will feature more prominently in future spaceborne programmes. The necessity of this for agricultural purposes has been demonstrated in the E.S.A.'s recent SAR-580 SLAR campaign (E.S.A./J.R.C. 1982; Haskell & Sorensen 1982), which was undertaken as part of the planning for the E.S.A.'s ERS-1 satellite. A U.K. example of the SAR-580 campaign (figure 2*a*) clearly demonstrates the usefulness of *both* X and L bands over land.

Another example reported by Sieber & Trevett (1982) demonstrates the potential that multifrequency and multipolarization imaging radars will have in crop discrimination. They produced co-registered images from the Oberpfaffenhofen test site of DFVLR. These considerably improved crop discrimination, the most effective image being the composite of the three frequencies X, C and L bands in VV polarization.

### 8. Discussion

The usable information content of commercial SLAR, Seasat and SIR-A (table 5) is large over land, especially at regional scales. The all-weather day and night capability of imaging radar, coupled with its sensitivity to surface slope, roughness and moisture, makes it an ideal complement to Landsat.

From approximately $18 \times 10^6$ km$^2$ of commercial SLAR imaging and many test-site investigations there has not seemed to be a firm consensus on the parameters most suitable for spaceborne radar. One of the problems in reaching consensus is that there are competing requirements between land and sea operations and between geology and land-use applications. Seasat and SIR-A confirmed the usefulness of L-band to applications *over land*. SIR-A showed that a moderate radar inspection angle (about 47° from vertical) produced image characteristics useful over a diversity of terrain, environment, vegetation and geology. Other wavebands, polarizations, depression angles, etc., remain to be tested in space.

Sir-a studies focus attention on five interpretational requirements for spaceborne radars *over land*.

1. *Repeat coverage*. Repetition of coverage is desirable for monitoring crops, for example, but for geological purposes a once-only cover or a repetition for different seasons, to permit multi-temporal comparisons, is all that is necessary. A long-cycle repetition, perhaps every 6–12 months, could also be useful to monitor desertification and hydrogeological changes and deforestation. Sir-a demonstrated that L-band could be particularly useful in desertification studies.

TABLE 5. SUMMARY COMPARISON OF PRINCIPAL CHARACTERISTICS
OF SEASAT AND SIR L-BAND (23 cm) IMAGING RADARS

|  | Seasat | SIR-A | SIR-B |
|---|---|---|---|
| orbit inclination/deg | polar, 108 | Shuttle, 38 | Shuttle, 57 |
| altitude/km | 794 | 268 | 225 |
| radar frequency/GHz | 1.275 | 1.272 | 1.282 |
| radar pulse bandwidth/MHz | 19 | 6 | 12 |
| swath width/km | 100 | 55 | variable, 30–60 |
| incidence angle/deg | 20 | 47 | variable, 15–60 |
| range resolution/m | 25 | 40 | variable, 55–17 |
| number of looks | 4 | 7 | 4 |
| processing | optical and digital | optical only | optical and digital |
| attributable cost/$M | 400 | 9 | ?13 (exclusive of launch costs) |

Global coverage could be achieved relatively quickly. Thome (1982), for example, calculates that the total *overland* Earth, about $145 \times 10^6$ km², could be sensed from polar orbit in 24 days, to provide image swaths 100 km wide with a 30 km side overlap (for 'stereo'; see Kaupp *et al.* (1982*b*)).

2. *Resolution*. For geological purposes the image interpretation matches the resolution available. The 40 m spatial resolution of sir-a produced images whose appearance were not markedly different from those of Seasat, even though the latter had 25 m resolution. For land-use purposes higher spatial resolution is needed for discrimination; although not as important for geology, geological interpretation would benefit from improved resolution.

3. *Image swath*. A swath of 50–100 km seems best to accommodate requirements for reasonably rapid regional coverage. The ideal imaging mode is probably a combination of simultaneous wide-swath moderate-resolution image with a narrow-swath (say 5–10 km) high-resolution (10 m or better) image.

4. *Radar frequency*. A combination of frequencies is probably the most important interpretational requirement. X and L bands have been widely tested, and experiments have also shown C-band to have merit. L-band with its lower power requirement proved itself on Seasat and sir-a. However, single-band images constrain interpretation, particularly in land-use applications. Multifrequency and multipolarization radar would permit better discrimination of lithological, vegetational and surface types. Experiments from space are needed to confirm the benefits to interpretation.

5. *Geometry*. Qualitative interpretation of radar images is dominated by recognition of

texture, shape and context. For geology there seems to be a side latitude of direction and inspection angle of radar illumination that is tolerable. Distortion rules out useful imagery over high-relief terrain, although the even illumination and consistent moderate inspection angles from space by SIR-A produced more acceptable images over high relief than those produced by SLAR. Varied inspection angles, especially over low relief and over large diverse areas rather than individual targets, need to be tested in conjunction with different frequencies.

N.A.S.A.'s SIR programme, and others, have been addressing the task of defining these requirements; the Shuttle offers an excellent vehicle to study them further. It maintains flexibility of payload, coverage and system parameters, allowing incorporation of technological development and providing extensive regional coverage over diverse terrain at relatively moderate cost. SIR-B and SAMEX will enable the testing and selection of the most appropriate system parameters (for both land and sea operations) before commitment to dedicated fully operational satellites.

The directors of Hunting Geology and Geophysics Limited authorized the publication of this paper. Hunting are joint official observers with the Royal Aircraft Establishment, Farnborough, to the SIR-A investigation team. SIR-A imagery reproduced here has been prepared by Hunting from optically correlated positive film transparencies supplied by the Jet Propulsion Laboratory, Pasadena.


REFERENCES AND SELECTED BIBLIOGRAPHY

Anon. (Various dates, 1970s and 1980s) *Projecto RADAM* (Multivolume reports covering geology, geomorphology, soils, vegetation and landuse). Rio de Janeiro: Ministerio das Minas e Energia. Summary maps at 1:1 000 000 scale, regional maps (in part) at 1:500 000 and 1:250 000 scales.

Arnold, H. J. P. 1981 *Br. J. Photogr.*, 9 Oct., pp. 1031–1041.

Azevedo, L. H. A. 1971 In *Proc. 7th Symp. Remote Sensing of Environment, ERIM, Univ. of Michigan, Ann Arbor*, vol. 3, pp. 2303–2306.

Black, R. *et al.* 1979 *Geol. Rdsch.* **68**, 543–564.

Blom, R. *et al.* 1982 In Keydel (1982), FA-6, pp. 9.1–9.6.

Blom, R. G. & Daily, M. 1982 *IEEE Trans. Geosci. Remote Sensing*, **GE–20**, 343–351.

Braun, H. M. 1982 In Keydel (1982), WP-9, pp. 1.1–1.6.

Bryan, M. L. 1973 *Radar remote sensing for geosciences: an annotated and tutorial bibliography.* ERIM, Rep. no. 193500 1-B. University of Michigan, Ann Arbor. (For Nat. Sci. Foundation, Contract no. G1-34089 X.)

Bryan, M. L. 1979 *Bibliography of geologic studies using imaging radar.* J.P.L. Pub. no. 79–53. N.A.S.A./J.P.L.

Carver, K. & Bush, T. F. 1979 *Airborne remote sensing of saline seeps: the 1978 Harding Co., S. Dakota experiment.* (Final Rep. N.A.S.A. Contract no. NAS 9-15421.) New Mexico, State Univ.

Carver, K. R. 1982 In Keydel (1982), TP-7, pp. 1.1–1.6.

Correa, A. C. 1980 In Harrison (1980), pp. 385–416.

Covault, C. 1982 *Aviation Week Space Technol.*, 2 August, pp. 17–18.

Curlander, J. C. 1982 *IEEE Trans. Geosci. Remote Sensing*, **GE–20**, 359–364.

Daily, M. *et al.* 1978 *Geophys. Res. Lett.* **5**, 889–892.

Dellwig, L. F. 1980 In Harrison (1980), pp. 351–364.

Diazgranados, D. A. 1979 *La Amazonia Colombiana y sus recursos. Proyecto Radar grametrico de las Amazonas* (*PRORADAM*). (Five volume report including memoir, and maps at 1:500 000 scale, for geology, soils, forestry/vegetation and landuse.) Bogota: Republica de Columbia.

Eberhart, J. 1982 *Science News* **121**, 419–420.

Elachi, C. 1980 *Science, Wash.* **209**, 1073–1082.

Elachi, C. 1982 In Keydel (1982), FA-6, pp. 5.1–5.6.

Elachi, C. *et al.* 1982*a* In Keydel (1982), FA-6, pp. 8.1–8.4.

Elachi, C. *et al.* 1982*b* *Science, Wash.* (In the press.)

E.S.A. 1981 *Synthetic aperture radar image quality: selected papers.* (Workshop ESRIN, Frascati, Dec. 1980.)

E.S.A./J.R.C. 1982 *SAR-580 Campaign 1981.* (Various reports by various authors for European Space Agency/Joint Research Centre.)

Estes, J. E. 1982 In Keydel (1982), FA-6, pp. 7.1–7.5.

Ford, J. P.　1982　In Keydel (1982), FA-6, pp. 6.1–6.6.
Ford, J. P. *et al.*　1980　*Seasat Views North America, the Caribbean, and Western Europe with imaging radar.* J.P.L. Pub. no. 80-67. N.A.S.A./J.P.L.
Froidevaux, C. M.　1980　In Harrison (1980), pp. 457–501.
Guyenne, T. D. & Lévy, G. (eds)　1981　*Coherent and incoherent radar scattering from rough surfaces and vegetated areas* (*Proceedings Workshop, Alpbach, Austria, March 1981*). E.S.A. Report no. SP-166. EARSeL/E.S.A.
Harrison, P. (chmn)　1980　*Radar geology: an assessment* (*Report of the radar geology workshop, Snowmass, Colorado, July 1978*). J.P.L. Pub. no. 80-61. N.A.S.A./J.P.L.
Haskell, A. & Sørensen, B. M.　1982　In Keydel (1982), WA-5, pp. 1.1–1.5.
Held, D. N. *et al.*　1982　In Keydel (1982), WP-8, pp. 2.1–2.6.
Honvault, C.　1981　In *Digest, 1981 International Geoscience and Remote Sensing Symposium* (*IGARSS '81*), *Washington June 1981*, pp. 1320–1326. Piscataway, N.J.: Institute of Electrical and Electronics Engineers Geoscience and Remote Sensing Society.
Hubbard, G.　1982　*Oilman*, October, pp. 19–25.
Hunting Geology and Geophysics Ltd.　1981　The evaluation of the data content of overland Seasat SAR imagery. Unpublished report prepared for Royal Aircraft Establishment, Farnborough. Final report. (Two volumes.)
Hunting Technical Services Ltd.　1977　NIRAD Project. Interpretation phase, progress report no. 3 (Digital analysis by Harwell). Federal Department of Forestry, Government of Nigeria.
Hunting Technical Services Ltd.　1978　NIRAD Project. Interpretation phase. Summary. Government of Nigeria.
Kaupp, V. H., Waite, W. P. & MacDonald, H. C.　1982*a*　*IEEE Trans. Geosci. Remote Sensing* **GE–20**, 383–390.
Kaupp, V. H. *et al.*　1982*b*　In Keydel (1982), TA-4, pp. 2.1–2.5.
Keydel, W. (ed.)　1982　In *Digest, 1982 International Geoscience and Remote Sensing Symposium* (*IGARSS '82*), *Munich, June 1982.* (Two volumes.) Piscataway, N.J.: Institute of Electrical and Electronics Engineers Geoscience and Remote Sensing Society.
Kobrick, J.　1982　*Astronomy*, August, pp. 18–22.
Lawrence, G. M. & Martin-Kaye, P. H. A.　1983　(In the press.)
Leberl, F.　1978　*ITC-Delft J.*, pt 1, pp. 167–190.
Leberl, F. W.　1980　In Harrison (1980), pp. 307–335.
de Loor, G. P.　1982　In Keydel (1982), TP-1, pp. 1.1–1.7.
Luther, C. A. *et al.*　1982　In Keydel (1982), TA-8, pp. 1.1–1.9.
McCandless, S. W. & Miller, B. P.　1975　*Acta astronaut.* **2**, 771–783.
MacDonald, H. C.　1980　In Harrison (1980), pp. 23–37.
MacDonald, H. C. & Waite, W. P.　1971　*Mod. Geol.* **2**, 179–193.
MacDonald, H. C. & Waite, W. P.　1973　*Mod. Geol.* **4**, 145–158.
McDonough, M. M. & Deane, G. D.　1979　In *Workshop Seasat SAR processor* (E.S.A. Rep. no. SP-154), pp. 9–18.
McDonough, M. M. & Martin-Kaye, P. H. A.　1983　*Int. J. Remote Sensing.* (In the press.)
McDonough, M. M., Martin-Kaye, P. H. A. & Deane, G. D.　1983　In *Proc. Symp. 'Seasat over Europe', Royal Society, April 1982.* EARSeL. (In the press.)
Martin-Kaye, P. H. A.　1973　In *Environmental remote sensing: applications and achievements.* (*Symp. Bristol, October 1972*) (ed. E. C. Barrett & L. F. Curtis), pp. 30–48. Edward Arnold.
Martin-Kaye, P. H. A.　1980　In Ackermann, F. (Ed.), *4th International Soc. Photogramm. Congress, Hamburg, July 1980* (*Int. Arch. Photogramm.* **23** (B8), Comm. 7) (ed. F. Ackermann), pp. 624–633.
Martin-Kaye, P. H. A. & Williams, A. K.　1973　In *Memoria IX Conferencia Inter-Guyanas, Caracas* (*Bol. Geol., Pub. especial* **6**), pp. 600–605.
Martin-Kaye, P. H. A. *et al.*　1980　In Harrison (1980), pp. 502–507.
Matsumoto, K. *et al.*　1982　In Keydel (1982), TP-7, pp. 2.1–2.5.
Matthews, R. E.　1975　*Active microwave workshop report.* N.A.S.A. Report no. SP-376, Washington, D.C.
Matthews, R. E.　1978　*Active microwave user workshop report.* N.A.S.A. Conference Pub. no. 2030. N.A.S.A./J.S.C.
Moore, R. K. *et al.*　1974　*Application of imaging radars: a bibliography.* Rep. Remote Sensing Lab. no. 265-2, University of Kansas, Lawrence.
Moreira, H. F.　1973　*Controle de qualidade de imagens de radar, Brazil.* (Dept. Nacional Producao Mineral, Projeto RADAM, tech. Rep. no. 57). Rio de Janeiro.
Nagler, R. G. & McCandless, S. W. Jr　1975　*Operational oceanographic satellites: potential for oceanography, climatology, coastal processes and ice.* Pasadena: J.P.L.
National Academy of Sciences　1977　*Microwave remote sensing from space for earth resources surveys.* N.A.S.A. publication no. CR-157891. Washington, D.C.: National Research Council.
Noel, J. & Pelleau, R. (eds)　1973　*Projet SARSAT. Synthese de l'étude des missions.* (Contributions from Sodeteg, Hunting Geology and Geophysics Ltd., B.R.G.M.)
Pala, D., Mussakowski, R. & Wedler, E.　1980　In *14th International Soc. Photogramm. Congress, Hamburg, July 1980* (*Int. Arch. Photogramm.* **23** (8), Comm. 7) (ed. F. Ackermann), pp. 754–771.
Paris, J. F.　1982　In Keydel (1982), FA-4, pp. 1.1–1.6.
Parry, D. & Trevett, J. W.　1979　*Geogrl J.* **145**, 265–281.
Raney, R. K.　1982　*The Canadian RADARSAT programme.* In Keydel (1982), TP-6, pp. 3.1–3.6.

Rouse, J. W. Jr (ed.)  1977  *Microwave remote sensing symposium, N.A.S.A./J.S.C., Houston, Texas, Dec. 1977* (and Workshop Rept).
Rouse, J. W. Jr  1982  In Keydel (1982), TP-1, pp. 3.1–3.5.
Sabins, F. F. Jr *et al.*  1982  *Bull. Am. Ass. Petrol. Geol.* **64**, 619–628.
Schaber, G. G. *et al.*  1976  *Bull. geol. Soc. Am.* **87**, 29–41.
Schlude, F.  1981  In Guyenne & Lévy (1981), pp. 29–30.
Sieber, A. J.  1982  In Keydel (1982), TA-1, pp. 5.1–5.3.
Sieber, A. & Trevett, J. W.  1982  *IEEE Geosci. Remote Sensing J.* (in the press.)
Taranik, J. V.  1982  In Keydel (1982), FA-6, pp. 1.1–1.5.
Taranik, J. V. & Settle, M.  1981  *Science, Wash.* **214**, 619–626.
Theile, B.  1982  In Keydel (1982), WA-7, pp. 1.1–1.6.
Thome, P. G.  1982  In Keydel (1982), WA-9, pp. 3.1–3.5.
le Toan, T.  1981  In Guyenne & Lévy (1981), pp. 99–110.
le Toan, T.  1982  In Keydel (1982), TP-1, pp. 3.1–3.5.
Ulaby, F. T., Moore, R. K. & Fung, A. K.  1982  *Microwave remote sensing – active and passive.* (Three volumes.) Remote Sensing Laboratory, University of Kansas.
Walters, R. L.  1968  *Radar bibliography for geoscientists.* C.R.E.S. Report no. 61-30, University of Kansas, Lawrence.
Wing, R. S.  1971  *Mod. Geol.* **2**, 1–21.
Wing, R. S. & Dellwig, L. F.  1970  *Tectonic development of the eastern Panamanian Isthmus as revealed through analysis of radar imagery.* (Annual Meeting, Geological Society of America, November 1970.)
Wing, R. S. & MacDonald, H. C.  1973  *Bull. Am. Ass. Petrol. Geol.* **57**, 825–840.

*Phil. Trans. R. Soc. Lond.* A **309**, 315–321 (1983)
*Printed in Great Britain*



# Synthetic aperture radar observations of ocean and land

By R. K. RANEY

*Canada Centre for Remote Sensing, Department of Energy, Mines and Resources,
2464 Sheffield Road, Ottawa, Canada K1A 0Y7*

[Plates 1–4]

A brief history of synthetic aperture radar (SAR) from early radar ideas to recent satellite
systems is presented. Properties of electromagnetic radar measurement that are
fundamental to SAR systems are presented and discussed by the use of example SAR
images derived from airborne and Seasat satellite sources. The next decade of
activity in radar satellites is outlined.

## 1. BACKGROUND

The first patent for a rudimentary radar was issued to Hulsmeyer (Germany) in 1904. There
were several technically successful but financially weak developments of special related devices
over the next 30 years, but it required the threat of World War II to motivate real progress.
Robert Watson-Watt proposed to the British Government in 1934 the principle of aircraft
detection by ground-based pulsed radar, and described in detail how it could be done. By the
spring of 1935 an experimental system proved to be successful, leading to the construction of the
British Home Chain, which saw five stations operational in 1938, a major technological
accomplishment.

The development of radar during World War II was explosively rapid, aided by high-level
technical exchange agreements amongst the Allies. The range and sensitivity of radars were
constantly improved, and smaller units were built for shipborne and airborne use.

These developments required higher power and shorter wavelengths than previously available. Whereas the work in 1935 was at $\lambda = 25$ m, sensitivity, size and performance requirements
led to wavelengths as short as $\lambda = 10$ cm by 1940, thus initiating microwave radar. The
pulsed high-power transmission requirement was met by the development of the cavity magne-
tron by John T. Randal & Henry A. Boot (University of Birmingham) during 1939–41,
achieving peak pulse powers on the order of 500 kW.

The development of radar by the Allies was very tightly classified during World War II.
For that reason, the several wavelengths in use received code names, in use to this day. Thus we
have inherited the following: L-band (25 cm), S-band, (10 cm), C-band (5 cm), X-band (3 cm),
and K-band (1 cm) in the popular microwave portion of the spectrum. Shorter waves are used
by 'millimetre radars'.

The close of World War II found aircraft equipped with radars that displayed on a 'plan
position indicator' (PPI) a 360° image of the terrain below the aircraft. Towards 1950 a modification of this approach led to a side-looking radar known as SLAR (side-looking airborne radar).
The radar antenna in this case, rather than spinning around through 360°, was kept pointing
to one side and the resulting imagery was that of the terrain parallel to the aircraft. The quality

of this imagery could be improved in resolution by shortening the range pulse, but the cross range or azimuth resolution was difficult to improve. Angular resolution is limited by the well known Rayleigh or diffraction limit, for which better performance requires a larger aperture. It is at this point that the logical development of synthetic aperture radar (SAR) begins.

An antenna carried on an aircraft or satellite cannot be made very large, thus apparently limiting the azimuth resolution. However, if the received energy could be recorded in memory for a set of successive scan lines, then perhaps the data in memory could be processed to achieve a much improved azimuth resolution, just as if those same scan lines had been simultaneously collected by a much larger antenna. This observation, first proposed by Carl Wiley of Goodyear Aerospace Corporation in 1954, lies at the heart of the synthetic aperture approach. The resulting theoretical azimuth resolution is $\rho = \frac{1}{2}d$ for an antenna of aperture size $d$.

Note that the diffraction-limited azimuth resolution of the synthetic aperture is not a function of wavelength and is improved as the aperture $d$ is made smaller. Both of these facts are in distinct contrast to the conventional diffraction-limited approach to a real aperture system. Note also that SAR azimuth resolution is not a function of range, and thus this radar technique is an ideal candidate for satellite applications.

There are four conditions that must be met by a side-looking radar sensor system for the resulting signal to yield successfully to synthetic resolution (see figure 1).

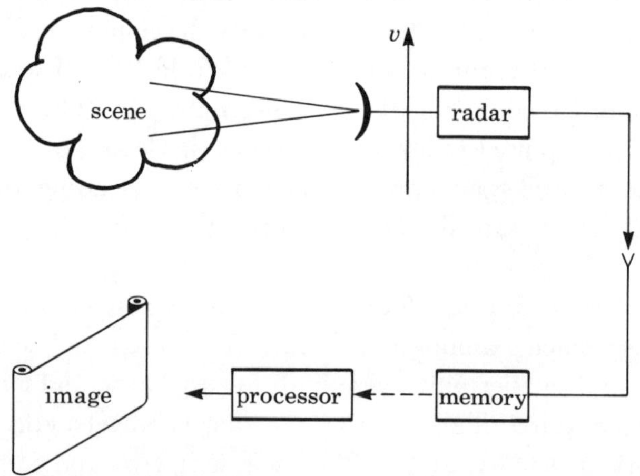



FIGURE 1. Conceptual diagram of a synthetic aperture radar.

1. Memory: the signal received at one point in time is of use in processing only in the context of all other signals received over the required synthetic aperture length. This requires a high-speed memory and in many applications a very large one at that. The memory accumulates data sequentially gathered by the radar as it traverses along the synthetic array length past the scene.

2. Processor: the set of signals must be processed to form an image. One differentiates between the signal history (input to processor) and the image itself (output from processor). For a SAR, the image processor is of importance equal to that of the radar itself, and is the most time-consuming task.

3. System coherence: individual frequencies may be processed only to the extent that they are present in memory. This requires that the wave front sampled by the radar as it traverses along the line of flight over the synthetic aperture length must have a consistent phase structure. If

there are across-track motion errors or timing errors in the system, they must be measured and compensated for, to maintain signal coherence. This usually imposes technological restraints on the radar itself, setting SAR systems apart from conventional side-looking real aperture radars.

4. Scene coherence: having a perfectly coherent radar is of no consequence if the scene or objects in it have their own random or deterministic motion. The scene itself must respect phase stability during the time over which the synthetic aperture is formed. If this condition is not satisfied then curious or disappointing results occur.

The first demonstration of an airborne SAR system was performed by the Willow Run Laboratories (now E.R.I.M.) of the University of Michigan in the late 1950s. The problem of maintaining coherence was met by the use of an inertial navigation and analogue phase-correction system. The memory and processor requirements were jointly achieved through recording the signals onto photographic film, which later was processed in an optical correlator.

A major improvement in the system, and a fundamental insight into its operation, was achieved by Emmett Leith. He applied the concept of an offset reference frequency (in common use in communication theory) to the wavefront reconstruction work of Dennis Gabor (first proposed by him in 1947) to arrive at the holographic view of the azimuth behaviour of a SAR. In this way, optical processing of SAR data was both simplified and established as the standard technique, essentially unchallenged until 1978. During that time, several experimental SAR systems were built at X, L, and S-bands; the military adopted X-band and developed production SAR for a variety of roles.

In 1978 the U.S.A. launched the first (and so far the only) satellite to carry a SAR system. Seasat as a proof-of-concept mission operated from July to October 1978, at which time there was a prime power system failure. The SAR was at L-band, with a resolution of 25 m × 25 m at four looks over a swath of 100 km, as seen by a nominal incidence angle of 20° from vertical.

Optical processing was the N.A.S.A. baseline plan for Seasat. However, there were independent developments of digital processing capability suitable for Seasat both in the U.S.A. and in Canada. The first digitally processed image from Seasat SAR was completed by MacDonald, Dettwiler and Associates (under funding from the Canada Sursat Project), and published in November 1978. Digital SAR processing proved to be outstanding in quality. The major remaining obstacle is speed. In 1978 it took 40 h to process one quarter frame of data, gathered by the satellite in a few seconds! The best rates currently are on the order of 2 h for one full frame of 100 km × 100 km.

## 2. IMAGE PROPERTIES

The image resulting from the processed data record collected by a SAR can be described in fundamental terms similar to more conventional passive imaging systems. The image brightness pattern is the result of a convolution of the system function (its response to an isolated point reflector) over the scene reflectivity function. Features of the image such as resolution, geometric properties, and background noise level are determined primarily by the SAR itself. Foundations for this description were set by P. M. Woodward.

Physical parameters of the surface that directly affect the image are surface roughness, dielectric constant and scene motion. Geometric parameters relating the radar and observed surface are polarization, radar wavelength and viewing angle.

*Surface roughness* is the most important scene characteristic determining relative reflectivity brightness, with rough surfaces appearing brighter in a side-looking radar image than smoother surfaces. The measure of roughness is wavelength $\lambda$ and incidence angle $\theta$ dependent according to the Rayleigh criterion, according to which the critical mean height variation $h_0 = \frac{1}{2}\lambda \cos \theta$, so that surfaces with variations less than $h_0$ appear 'smooth', and surfaces with height variations larger than $h_0$ appear 'rough'. As a general rule, one sees in images this effect at work, so that smooth water appears dark, most agricultural features show various shades of grey, whereas forests or major geological features are brightly represented.

Figure 2, plate 1, an image from the Seasat SAR processed digitally by M.D.A. (Canada), illustrates the differential brightness response to surface roughness. Radar data have a very wide range of intensity, even for natural surfaces, which for a scene such as this is typically two orders of magnitude.

The *complex dielectric constant* describes the electromagnetic properties of a material in quantitative terms. To first order, good conductors are good reflectors, and good insulators are poor reflectors. Most natural surfaces such as ice or vegetation are partly conducting. For these materials, microwave radiation penetrates the volume, is attenuated, and is reflected from within the material. These properties allow the possibility of measurement by SAR of certain material properties such as soil moisture, although the variable of interest is easily masked by system or roughness variations. Since for many interesting surfaces (e.g. ice) the dielectric constant is in fact a function of temperature and salt content, it is a source of unknown reflectivity variation as well as latent information. These are areas of active research in microwave remote sensing.

Figure 3, plate 2, shows the impact of rain on the reflectivity of a nominally dry region of Wyoming, as seen by Seasat in August 1978. The rain (brighter streaks) caused the dielectric constant of the surface to increase and hence the increased reflectivity.

*Scene motion* has impacts on imagery that are unique to SAR systems, a consequence of the time required to form the synthetic aperture along the flight track. Two classes of motion are of interest, deterministic and random.

For a discrete object with a (slowly varying) velocity, its image will be out of focus in azimuth in response to its along-track velocity component and any accelerations, and will be shifted in azimuth from its true position in proportion to its velocity towards the radar track. Figure 4, plate 3, shows the azimuthal shift of a ship's image relative to the wake of the ship. The amount of shift can be measured and reduced through the known radar viewing geometry to estimate the ship's speed accurately. The lack of such shift for waves accompanying the ship's progress is illustrative of the more subtle mechanisms that govern the formation of their images.

In Earth observation, random motions are of frequent occurrence, either on the surface of water bodies or wind-blown vegetation. It can be shown that the average reflectivity of such scenes is transferred to the image of a SAR unspoilt by the random motion, but that the ability of the SAR to maintain resolution is adversely affected if the scene complex correlation time is shorter than the time required by the SAR to form the aperture. This is a characteristic that frequently restricts the imaging response of a SAR to oceanic wave patterns.

One problem of oceanic wave imaging is illustrated in figure 5, plate 3, in which a swell system is propagating into a field of floating ice off the coast of Labrador. The waves are visible on the ice, but not on the water. This is due primarily to the fact that the relative time-varying complexity of the water surface reduces the ability of the SAR to achieve spatial resolution, in contrast to the conditions for the same dominant wave system in ice. This is an area of active research. (Data are from the Canadian X-band airborne SAR, and were processed optically.)

FIGURE 2. A Seasat SAR image of Vancouver, B.C.

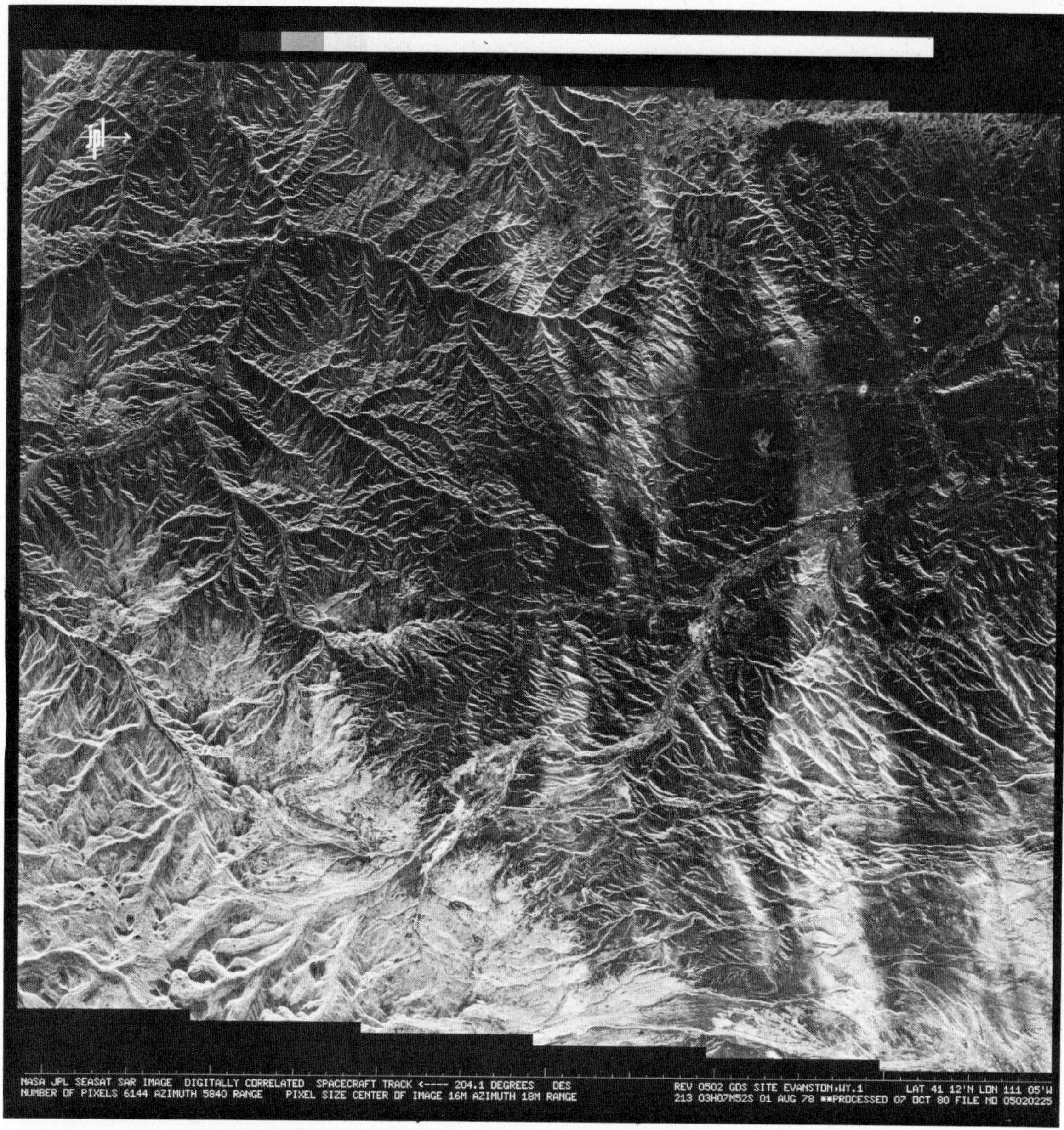



FIGURE 3. Effect of recent rain on dry terrain in Wyoming: the streaks are as a result of rain (verified by A. Loomis). Pixel size 16 m × 18 m. (Courtesy of N.A.S.A.)

   

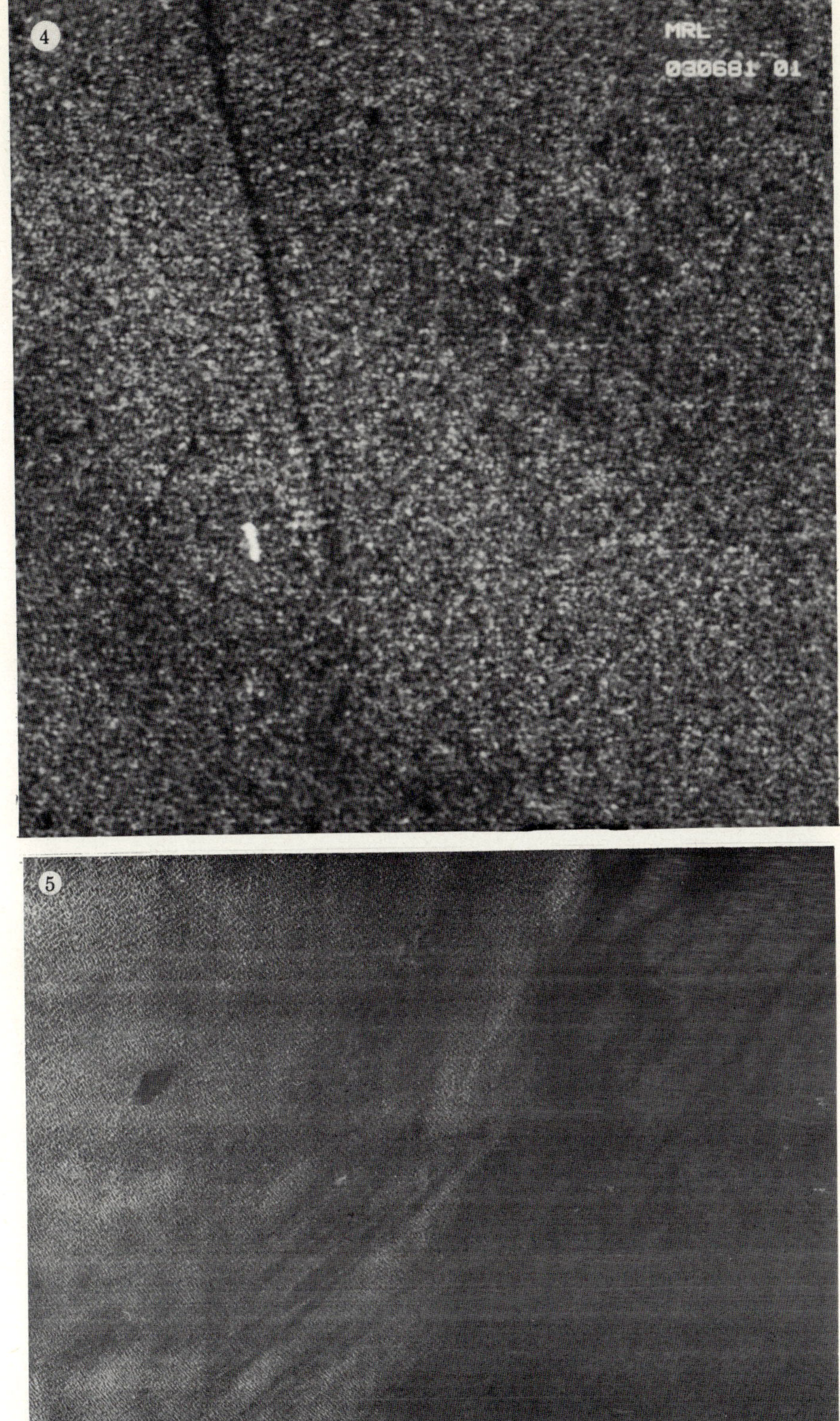

FIGURE 4. Image displacement of a ship caused by its motion (enlarged image, courtesy of Marconi Ltd).

FIGURE 5. Waves in an ice and water environment: waves are visible in
the floating ice only (bottom of figure is radar near range).

  

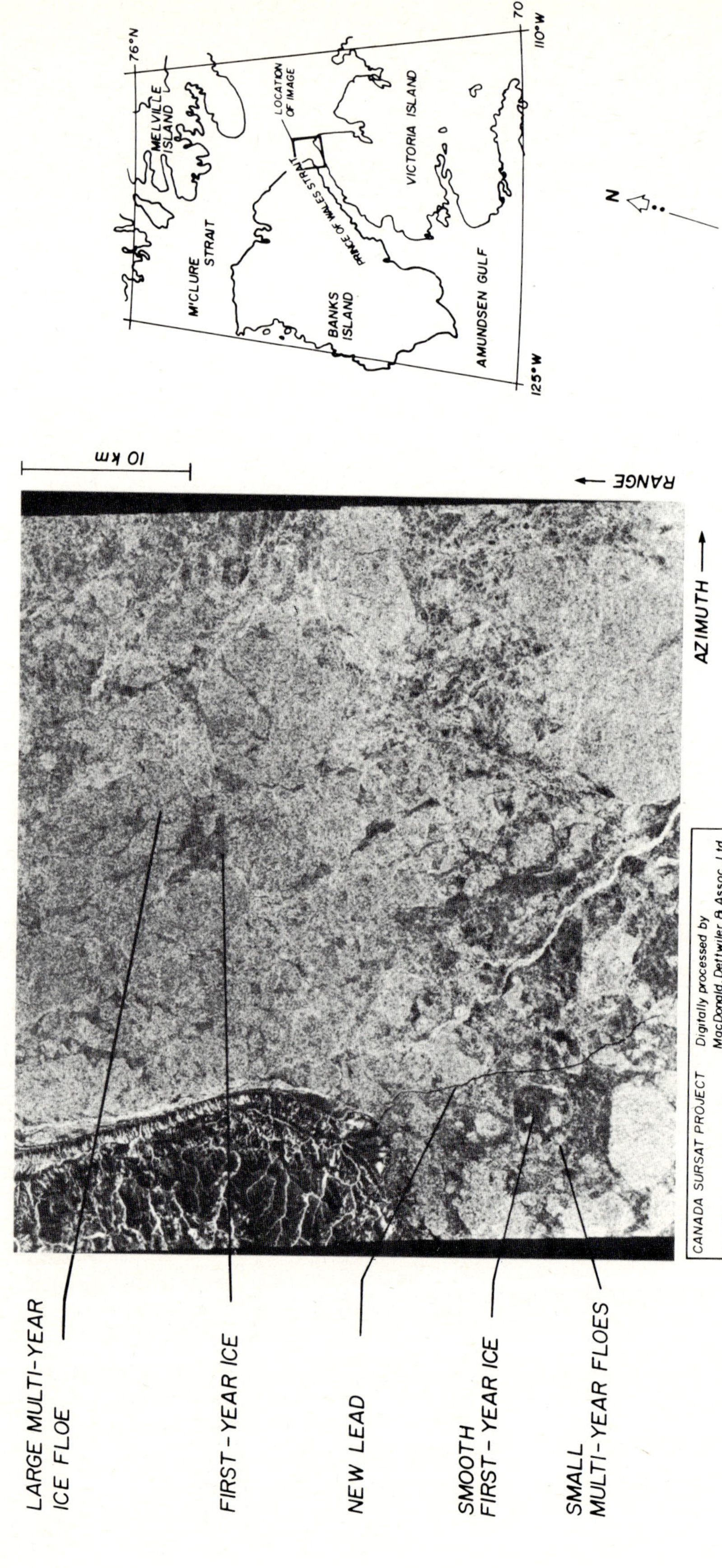


FIGURE 7. Seaset SAR image of Peel Point, N.W.T. (25 m resolution, 4 looks).

### 3. THE DECADE AHEAD

Worldwide, there are three free-flying Earth observation satellites planned for the next 10 years that are to carry SAR systems: ERS-1 (European Remote Sensing Satellite), Radarsat (Canada) and J-ERS-1 (Japanese Environmental Remote Sensing Satellite). The capabilities of these are in outlined table 1. The prime contractor for the ERS-1 radar system (whose acronym is AMI) is Marconi Ltd.

TABLE 1. COMPARISON OF FUTURE SATELLITE SARS

|  | Radarsat | ERS-1 | J-ERS-1 |
| --- | --- | --- | --- |
| launch | 1990 | 1988 | 1988 |
| radar band | C | C | L |
| processing | Par | Easier | Harder |
| resolution/m (looks) | 25 (3) | 30 (4) | 25 (4) |
| swath width/km | *ca*. 150 | 80 | 75 |
| accessible swath/km | 500 | 80 | 75 |
| incidence angle/deg | 20–45 | 22 | 33 |
| sensitivity/dB | −25 | −18 | ? |
| coverage | <76° N | <88° N | <80° N? |
| coverage interval at 73° N | <1 day | >4 days | ? |
| downlink | digital compatible | digital compatible | digital compatible |
| on-board recorder | yes | no | yes |

In addition, there are three shuttle imaging-radar missions planned, MRSE (Microwave Remote Sensing Experiment, the German contribution to Space lab), SIR-B (1984, N.A.S.A. approved, duration 1 week, maximum northern coverage 48° N), and an experimental SAR (1986, not approved by N.A.S.A., duration one week). This discussion does not consider these shuttle-based radars, since for operational requirements these systems are of little interest. (Although Radarsat, ERS-1 and J-ERS-1 are planned to be launched in the 1988–1990 time frame, none of these programmes are fully approved at the final build to completion commitment level.)

The advance of ERS-1 over Radarsat by 2 years is an advantage to both programmes. For Canada, a limited amount of data will be available to serve as test, validation and qualification input for ground stations, signal processors and product dissemination facilities. There will be an opportunity for user feedback and pre-operational experience. For E.S.A., there will be a genuine user 'on-line', ready to take advantage of the satellite information, thus serving as a pilot project for the validation of the satellite SAR system.

The radar system frequency has been selected to be C-band (5.3 GHz) for both the Canadian and E.S.A. systems. The Japanese system, essentially a copy of the Seasat SAR, is at L-band. The higher frequency of C-band is preferred for ice applications, based on Seasat and airborne experiments.

From an operational point of view, the two main issues are coverage and the availability of data to the user. The potential data-flow bottleneck is SAR data processing. Once processing speed is planned that is adequate to support Radarsat and Canadian users, then Canadian processing of ERS-1 data is a simple task. On the other hand, some compromises would be required to accommodate J-ERS-1 data, because L-band is more difficult to process accurately and rapidly.

The nominal swath of Radarsat is much wider than that of ERS-1 (150 km as against 80 km).

Furthermore, the Radarsat swath actually imaged can be selected (during a pass or mission if need be) anywhere within an accessibility zone 500 km wide. This has two major advantages.

1. It is possible to image any given location north of 51° N at least once every 72 h. This is of critical importance for site-specific or route-specific applications, such as shipping. This frequency of coverage, if met by a fixed swath system such as Seasat or ERS-1, would require a network of at least three satellites.

2. It is possible to gather data on subsequent orbits from one Earth location at different incidence angles, such as near 22° and near 43°. This provides a basis for an all-weather stereoscopic data set worldwide, which would be of great potential value in economic geology.

Thus it is that south of 75° N, Radarsat has substantial coverage advantages over ERS-1. This is achieved by use of swath steering and by use of a south-looking radar, chosen to maximize coverage of the Northwest Passage. But, as a consequence, the extreme northern regions are not covered by Radarsat; thus, north of 75 °N, ERS-1 has substantial coverage advantage (see figure 6), the Northern Beaufort Sea for example. Data from ERS-1 (out of range of

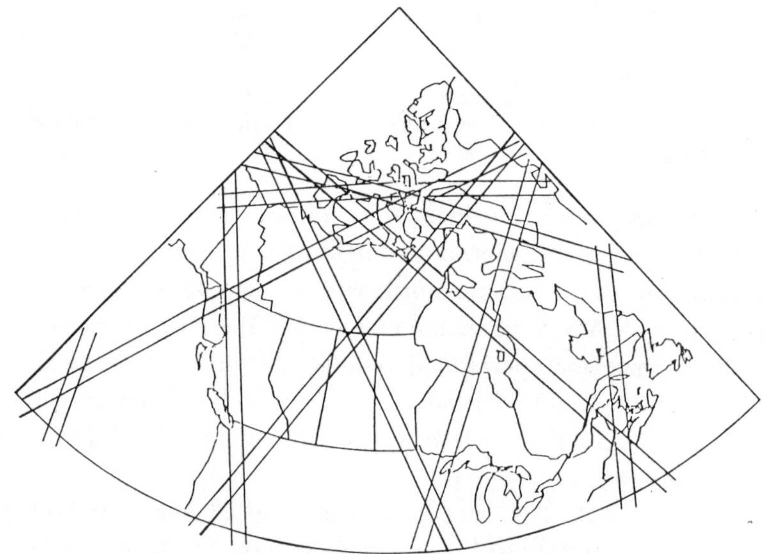

FIGURE 6. Arctic swath coverage for Radarsat SAR. Altitude, 1001.1 km; inclination, 99.48°; swath, 150 km (l.h.s.). Elevated Sun-synchronous orbit; 1 day of ascending and descending passes in a 3-day subcycle.

Radarsat) will be of importance as input to the ice-generating models being developed for the western Arctic. For this purpose the less frequent coverage of ERS-1 is satisfactory. Therefore, if Radarsat and ERS-1 lifetimes should overlap, these two radar sensors would be complementary.

Radarsat is designed to be more sensitive to weak signal variations than ERS-1. This means that distinction between smooth water and first-year ice will be more reliable from Radarsat. However, for the multiyear ice expected in the far North, ERS-1 performance should be satisfactory. It will be acceptable as a pilot project data source for most mid-Arctic conditions before the availability of Radarsat data. ERS-1 sensitivity should be adequate for most Canadian land applications.

The sensitivity parameter is closely related to the incidence angle parameter. For certain applications, the nominal angle of 22° (ERS-1) is nearly optimal for work such as on soil moisture, yet at just these angles, other applications, such as crop monitoring or iceberg detection, are poorly served. No one incidence angle can satisfy all users.

## Conclusions

Synthetic aperture radar has proved itself to be of value in scientific applications of land and sea observation. Future satellite systems are being planned that would employ SAR systems for experimental and operational Earth remote sensing. Both system technology and application methodology are under active development as we begin this decade.

## Bibliography

Beal, R. C., DeLeonibus, P. S. & Katz, I. (eds) 1981 *Spaceborne synthetic aperture radar for oceanography.* Johns Hopkins University Press.

Ford, J. P. *et al.* 1920 *Seasat views North America, the Caribbean, and western Europe with imaging radar (N.A.S.A. Jet Propulsion Laboratory Report* no. 80-67).

Fu, L. L. & Holt, B. 1982 *Seasat views oceans and sea ice with synthetic aperture radar (N.A.S.A. Jet Propulsion Laboratory Report* no. 81-120).

Long, W. L. 1975 *Radar reflectivity of land and sea.* Lexington Books.

Raney, R. K. 1982 *Int. J. Remote Sensing* **3**, 243–258.

*Phil. Trans. R. Soc. Lond.* A **309**, 323–335 (1983)
*Printed in Great Britain*



# Image processing for remote sensing

By J. Kittler

*S.E.R.C. Rutherford Appleton Laboratory, Chilton, Didcot, Oxfordshire OX11 0QX, U.K.*

In this paper a number of approaches to multispectral image segmentation and classification are considered. The methods range from the simple Bayesian decision rule for classification of image data on pixel-by-pixel basis, to sophisticated algorithms using contextual information. Both the spatial pixel category dependencies and the two-dimensional correlation-type contextual information have been incorporated in decision-making schemes. The aim of these algorithms is to achieve a greater reliability in the process of interpretation of remote-sensing data.

## 1. Introduction

Remote sensing is concerned with the problem of detecting the nature of an object and of its monitoring by means of external non-contact observations. In the study of Earth resources the observations are made by multispectral cameras and other passive and active sensors (infrared scanner, scanning radiometer, gamma-ray spectrometer, radar scatterometer, radar imager). The acquired multispectral–multisensor image data are then analysed by electronic means.

Image processing of remotely sensed data involves image correction, image preprocessing, image segmentation and, finally, image interpretation. These processing stages are depicted in figure 1.

The role of the first stage is to correct the recorded data for geometric distortions, the effects of the propagation medium, atmospheric variations, etc. In general, image data to be processed will be multichannel. If the data in individual channels are acquired by using different sensors, or at different instants, it is essential to massage it so that all picture elements (pixels) representing a particular point on the ground are brought into exact correspondence. This process of image alignment involves congruencing, rectification or simple image registration depending on the configuration of the sensing system. For detailed discussion of these topics the reader is referred to Bernstein (1978) and Haralick (1976).

Image preprocessing techniques have been developed to minimize variability in recorded data due to noise and various artefacts such as viewing-angle variations. A detailed treatment of digital filtering of two-dimensional data can be found in Huang (1979). Grey tone normalization for atmospheric, viewing-angle and intensity variations is reviewed in Haralick (1976). Image compression and coding for storage and transmission purposes also constitute an integral part of the preprocessing stage. Pratt (1978) provides a broad coverage of various approaches to this problem.

The ultimate goal of remote-sensing data processing is the interpretation of image segments that exhibit similar statistical properties. The lowest level of interpretation is the segment classification obtained at the output of the third stage of the model. However, this form of output rarely suffices because we are usually interested in extracting additional information from the data, such as the shape and size of objects that these segments represent and their spatial relation with other objects. Methodology for these higher levels of interpretation can be found in Fu (1977, 1982 *a, b*), Pavlidis (1977), Devijver & Kittler (1982) and Kittler (1983). Any ambiguity



25-2

in interpretation usually has to be resolved by using ancillary data such as topographic, soil-type and geopolitical maps, general knowledge about cultivation practices, rainfall conditions, seasonal characteristics – briefly, world models.

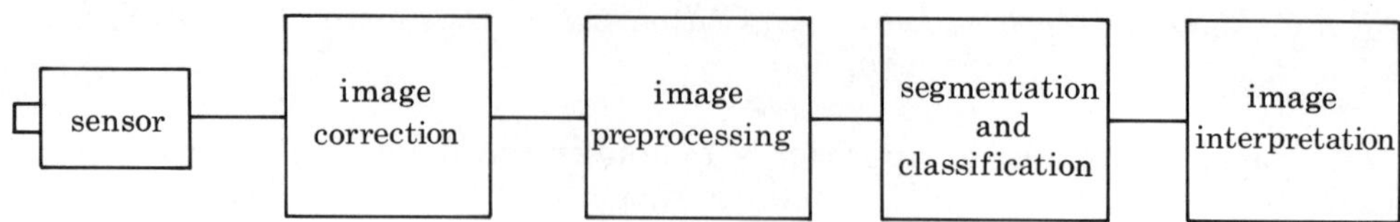


FIGURE 1. Image-processing system.

This paper concentrates on the problem of image segmentation and classification. In §2 the problem of classification of multispectral pixel data will be formulated and a simple solution based on the Bayes classification rule introduced. Section 3 deals with object classification by using spectral and textural features. The effect of one-dimensional spatial correlation between neighbouring pixels on the classification performance is studied in §4. Classification algorithms exploiting two-dimensional spatial dependences of pixel data are the subject of §5. Finally, §6 discusses probabilistic relaxation labelling algorithms that incorporate contextual information.

## 2. PIXEL-BY-PIXEL CLASSIFICATION

The first step in the analysis of remotely sensed data is to identify homogeneous segments in the image. Each homogeneous segment is then associated with one of the possible classes $\omega_i (i = 1, 2, ..., m)$ of land cover. This image segmentation and segment association process can be carried out in one of the following ways. Either we can first classify all the data on a pixel-by-pixel basis and then link identically labelled pixels to form connected segments. These segments are then associated with the class corresponding to the pixel label. Alternatively, we can first detect segments of pixels exhibiting similar properties. These homogeneous segments are then classified on a segment basis to the appropriate categories. Because of this duality, I shall consider the image segmentation and classification problems together.

I shall now describe a model for the multispectral remotely sensed data for analysis. For each point (or site) in the image we have a multivariate observation

$$x = [x_1, x_2, ..., x_d]^T, \tag{1}$$

where $x_i$ is the measurement obtained by the $i$th spectral channel of the sensor. We shall assume that the data generation process can be modelled statistically. In other words $x$ is considered to be a random variable having a unique conditional probability distribution for each class $\omega_i$. It will be convenient to characterize the probability model for $x$ in terms of the set of class conditional probability density functions $p(x|\omega_i)$ and the *a priori* class probabilities $P(\omega_i), \forall i$.

The above probabilities constitute all the essential ingredients for making decisions about class membership of patterns $x$, provided that it is appropriate to weigh all classification errors equally. The corresponding optimal decision rule, known as the Bayes minimum error rule, states (Devijver & Kittler 1982)

$$\text{assign} \quad x \quad \text{to} \quad \omega_i \quad \text{if} \quad P(\omega_i|x) = \max_j P(\omega_j|x). \tag{2}$$

Thus pattern $x$ is assigned to class $\omega_i$ if the *a posteriori* probability $P(\omega_i|x)$ of class $\omega_i$ given $x$ is greater than the *a posteriori* probability of any other class. The functions $P(\omega_j | x)$ can be computed by using the Bayes formula relating the conditional probabilities as

$$P(\omega_j | x) = \{P(\omega_j)\, p(x | \omega_j)\}/p(x), \tag{3}$$

where $p(x)$ is the mixture density given as

$$p(x) = \sum_{j=1}^{m} P(\omega_j)\, p(x | \omega_j). \tag{4}$$

The Bayes decision rule (2) has been in existence in one form or another for more than two centuries. The recent and current research in statistical pattern recognition relating to this topic is concerned with the problems of its implementation. The problems lie first of all in the fact that we do not know the probability functions $p(x | \omega_j)$. The only information assumed to be available for the recognition system design is a set of training patterns of which the class membership is known, that is ground truth data.

Second, the enormous amount of data that must be processed in one image alone has debarred the use of computationally involved classification schemes. Very simple decision-making algorithms are frequently found in commercially available systems. A typical example is the slicer classifier where a pixel is assigned to class $\omega_i$ if each component of vector $x$ falls within a specific interval corresponding to this class.

Alternatively we can assume that the probability distribution of vector $x$ has a parametric form. Under such an assumption the decision rule in (2) also becomes parametric. Apart from the resulting computational simplicity of the decision rule, this has the additional benefit that the probability distribution functions can be estimated more accurately because we need to infer only the parameters of these distributions.

The popular assumption of normality (Devijver & Kittler 1982), i.e.

$$p(x | \omega_i) = [(2\pi)^d |\Sigma_i|]^{-\frac{1}{2}} \exp\{-\tfrac{1}{2}(x - \mu_i)^{\mathrm{T}} \Sigma_i^{-1}(x - \mu_i)\}, \tag{5}$$

where $\Sigma_i$ is the covariance matrix of the $i$th class and $\mu_i$ is the mean vector, leads to the following parametric rule:

$$\text{assign} \quad x \quad \text{to} \quad \omega_i \quad \text{if} \quad (x - \mu_i)^{\mathrm{T}} \Sigma_i^{-1}(x - \mu_i) + K_i = \min_j \{(x - \mu_j)^{\mathrm{T}} \Sigma_j^{-1}(x - \mu_j) + K_j\}. \tag{6}$$

In (6) $K_i$ denotes

$$K_i = \log[(2\pi)^d |\Sigma_i|] - \log P^2(\omega_i). \tag{7}$$

Further, if the covariance matrices $\Sigma_i$ and class *a priori* probabilities $P(\omega_i)$ are identical, i.e. $\Sigma_i = \Sigma$, $P(\omega_i) = 1/m$, $\forall i$, then we can achieve even greater simplification, for now $K_i = K_j$, $\forall_j$ and (6) can be rewritten as

$$\text{assign} \quad x \quad \text{to} \quad \omega_i \quad \text{if} \quad (2x - \mu_i)^{\mathrm{T}} \Sigma^{-1} \mu_i = \max_j (2x - \mu_j)^{\mathrm{T}} \Sigma^{-1} \mu_j. \tag{8}$$

It follows that the decision rule in (8) is linear in $x$.

Finally when $\Sigma_i = I$ for all classes, the resulting decision rule is the well known nearest mean (minimum distance) classifier, i.e.

$$\text{assign} \quad x \quad \text{to} \quad \omega_i \quad \text{if} \quad \delta(x, \mu_i) = \min_j \delta(x, \mu_j), \tag{9}$$

where $\delta(x, \mu_j)$ is the Euclidean distance between vectors $x$ and $\mu_j$, defined as

$$\delta(x, \mu_j) = \{[x - \mu_j]^{\mathrm{T}}[x - \mu_j]\}^{\frac{1}{2}}. \tag{10}$$

The various simplifying assumptions can often be justified on the basis of computational involvement only. Although this will inevitably introduce approximation errors, some consolation can be drawn from the fact that with a given data base, simpler models can be estimated more accurately than more complicated ones.

In this paper I shall not impose any engineering constraints that would preclude the discussion of more sophisticated classification schemes. I shall adopt the view that the rapidly advancing technology will make it possible to implement such schemes in the near future. The recent commercial availability of parallel processors such as CLIP4 provides supporting evidence for this view. Moreover, the greatly improved data quality achieved with the launch of Landsat D/D makes it more important than ever before to use algorithms that can fully exploit the information content of the remotely sensed imagery.

The additional information that can be used is conveyed by spatial characteristics of the data. In particular, in classification of groups of pixels rather than one pixel at a time, we can take advantage of class homogeneity of land surface covers. Other useful information is contained in texture, shape, structural relations between land cover objects, context, and general dependences between pixels due to instrumental scanning errors, atmospheric turbulence, noise, weather conditions, etc. Algorithms that take some of these factors into consideration will be the subject of discussion in the following sections.

## 3. Object classification

### (a) Spectral features

The classification of data on a pixel-by-pixel basis discussed in the previous section can be affected by noise. If we consider that the size of land cover objects is usually considerably larger than the area corresponding to a single pixel, then the neighbouring pixels are likely to have similar properties. We can take advantage of this observation and identify homogeneous segments in the image that can be subsequently classified on a pixel-group (object) basis. In the context of the statistical estimation theory the expected improvement in performance is based on the premise that the larger the sample from which a statistic is inferred, the better its estimate.

Before we can proceed with object classification we need a procedure for determining homogeneous segments in the image. Kettig & Landgrebe (1976) and Landgrebe (1980) suggest partitioning the image into small regions. Those that have similar properties are then merged to create homogeneous segments.

*Segmentation procedure*

1. Group pixels into cells of $2 \times 2$ pixels (see figure 2).
2. Test statistical homogeneity. If the test fails then classify each singular pixel individually.
3. Check adjacent cells of non-singular pixels for statistical similarity. If similar, then merge the cells into one segment.

By using this procedure homogeneous segments will grow to their natural boundaries.

I now denote the set of pixels belonging to one of the segments by $X$, i.e.

$$X = \{x_1, x_2, ..., x_n\}. \tag{11}$$

By analogy with (2), I shall assign all the pixels in the set to class $\omega_i$ satisfying

$$P(\omega_i|X) = \max_j P(\omega_j|X). \tag{12}$$

The *a posteriori* probability $P(\omega_j|X)$ can be expressed in terms of the joint conditional and mixture probability density functions, i.e.

$$P(\omega_j|X) = \frac{p(x_1, ..., x_n|\omega_j)\,P(\omega_j)}{p(x_1, ..., x_n)}. \tag{13}$$

<table>
<tr><td>$k,l$</td><td>$k,l+1$</td></tr>
<tr><td>$k+1,l$</td><td>$k+1,\ l+1$</td></tr>
</table>

FIGURE 2. A pixel cell.

If the variables $x_k$ ($k = 1, 2, ..., n$) are conditionally statistically independent then the numerator in (13) can be evaluated as a product of pixel densities, i.e.

$$p(x_1, ..., x_n|\omega_j) = \prod_{k=1}^{n} p(x_k|\omega_j). \tag{14}$$

In general, conditionally independent variables will be unconditionally dependent and the denominator cannot therefore be simplified in a similar manner. However, as the denominator is identical for all classes and only finding the most probable class is of interest, the following rule can be used:

$$\text{assign} \quad X \quad \text{to} \quad \omega_i \quad \text{if} \quad P(\omega_i) \prod_{k=1}^{n} p(x_k|\omega_i) = \max_j P(\omega_j) \prod_{k=1}^{n} p(x_k|\omega_j). \tag{15}$$

### (b) Texture measures

In many cases the spectral characteristics of data varies from pixel to pixel. If these spatial variations are (locally) stationary, they will give rise to an apparent regular spatial pattern, which is referred to as texture. Texture can be characterized by quantitative textural measures. Many such measures have been suggested in the literature but it is beyond the scope of this paper to discuss them in detail. Broadly speaking, the following categories of texture measures have been proposed (Haralick 1979; Pratt 1978; Niemann 1981; Weszka *et al.* 1976): first-order statistics of grey level differences; second-order statistics of the grey tone spatial dependence matrix; run length statistics; two-dimensional power spectrum measurements; edgeness per unit area. However, this list is by no means exhaustive.

In general there are no guidelines as to what texture measures are best suited in given circumstances. For given data, answers to these questions can be found by using feature-selection methods. More specifically, each texture measure is considered as a candidate feature; suppose we extract $N$ such measures. Then feature selection is concerned with the problem of selecting a subset, $d < N$, of these candidate measures that allow for best discrimination between various classes of textures. The selected subset of features forms a feature vector $x$, which can be classified by using the decision rule in (2). A recent discussion of this topic can be found in Devijver & Kittler (1982).

It is apparent that texture measures can be computed only for groups of neighbouring pixels (not for a single pixel). The most appropriate procedure is to specify a window of $k_x \times k_y$ pixels and base the texture measure computation on the resulting group of $n = k_x k_y$ pixels. Again there are two options. Either this window can be centred at successive pixels and for each a vector of texture measurement can be obtained. This vector can then be assigned to an appropriate land cover category.

Alternatively the image can be segmented first by using the procedure described in the previous subsection. Note that here the initial cells are $k_x \times k_y$ instead of $2 \times 2$ pixels. Once homogeneous segments are detected, the texture statistics can be recomputed for the complete segments and the segments then classified.

### 4. One-dimensional spatial dependences

With the exception of texture classification discussed in the previous subsection, so far I have assumed that pixel data are conditionally independent. In other words, observation $x_{kl}$ corresponding to the $(k, l)$th pixel is independent of the measurements on pixels in its neighbourhood. In practice the value of $x_{kl}$ is likely to be correlated with those of the surrounding pixels, $x_{k+r, l+s}$, where $r$ and $s$ are small positive and negative integers. In general, these correlations should be modelled by two-dimensional spatial random processes and I shall consider such models in the next sections. Here I shall assume that there are correlations only between successive variables in each scan line. This assumption has been shown to be realistic in a number of experimental studies. In particular, low-order autoregressive and autoregressive moving-average models appear to fit the Landsat data reasonably well (Craig 1979; Tubbs & Coberly 1978; Tubbs 1980). It allows us to model the data by using simple one-dimensional spatial models, which have an analogy in the time-series analysis.

For simplicity I shall further assume that observations in one spectral channel are not spatially dependent on observations in other channels. Thus for modelling purposes I consider only one spectral channel at a time. Now suppose the sensor is scanning the $k$th line of the current image frame in some spectral channel. According to the ARMA model the $l$th observation $x_l$ is related to the previous observations as

$$\sum_{t=0}^{p} \psi_t x_{l-t} = \sum_{t=0}^{q} \varphi_t (\epsilon_{l-t} + \nu_{l-t}), \tag{16}$$

where $p$ and $q$ denote the autoregressive and moving-average processes respectively, $\psi_t$ and $\varphi_t$ are the parameters of the process and $\epsilon_l$ is a normally distributed independent random variable with zero mean and variance $\sigma^2$; $\nu_l$ is the true (uncorrupted) intensity at the $l$th pixel, characteristic of a particular class of land cover. Note that $\psi_0 = \varphi_0 = 1$.

The model in (16) depicts the situation where due to crosstalk or a cell–spot overlap, the measured output $x_l$ at the $l$th pixel in each scan line depends on the outputs at pixels $x_{l-t}$ ($t = 1$, $2, \dots, p$) and the history of the noise.

This general type of model may also be found useful in describing correlation between the noise variables. For instance, atmospheric turbulence will give rise to spatial dependence of the noise, which strictly speaking is two-dimensional. However, because the data are acquired in a scanning fashion and the atmospheric conditions change as a function of time, the noise components in one scan line can be considered independent of those in the previous line.

I shall now analyse the effect on pixel classification of a particular type of model with $p = 1$ and $q = 0$. A model of this order has been suggested for remotely sensed data by Tubbs & Coberly (1978). Ignoring the transient effect in the border regions between homogeneous segments, from (16)

$$y_l + \psi_1 y_{l-1} = \epsilon_l, \tag{17}$$

where $y_l$,

$$y_l = x_l - \mu_l, \tag{18}$$

is the centralized observation on the $l$th pixel and $\mu_l$ is the corresponding mean. From (17) the variance, $\tau^2$, of $y_l$ satisfies

$$\tau^2 = E\{y_l^2\} = \sigma^2 + \psi_1^2 \tau^2, \tag{19}$$

which can be rearranged as

$$\tau^2 = \sigma^2/(1 - \psi_1^2). \tag{20}$$

From the stability point of view parameter $\psi_1$ must satisfy $|\psi_1| < 1$. Thus the variance of $x_l$ is larger than that of $\epsilon_l$. It follows that the AR(1) autoregressive process introduces between-variable dependences that give rise to observations with a considerably larger variance than the noise process $\epsilon_l$.

Usually the larger the variance, the more difficult it is to achieve a low-error pixel classification performance. It is therefore desirable to base the classification process on a filtered variable that exhibits a lower variance than $\tau^2$.

In the context of model (17) this filtered observation $z_l$ can be computed as

$$z_l = x_l + \psi_1 x_{l-1}. \tag{21}$$

Variable $z_l$ in (20) has class-conditional mean $v_{li}$ $(i = 1, 2, ..., m)$ and variance $\sigma^2$. It is independent of previous outcomes $z_{l-t}$, $t = 1, 2, ...$, and, therefore, it should afford a more reliable classification of pixel data. However, this conjecture will now be scrutinized to verify its validity.

Assume that the pixel data is normally distributed with the class mean $\mu_{li}$ and equal variance $\tau^2$. Further, suppose that *a priori* probabilities of classes $\omega_i$ are equal. Then our ability to discriminate between classes $\omega_i$ and $\omega_j$ on the basis of observation $x_l$ can be measured by using the Mahalanobis distance (Devijver & Kittler 1982), defined as

$$J(\omega_i, \omega_j) = (\mu_{li} - \mu_{lj})^2/\tau^2. \tag{22}$$

Suppose that pixel $x_l$ is classified on the basis of the filtered measurement $z_l$. Then

$$J'(\omega_i, \omega_j) = (v_{li} - v_{lj})^2/\sigma^2. \tag{23}$$

I shall now compare the criteria $J$ and $J'$ for model (17). From (16), (17) and (18),

$$\mu_l = -\psi_1 \mu_{l-1} + v_l, \tag{24}$$

which for a homogeneous region where $\mu_l = \mu_{l-1}$ leads to

$$\mu_l = v_l/(1 + \psi_1). \tag{25}$$

Substituting from (20) and (25) into (23),

$$J'(\omega_i, \omega_j) = \{(1 + \psi_1)/(1 - \psi_1)\} J(\omega_i, \omega_j). \tag{26}$$

Thus in this case the classification performance for the filtered image will improve or deteriorate depending on the sign of parameter $\psi_1$, i.e. whether the model is basically a low-pass or high-pass filter.

In view of these results it is important to study the model effects very carefully before a commitment to a specific classification strategy is made. It is possible that the combined data generation and acquisition process is such that better results can be achieved in the original observation space. On the other hand, if the model of the process indicates that it is advantageous to classify pixel data by using filtered observations, an appropriate filtering scheme must be derived from the model. The extension of these results to multivariate observations is straightforward.

[ 87 ]

## 5. Two-dimensional spatial dependences

In this section I shall consider how the contextual information imbedded both in pixel category relations and in two-dimensional correlations can be exploited to improve classification performance. In the preceding sections I made use of both types of information either explicitly or implicitly. The particular contextual information exploited is that neighbouring pixels are likely to belong to the same category. This relation, which I have termed homogeneity, led to the object classification rule (15) which has been shown to yield a lower classification error than the basic rule in (2) (Landgrebe 1980).

FIGURE 3. Neighbouring pixels.  FIGURE 4. The four-neighbourhood.

Before any observations are made, the pixel category relation that gives rise to homogeneity can be characterized in terms of joint probabilities. Adopting the notation in figure 3, the dependence of the class membership $\theta_k$ of the $k$th pixel on the class membership $\theta_{k+1}$ of its neighbour $k+1$ can be quantified by the probability of joint occurrence of $\theta_k$ and $\theta_{k+1}$, $P(\theta_k, \theta_{k+1})$. For homogeneity this is $P(\theta_k = \omega_i, \theta_{k+1} = \omega_i)$. (Here I am using the same notation for the random variable $\theta_k$ and its realization.)

In principle, the probability that label $\theta_k$ will take a particular value could depend on all the pixels in the image, and certainly on more than one of the neighbours of pixel $k$. For simplicity and not without justification I shall, however, assume that the region of mutual influence extends only to the four pixels 1, 2, 3 and 4 as shown in figure 4. (Index $k$ has been dropped for convenience.) In other words the effect of more distant pixels and of the pixels in the diagonal directions will be ignored. Then formally homogeneity can be characterized by the joint probability $P(\theta_0 = \omega_i, \theta_1 = \omega_i, \theta_2 = \omega_i, \theta_3 = \omega_i, \theta_4 = \omega_i)$.

Homogeneity, of course, is not the only pixel category relation that can occur and is of interest in remotely sensed data. A particular land cover at one pixel may provide strong contextual evidence for another class at a neighbouring pixel; for instance, the land cover on each side of a road is likely to be similar and in any case its special signature will be different from that of the road. Objects on a river are more likely to be boats rather than cars. In general, therefore, what is of interest is the probability of particular spatial relations between pixel categories. The contextual information provided by these relations is embodied in the joint *a priori* class probabilities $P(\theta_0, \theta_1, ..., \theta_4)$. Introducing a vector notation $\boldsymbol{\theta}$ for the class labels, i.e.

$$\boldsymbol{\theta} = (\theta_0, \theta_1, \theta_2, ..., \theta_4)^{\mathrm{T}}, \tag{27}$$

this joint probability can be written as $P(\boldsymbol{\theta})$.

In the following subsections I shall consider how these joint probabilities can be incorporated in the decision-making process.

*(a) Decision rules based on contextual information*

Similarly to object classification discussed earlier, decisions are to be based on a set of observations rather than on one pixel at a time. As distant pixels are assumed to have no bearing on the classification of pixel $x_0$, the set of interest, $X$, is defined by

$$X = \{x_0, x_1, ..., x_4\}. \tag{28}$$

Note that here we use $X$ to classify only $x_0$, not all the elements of the set as in § 3 $a$.
By analogy with (12) $x_0$ should be assigned to the most likely class, i.e.

$$\text{assign} \quad x_0 \quad \text{to} \quad \omega_l \quad \text{if} \quad P(\omega_l|X) = \max_i P(\omega_i|X). \tag{29}$$

The application of the Bayes formula for calculating conditional probabilities would transform the decision rule (29) into a scheme defined in terms of probability density functions $p(X|\omega_i)$. However, because of the complex dependences assumed between the variables of the four-neighbourhood, this probability density function cannot in general be simplified. Inference of the joint density function $p(X|\omega_i)$ would lead to estimation and computational problems, because working with $X$ amounts to increasing the dimensionality of the random variable by a factor of five. I shall therefore attempt to simplify the decision rule by considering the original *a posteriori* functions $P(\omega_i|X)$ direct.

Expand $P(\omega_i|X)$ as

$$P(\omega_i|X) = \sum_{S_i} P(\theta|X) = \{1/p(X)\} \sum_{S_i} p(X|\theta)\, P(\theta), \tag{30}$$

where $S_i$ is the set of $\theta$ such that $\theta_0 = \omega_i$. Because the denominator term $p(X)$ is identical for all the classes it can be ignored, thus leading to the decision rule

$$\text{assign} \quad x_0 \quad \text{to} \quad \omega_l \quad \text{if} \quad \sum_{S_l} p(X|\theta)\, P(\theta) = \max_i \sum_{S_i} p(X|\theta)\, P(\theta). \tag{31}$$

Moreover, if the class labels are independent, i.e. the spatial configuration of pixel classes conveys negligible contextual information, the rule becomes

$$\text{assign} \quad x_0 \quad \text{to} \quad \omega_l \quad \text{if} \quad \sum_{S_l} p(X|\theta)\left\{\prod_{j=1}^{4} P(\theta_j)\right\} P(\omega_l) = \max_i P(\omega_i) \sum_{S_i} p(X|\theta) \prod_{j=1}^{4} P(\theta_j). \tag{32}$$

These classification schemes involve estimation of the joint probability density functions $p(X|\theta)$. In general these functions will be very difficult to obtain. A notable exception is when the data can be adequately modelled as an autonormal scheme discussed by Besag (1974), Bartlett (1975) and Fu & Yu (1980). The joint density function of such a scheme for univariate observations is defined as

$$p(X|\theta) = (2\pi\sigma^2)^{-\frac{5}{2}} |B|^{\frac{1}{2}} \exp\left\{-\tfrac{1}{2}\sigma^{-2}(X-u)^{\mathrm{T}} B(X-u)\right\}, \tag{33}$$

where $\sigma^2 B^{-1}$ is the variance–covariance matrix, $u$ is the vector of means of classes $\theta_0, \theta_1, ..., \theta_4$ and $X$ is a vector of univariate observations corresponding to the four-neighbourhood in figure 4. Methods for estimating $\sigma^2$ and $B$ are documented in Fu & Yu (1980).

Under the assumption that the appearance of $x_j$ is a function of $\theta_j$ only, which implies that

$$p(x_j|\theta_j, x_l, \theta_l, l \neq j) = p(x_j|\theta_j), \tag{34}$$

$p(X|\theta)$ in (31) can be simplified as

$$p(X|\theta) = p(x_0|\omega_i) \prod_{j=1}^{4} p(x_j|\theta_j). \tag{35}$$

From (31) and (35) the joint class probability decision rule becomes

$$\text{assign} \quad x_0 \quad \text{to} \quad \omega_l \quad \text{if} \quad p(x_0|\omega_i) \sum_{S_l} P(\theta) \prod_{j=1}^{4} p(x_j|\theta_j) = \max_i p(x_0|\omega_i) \sum_{S_i} P(\theta) \prod_{j=1}^{4} p(x_j|\theta_j). \tag{36}$$

In addition to the density function $p(x|\omega_i)$, the decision rule (36) requires the estimation of *a priori* joint probability function $P(\theta)$. This will require the availability of a large quantity of ground truth data.

Finally, it should be noted that a pixel can be classified by determining the most probable combination of class labels $\theta$ for the appropriate neighbourhood. Accordingly,

$$\text{assign} \quad x_0 \quad \text{to} \quad \omega_l \quad \text{if} \quad p(X|\theta^*) P(\theta^*) = \max_{S_i} p(X|\theta) P(\theta)$$
$$\text{and} \quad \theta_0^* = \omega_l. \tag{37}$$

This scheme has been proposed in Fu & Yu (1980) but in my view it is associated with some conceptual problems. In particular each pixel ends up with five labels that may differ. Even if all but one are ignored, potential inconsistencies in the resulting labelling cannot be eliminated.

### (b) *The compound decision rule*

In addition to assumption (34) I shall further assume that the contextual classification between any non-adjacent cells is negligible. For $l > 0$ this implies that

$$p(x_j| x_l, \theta_l, l \neq j) = p(x_j|\theta_0), \tag{38}$$

because pixel 0 is the only adjacent cell to pixels $x_j (j = 1, 2, ..., 4)$. Under these two assumptions $P(\omega_i|X)$ in (29) can be expressed as

$$P(\omega_i|X) = p(X|\omega_i)\frac{P(\omega_i)}{p(X)} = \frac{P(\omega_i)}{p(X)} \prod_{j=0}^{4} p(x_j|\theta_0 = \omega_i). \tag{39}$$

To avoid the need for estimating conditional probability functions $p(x_j|\theta_0)$ $(j \geq 1)$, I shall expand $p(x_j|\theta_0)$ as

$$p(x_j|\theta_0) = \sum_{r=1}^{m} p(x_j, \theta_j = \omega_r|\theta_0) = \sum_{r=1}^{m} p(x_j|\omega_r) P(\omega_r|\theta_0), \tag{40}$$

provided that the probabilities $P(\omega_r|\theta_0)$ are isotropic. Substituting (40) into (39) the compound decision rule is obtained (Welch & Salter 1971):

$$\text{assign} \quad x_0 \quad \text{to} \quad \omega_l \quad \text{if} \quad P(\omega_l) p(x_0|\omega_l) \prod_{j=1}^{4} \sum_{r=1}^{m} p(x_j|\omega_r) P(\omega_r|\omega_l)$$
$$= \max_i P(\omega_i) p(x_0|\omega_i) \prod_{j=1}^{4} \sum_{r=1}^{m} p(x_j|\omega_r) P(\omega_r|\omega_i). \tag{41}$$

$P(\omega_r|\omega_i)$ are called transition probabilities. If these probabilities exhibit directional dependence, $P(\omega_r|\omega_i)$ must be replaced by $P(\theta_j = \omega_r|\omega_i)$. For brevity I shall denote these directional probabilities by $P_j(\omega_r|\omega_i)$.

Thus in comparison with (2) the only additional information needed to implement decision rule (41) are the class transition probabilities $P(\omega_r|\omega_i), \forall i, r$, or directional transition probabilities $P_j(\omega_r|\omega_i)$ $(j = 1, 2, ..., 4)$. This is computationally less involving than the estimation of the joint probability function $P(\theta)$ of the method in (36).

### 6. Relaxation labelling

Up to now I have discussed non-iterative decision-making schemes incorporating contextual information. In all cases the aim has been to determine the class membership of a pixel by computing the *a posteriori* probabilities $P(\omega_i|X)$ of class $\omega_i$, given observations on the pixel and its neighbours. An alternative approach is to compute these probabilities recursively. The basic updating formula for the probability $P(\omega_i, x_0)$ of joint occurrence of $\omega_i$ and $x_0$ is

$$P^n(\omega_i, x_0) \Rightarrow P^{n-1}(\omega_i, X). \tag{42}$$

The functions $P^n(\omega_i, x_0)$ appear as arguments in $P^n(\omega_i, X)$ and so on.

Note that strictly speaking $P^n(\omega_i, x_0)$ is no longer the joint probability of class $\omega_i$ and pixel $x_0$. It is a variable that should eventually be equivalent to $P(\omega_i, X)$. Initially $P^0(\omega_i, x_0)$ is set equal to

$$P^0(\omega_i, x_0) = P(\omega_i|x_0)\,p(x_0). \tag{43}$$

As before, the actual form of the function $P(\omega_i, X)$ will depend on the assumptions made for the dependence between variables in $X$ and their classes. Suppose that the spatial dependence of pixels can be characterized by the joint probability $P(\omega_i, X)$ in (30) with the *a priori* class probabilities being independent. Then, from (32),

$$P(\omega_i, X) = \sum_{S_i} p(X|\boldsymbol{\theta})\,P(\omega_i) \prod_{j=1}^{4} P(\theta_j). \tag{44}$$

Under the assumption of known $p(X|\boldsymbol{\theta})$, the expression (44) can be used to drive the relaxation process (42). For univariate observations the density function in (33) is of practical importance in many applications. After the initialization stage the *a priori* probabilities in (44) lose their identities and are replaced by the current estimates of the conditional probabilities $P^{n-1}(\omega_j|x_0)$, i.e. (Kittler & Föglein 1983 $a$)

$$P^n(\omega_i, x_0) \Rightarrow P^{n-1}(\omega_i|x_0) \sum_{S_i} p(X|\boldsymbol{\theta}) \prod_{j=1}^{4} P^{n-1}(\theta_j|x_j). \tag{45}$$

The probabilities $P^n(\omega_i, x_0)$ may have to be normalized to ensure that

$$\sum_{i=1}^{m} P^n(\omega_i|x_0) = 1. \tag{46}$$

Yu & Fu (1983) proposed an algorithm that can be interpreted as a simplification of the relaxation process in (45). I shall approximate the probabilities $P^n(\theta_j|x_j)$ $(j = 1, 2, ..., 4)$ as follows:

$$P^n(\theta_j = \omega_r|x_j) = 1 \quad \text{if} \quad P^n(\theta_j = \omega_r|x_j) = \max_l P^n(\theta_j = \omega_l|x_j); \tag{47}$$

$$P^n(\theta_j = \omega_r|x_j) = 0 \quad \text{otherwise.} \tag{48}$$

I shall denote the vector $\boldsymbol{\theta}$ at the $n$th iteration with $\theta_0 = \omega_i$ and $\theta_j$ such that (47) holds, by $\tilde{\boldsymbol{\theta}}_n$. Then the summation in (45) becomes

$$\sum_{S_i} p(X|\boldsymbol{\theta}) \prod_{j=1}^{4} P^{n-1}(\theta_j|x_j) = p(X|\tilde{\boldsymbol{\theta}}_n) \tag{49}$$

and the recursive formula simplifies to

$$P^n(\omega_i, x_0) = P^{n-1}(\omega_i|x_0)\,p(X|\tilde{\boldsymbol{\theta}}_{n-1}). \tag{50}$$

The corresponding algorithm can now be stated as follows.

(i) Classify pixels by using the minimum error Bayes rule and set

$$P^0(\omega_i|\pmb{x}_0) = P(\omega_i|\pmb{x}_0).$$

(ii) For every pixel $\pmb{x}_0$ at the stage $n$ form vector $\tilde{\pmb{\theta}}_{n-1}$ and compute $p(X|\tilde{\pmb{\theta}}_{n-1})$.

(iii) Use (50) to update class probabilities, i.e.

$$P^n(\omega_i|\pmb{x}_0) \Rightarrow P^{n-1}(\omega_i|\pmb{x}_0)\, p(X|\tilde{\pmb{\theta}}_{n-1})/p(X),$$

and normalize them to satisfy condition (46).

(iv) Reclassify pixels according to the minimum error Bayes rule, i.e.

$$\text{assign} \quad \pmb{x}_0 \quad \text{to} \quad \omega_l \quad \text{if} \quad P^n(\omega_l|\pmb{x}_0) = \max_i P^n(\omega_i|\pmb{x}_0).$$

(v) If no pixel is reassigned then terminate the algorithm, else return to step 2.

If the spatial dependence of pixels satisfies the assumptions made in §5$b$, it can be shown (Kittler & Föglein 1983$b$) that the corresponding algorithm for updating $P^n(\omega_i|\pmb{x}_0)$ is the conventional relaxation algorithm (Zucker *et al.* 1978; Rosenfeld *et al.* 1976). It permits the determination of pixel labels that are compatible in the sense of the transition probabilities.

I shall denote the modified *a posteriori* probability of class $\omega_i$ given an arbitrary pixel $\pmb{x}_0$ after $n$ iterations by $P^n(\omega_i|\pmb{x}_0)$. Then at the next stage this probability is updated according to

$$P^{n+1}(\omega_i|\pmb{x}_0) = \frac{P^n(\omega_i|\pmb{x}_0)\, Q^n(\omega_i)}{\sum\limits_{r=1}^{m} P^n(\omega_r|\pmb{x}_0)\, Q^n(\omega_r)}, \tag{51}$$

where $Q^n(\omega_r)$ is the so-called neighbourhood function. It is defined as

$$Q^n(\omega_r) = \sum_{j=1}^{4} d_j \sum_{l=1}^{m} P_j(\omega_r|\omega_l)\, P^n(\omega_l|\pmb{x}_j), \tag{52}$$

where $d_j$ is a neighbourhood weight.

The stability and convergence of this relaxation algorithm has been studied by Zucker *et al.* (1978). The effect of weights $d_j$ and of the transition probabilities on the algorithm performance has been investigated by Richards *et al.* (1980). It has been shown that a suitable choice of $d_j$ can prevent the loss of corners, lines, endpoints and other degradations in pixel labelling.

## 7. Conclusions

In the paper a number of approaches to multispectral image segmentation and classification have been considered. The methods range from the simple Bayesian decision rule for classification of image data on pixel-by-pixel basis, to sophisticated algorithms using contextual information. Both the spatial pixel category dependences and the two-dimensional correlation-type contextual information have been incorporated in decision-making schemes. The algorithms have been developed to achieve a greater reliability in the process of the interpretation of remote-sensing data.

## References

Bartlett, M. S. 1975 *The statistical analysis of spatial pattern*. London: Chapman & Hall.
Bernstein, R. (ed.) 1978 *Digital image processing for remote sensing*. New York: IEEE Press.
Besag, J. E. 1974 *Jl R. stat. Soc.* B **36**, 192–236.
Craig, R. G. 1979 In *Proc. 13th Int. Symp. Remote Sensing of Environment, Ann Arbor, Michigan*, pp. 1517–1524.
Devijver, P. & Kittler, J. 1982 *Pattern recognition: a statistical approach*. Englewood Cliffs: Prentice-Hall.
Fu, K. S. (ed.) 1977 *Syntactic pattern recognition, applications*. Berlin: Springer-Verlag.
Fu, K. S. 1982$a$ *Syntactic pattern recognition and applications*. Englewood Cliffs: Prentice-Hall International.

Fu, K. S. 1982*b* In *Pattern recognition theory and applications* (ed. J. Kittler, K. S. Fu & L. F. Pau), pp. 139–155. Dordrecht: D. Reidel.

Fu, K. S. & Yu, T. S. 1980 *Statistical pattern classification using contextual information*. Chichester: John Wiley & Sons.

Haralick, R. M. 1976 In *Digital picture analysis* (ed. A. Rosenfeld), pp. 5–63. Berlin: Springer-Verlag.

Haralick, R. M. 1979 *Proc. IEEE* **67**, 786–804.

Huang, T. S. (ed.) 1979 *Picture processing and digital filtering*. Berlin: Springer-Verlag.

Kettig, R. L. & Landgrebe, D. A. 1976 *IEEE Trans. Geosci. Electron.* **GE-4**, 19–26.

Kittler, J. 1983 In *Physical and biological processing of images* (ed. O. J. Braddick and A. C. Sleigh), pp. 232–243. Berlin: Springer-Verlag.

Kittler, J. & Föglein, J. 1983*a* In *Proc. 3rd Scandinavian Conf. Image Analysis, Copenhagen*.

Kittler, J. & Föglein, J. 1983*b* (Submitted.)

Landgrebe, D. A. 1980 *Pattern Recogn.* **12**, 165–175.

Niemann, H. 1981 *Pattern analysis*. Berlin: Springer-Verlag.

Pavlidis, T. 1977 *Structural pattern recognition*. Berlin: Springer-Verlag.

Pratt, W. K. 1978 *Digital image processing*. New York: John Wiley and Sons.

Richards, J. A., Landgrebe, D. A. & Swain, P. H. 1980 In *Proc. 5th Int. Conf. Pattern Recognition, Miami*, pp. 61–65.

Rosenfeld, A., Hummel, R. & Zucker, S. 1976 *IEEE Trans. Syst. Mgmt Cybern.* **SMC-6**, 420–433.

Tubbs, J. D. 1980 *IEEE Trans. Syst. Mgmt Cybern.* **SMC-10**, 177–180.

Tubbs, J. D. & Coberly, W. A. 1978 In *Proc. 12th Int. Symp. Remote Sensing of Environment, Manila, Philippines*.

Welch, J. R. & Salter, K. G. 1971 *IEEE Trans. Syst. Mgmt Cybern.* **SMC-1**, 24–30.

Weszka, J., Dyer, C. & Rosenfeld, A. 1976 *IEEE Trans. Syst. Mgmt Cybern.* **SMC-6**, 269–285.

Yu, T. S. & Fu, K. S. 1983 *Pattern Recogn.* **16**, 89–108.

Zucker, S., Krishnamurthy, E. & Haar, R. 1978 *IEEE Trans. Syst. Mgmt Cybern.* **SMC-8**, 41–48.

*Phil. Trans. R. Soc. Lond.* A **309**, 337–359 (1983)
*Printed in Great Britain*



# Satellite monitoring of the ocean for global climate research

By J. D. Woods
*Institut für Meereskunde an der Universität Kiel,*
*Dusternbrooker Weg 20, D-2300 Kiel 1, F.R.G.*

The study of climate embraces a broad range of timescales, from weeks to millions of years. This paper concentrates on the narrow spectral band 'several weeks to several decades' chosen for study in the World Climate Research Programme (WCRP). The programme is divided into three streams, concerned with climate variation on timescales of several weeks, several years and several decades, respectively. The aim is to discover how far it is possible to predict natural climate variation and man's influence on climate in each of these spectral bands. It is believed that an improved understanding of, and an ability to monitor and model, the World Ocean will be critical to the success of the WCRP in each stream. International experiments are now being planned to achieve these improvements. Satellite monitoring of the ocean offers a number of advantages for these experiments, including the following: global coverage, accuracy and consistency, novel products, tracking and communication. The paper reviews specifications for monitoring the ocean in the context of the WCRP and assesses the extent to which satellite monitoring will help towards meeting these requirements. The special needs of two major projects, the World Ocean Circulation Experiment (WOCE) and the Tropical Ocean and Global Atmosphere (TOGA), are described. The prospects seem good for a new generation of ocean-observing satellites suitable for climate research in the next decade. They will be crucial to the success of the World Climate Research Programme.

## Introduction

The climate of a planet is its atmospheric response to solar and internal heating. The response mechanism involves not only the atmosphere, but also the other elements of the planetary climate system. The existence of the World Ocean makes Earth climatologically unique among the planets of the solar system. It is difficult to think of any problem in global climatology, regardless of the timescale involved, that does not centrally involve the ocean. Examples of ocean–climate interaction, based on the review by Woods (1983), are given in table 1. In this paper I shall concentrate on the shorter timescales commensurate with human activities and therefore offering potential benefits if they could be predicted. The World Climate Research Programme (WRCP 1980–2000) has been established by the World Meteorological Organization (WMO) and the International Council of Scientific Unions (ICSU) with the following aims:

(1) to determine to what extent climate can be predicted;

(2) to determine the extent of man's influence on climate.

Highest priority is being given to climate variation on timescales of 'several weeks to several decades'. The underlying assumption is that it might be possible one day to achieve some kind of prediction of climate changes with sufficient lead time for industry and public services to prepare for them. Meteorologists have had considerable success in forecasting the weather by systematic global observations fed into computer models based on the laws of dynamics and

[ 95 ]

physics. It is hoped that an appropriate combination of global observations and dynamic-physical models can also provide the basis for climate prediction. But it is not yet certain whether that is going to be possible. An intense research programme including model development and data collection is being planned for the remaining years of this century (Houghton & Morel 1983).

TABLE 1. THE INFLUENCE OF THE OCEAN ON GLOBAL CLIMATE CHANGES
WITH DIFFERENT TIMESCALES

| timescale | climatological phenomenon | ocean contribution |
|---|---|---|
| order $10^9$ years | evolution of the atmosphere | chemistry and biology |
| order $10^9$ years | evolution of the continents | sedimentation and coastal erosion |
| $10^8$–$10^9$ years | evolution of the biosphere | spreading from the ocean to the land |
| $10^7$–$10^8$ years | continental drift opens and closes narrow sills between ocean basins | global energy cycle modified by the interchange of water masses between ocean basins via narrow sills; for example: opening of the Drake Passage, closing of the Meso-American gap between Pacific and Atlantic, opening of the overflow from Arctic Ocean to Atlantic Ocean, temporary closing of the Straits of Gibraltar (leading to desiccation of the Mediterranean) |
| order $10^8$ years | interval between ice ages | evaporation supplies water for the polar glaciers; |
| order $10^7$ years | duration of ice ages (due to continents drifting near the poles) | biology changes atmospheric $CO_2$; deep convection in polar seas creates the ocean cold water sphere, thermally isolating the upper ocean |
| $10^4$–$10^5$ years | fluctuations in ice ages due to variations of the Earth's orbit | sea level changes affect biological processes modifying atmospheric $CO_2$<br>sea surface a few degrees cooler 18 000 years ago at the maximum glaciation |
| centuries | little ice age | unknown |
| decades | heating by $CO_2$ pollution | accommodation of $CO_2$ by ocean chemistry and biology halves the greenhouse effect<br>ocean thermal lag delays by several decades the temperature rise in the atmosphere |
| years | inter-annual variability including the Southern Oscillation | teleconnections in ocean due to El Niño, Gulf Stream, etc.<br>sea ice variation<br>mid-latitude heat content anomalies in the oceanic boundary layer |
| months | the regular seasonal cycle, including: the meridional migration of the Westerlies and Trades; the monsoons | seasonal heat storage in the upper ocean<br>global circulation of about 1 % of the solar heat input<br>land–sea difference of annual temperature range<br>waxing and waning of sea ice cover |
| days | weather | sea-surface temperature distribution; sea ice distribution |

The programme strategy for the World Climate Research Programme (WCRP 1983) is divided into three streams, each concerned with a particular range of timescales:

stream 1, timescales of order months;

stream 2, timescales of order years;

stream 3, timescales of order decades.

The ocean is important for all three streams. An oceanographic programme (WCRP-O, where the O stands for oceanography) is being planned with specific activities devoted to each of the three WCRP streams. Many of the activities are being organized by research groups in a single institute, or at several institutes in one country, or involve informal collaboration between groups in several countries. International organizations like ICES (the Inter-governmental

Council for Exploration of the Sea), Unesco and the Nato Science Committee, are actively supporting the international aspects of these activities.

Experience in GARP (the Global Atmospheric Research Programme) showed that more formal co-ordination is needed in the case of major international experiments such as GATE (the GARP Atlantic Tropical Experiment, in 1974) and FGGE (the First GARP Global Experiment, in 1979). One of the critical factors favouring formal co-ordination was the need for satellite observations as an essential ingredient of the overall observing programme in GATE and FGGE. Although the specification and design of each satellite observing system, and the scientific analysis of the resulting data, all depend critically on the contributions of individual scientists, the need to create a case sufficiently powerful to convince governments and space agencies to allocate resources to meteorological rather than, say, astronomical satellites demanded a strong central lobby. The case presented by the lobby would be convincing only if its scientific objectives were simple and supported by the meteorological community as a whole. The task of the Joint Organizing Committee (JOC) for GARP was to help crystallize scientific objectives and to translate them into simple experimental plans, involving satellite observations, that would be widely accepted by the international meteorological community. Then, having prepared a strong case for satellites for GARP, the JOC had to act as the lobby to make sure that they were launched in time for the major experiments. The successful accomplishment of these tasks by the JOC for GARP sets a precedent for the new Joint Scientific Committee (JSC) for the WCRP and, for the oceanographic aspects, their partner, the Committee for Climate Change and the Ocean (CCCO), which is sponsored jointly by ICSU (through SCOR, the Scientific Committee for Ocean Research) and the Intergovernmental Oceanographic Committee (IOC, of Unesco).

If it becomes feasible, climate prediction will involve not only the atmosphere but also the other elements of the planetary climate system. The WCRP requirements for satellites will therefore involve missions dedicated to observing the atmosphere, the land surface, the cryosphere and the oceans. Progress is being made in all these elements. The successors of the GARP generation of meteorological satellites, together with novel meteorological missions designed to support the Earth Radiation Budget Experiment (ERBE) and the International Satellite Cloud Climatology Project (ISCCP), are now established. So is a new generation of satellites to observe the land surface: the successors of the Landsat series, and the new SPOT series (Eagleson 1982). The cryosphere (in particular the sea ice, which exhibits large seasonal and inter-annual variation) has been routinely monitored for many years by satellites carrying microwave radiometers (Untersteiner 1983). The case for meteorological, land surface and sea-ice missions for the WCRP rests securely on a great deal of experience gained in the 1970s. By contrast, the case for satellite missions to observe the oceans is newer, and therefore less secure. Nevertheless, oceanographers have made considerable progress during recent years, and a strong case can now be made for dedicated WCRP-O satellites. The experience comes from three main sources:

(1) analysis of data from meteorological satellites to estimate sea surface temperature, insolation and surface wind stress;

(2) the Coastal Zone Colour Scanner on the Nimbus-7 mission;

(3) the first dedicated ocean-observing mission, Seasat, which carried a variety of novel instruments, including radars and passive radiometers in the microwave, infrared and visible wavebands.

[ 97 ]

Cracknell (ed.) (1981), Stewart (1982) and Gower (ed.) (1982) have reviewed the state of the art of satellite oceanography at the start of the 1980s. My aim in this paper is to summarize the case that is being developed for a new generation of dedicated ocean-observing satellites to be used as central elements of WCRP oceanographic experiments during the remaining years of this century. The next section discusses why satellites will be crucial to the success of WCRP-O experiments. The subsequent sections look in more detail at the experiments planned for the three streams of the WCRP, identifying the role of satellite observations in each.

### THE IMPORTANCE OF SATELLITES FOR WCRP OCEANOGRAPHIC EXPERIMENTS

Traditionally the ocean has been observed from ships, and ship measurements will play a key role in the WCRP-O experiments, especially for the establishment of the three-dimensional distributions of temperature, salinity and chemicals used as tracers in circulation studies. Satellites can only observe the surface of the ocean, and therefore only compete with surface observations from ships. But the WCRP-O specifications for surface observations cannot be satisfied by ship data alone. Satellites do not satisfy all the requirements either, but they offer a number of crucial advantages. It is doubtful whether the JSC and CCCO would proceed with the experiments if satellite observations were not going to be available. The advantages can be grouped into four categories: (1) global coverage, (2) accuracy and consistency, (3) novel products, and (4) tracking and communications. These will now be considered in turn.

### Global coverage

Whereas it is possible to make quite good weather forecasts by using observations and a model that covers only part of the globe, that will not be true for climate prediction, even on the shortest timescales (stream 1). Empirical studies (Horel & Wallace 1981; Wallace & Gutzler 1981; Keshavarmurty 1982; Shukla & Wallace 1983) and modelling studies (Pan & Oort 1983) both show that the atmospheric climate responds remotely to anomalies in the surface fluxes of energy and water from the ocean. It is only meaningful to discuss the climate in terms of the *global* pattern of fluxes from the ocean (g.p.f.o.). Recent theoretical studies by Webster (1981) and Hoskins (1983) have helped to explain the mechanism by which the atmosphere responds remotely to anomalies in the g.p.f.o. Their theories make interesting predictions about seasonal and regional variation in the sensitivity of the response, predictions that have influenced the design of WCRP-O experiments, as we shall see later. Nevertheless, it is essential to measure the *global* pattern of ocean fluxes. They also serve as upper boundary conditions for ocean circulation models, which are needed to predict changes in the contribution of the ocean to the global circulations of energy and water in the planetary climate system. Global observations are a *sine qua non* for the WCRP. The sampling density must be sufficient to give statistically significant data with a spatial resolution of one megametre† and temporal resolution of one month. The distribution of ship observations (figure 1) cannot meet that specification, especially in the Southern Hemisphere. But, given appropriate orbits, satellites sample the whole globe systematically and predictably. They do not suffer from the kind of biases found in data bases derived from ships, which are likely to have avoided typhoons, for example. On the other hand, satellite-derived data may have biases of a different kind, for example in favour of cloud-free regions; so care must be taken in using them, too. Nevertheless one of the great attractions of

† A higher meridional resolution is needed in the Tropics (see table 5).

satellite observations is their potential for complete, systematic, global coverage. A recent example is shown in figure 2.

### Accuracy and consistency

Climate prediction will demand a very high order of accuracy in ocean surface data. The specification depends on the aims of the forecast; during the research phase it depends on

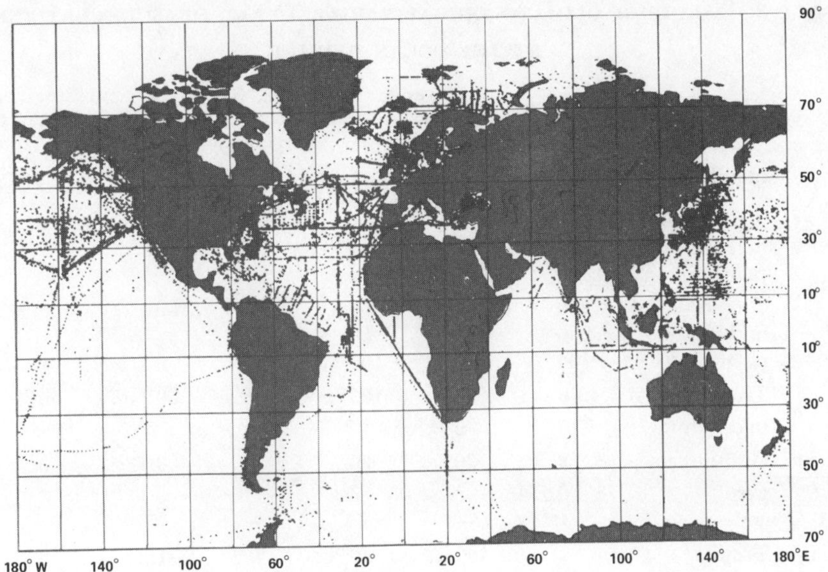

FIGURE 1. Distribution of ship observations reported during 1978 through IGOSS.

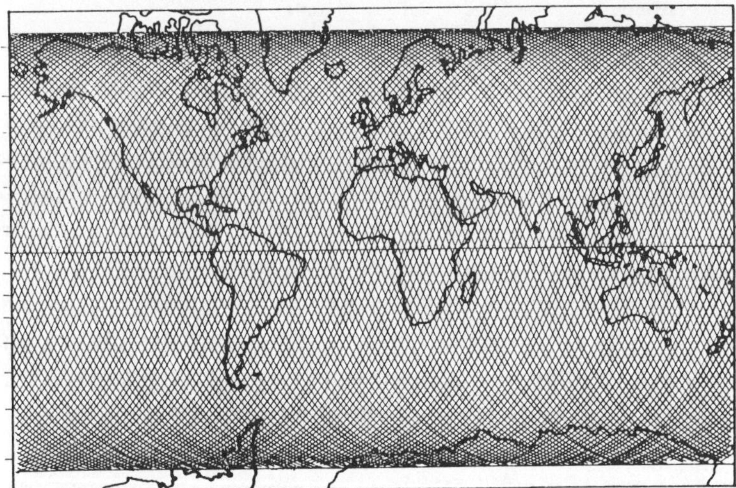

FIGURE 2. Coverage of the ocean by Seasat in an 8 day repeat orbit.

which stream of the WCRP is being pursued. As a rule of thumb we can specify the required accuracy as, say, 20% of the relevant signal, which might be the synoptic global range, the annual range at one place, the range of short-term inter-annual variability, or the secular change expected over a given period. The existing data base and predictions based on simple climate models give some idea of what these ranges are (table 2), but it must be admitted that there are large uncertainties due to measurement error and inadequate sampling, especially in the Southern Hemisphere. Examining table 2, we note that the uncertainty in existing data severely limits the areas of the World Ocean in which our climatological knowledge meets the 20% criterion. In many places the error is greater than 100% of the climatological signal, making it difficult to be sure even of the sign of the variation. One of the most urgent tasks of the WCRP-O is to collect a new data base that will meet the 20% specification for the mega-metre-monthly average values of the variables listed in table 2, everywhere around the globe.

TABLE 2. RANGES OF CLIMATOLOGICAL VARIABLES AND MEASUREMENT ERRORS AT THE OCEAN SURFACE

| | | range | | | | measurement error | | |
| variable | qualification | regional<br>Mm | seasonal<br>month | inter-<br>annual | secular<br>$(2 \times CO_2)$ | ship | satellite<br>1985 | 1990 |
|---|---|---|---|---|---|---|---|---|
| temperature | Tropics | 10 | 3 | 6 | 2 | 0.5 | 1 | 0.3 |
| K | mid-latitude | 20 | 10 | 1 | 8 | | | |
| | polar | 5 | 5 | 1 | 10 | | | |
| energy flux | Tropics | 150 | 100 | 90 | 15 | 50 | 50 | 50 |
| W m$^{-2}$ | mid-latitude | 200 | 100 | 30 | 15 | | | |
| | polar | 400 | 100% | 100% | ? | | | |
| precipitation | Tropics | 50 | 100% | 100% | ? | 50%(?) | 50%(?) | 10%(?) |
| cm month$^{-1}$ | mid-latitude | 10 | 30% | ? | ? | | | |
| pressure gradient | Gulf Stream | 2 | 30 | 30 | ? | 0.1 | 0.1 | 0.02 |
| m (100 km)$^{-1}$ | gyre | 0.1 | ? | ? | ? | | | |
| | eddies | 1 | — | — | — | | | |
| Ekman pumping | Tropics | 1 | 100% | 50% | ? | 1(?) | 0.2 | 0.2 |
| m day$^{-1}$ | mid-latitude | 0.1 | 50% | 30% | ? | | | |
| heat content | Tropics | 0.5 | 0.1 | 0.3 | ? | 0.02† | 0.1 | 0.02 |
| m elevation | mid-latitude | — | 0.1 | 0.01 | ? | | | |

† Tide gauge.

The contribution of satellite measurements can be assessed by comparing their accuracies with these specifications. Some recent estimates of the uncertainties in existing satellite data are listed in table 3. In many cases the data are so new that algorithms used to produce the required variables are still under development; we can look forward to further hardware improvements in the new 'WCRP-generation' of ocean-observing satellites.

Although it may take some time before the accuracy of satellite observations exceeds that of the best ship observations, they will from the start offer consistency of data quality, which has always been a worry with routine observations from ships of opportunity. For example, there is evidence of systematic differences between sea-surface temperature measurements made by different methods (bucket, plate and engine intake thermometers), as discussed by Tabata (1978). Subjective estimates of rainfall rate from ships and islands present special problems of consistency (Mintz 1981). There are obvious advantages of a data set collected from carefully inter-calibrated sensors on a small number of satellites, each sampling the whole globe (Tabata & Gower 1980).

*Novel products*

Novel instruments on satellites and their systematic high-resolution scanning of the ocean opens the door to monitoring variables that are of great importance to modellers, but whose global patterns are known only sketchily from the existing ship observations. The highest priority for any dynamicist is the vertical motion due to Ekman convergence and synoptic-scale development. Saunders (1976) has discussed the problems of calculating the mean wind-stress curl over various averaging areas and intervals from ship data that are incoherent with respect to the atmospheric weather events controlling wind stress. Guymer *et al.* (1983) have produced the first synoptic map of wind-stress curl from Seasat radar scatterometer measurements, which are coherent with respect to the weather (figure 3). The annual vertical displacement due to

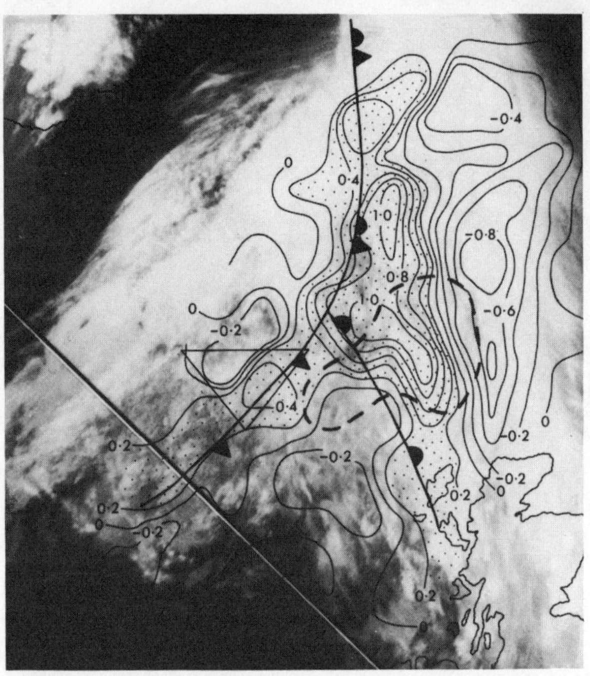

FIGURE 3. The first map of wind stress curl that resolves the variation with atmospheric weather. From the Seasat scatterometer during JASIN analysed by Guymer *et al.* (1983). The weather is shown in three ways: (i) by conventional synoptic analysis (the lines marking an occluded front), (ii) by contours of rainfall rate estimated from the Seasat scanning multichannel microwave radiometer, and (iii) by a cloud image from the NOAA-5 weather satellite which passed overhead 5 h earlier. (Reproduced by kind permission of the Director, Institute of Oceanographic Sciences, Wormley, and Dundee University.)

Ekman pumping lies in the range $\pm 100$ m. Similar vertical displacement is achieved in a month by transient 'synoptic-scale' motions (eddies & Rossby waves) in the ocean. The energy of the synoptic-scale motions exhibits considerable regional (Dantzler 1976) and seasonal (Dickson *et al.* 1982) variability. The dimpling of the sea surface by these (order 100 km) ocean synoptic-scale motions can be measured by satellite altimeter (figure 4). The sampling is unlikely to be coherent, so it will not be possible to measure the vertical velocity associated with these transient motions; but monitoring the regional and seasonal variation of the statistics of the surface dimpling will provide an invaluable product for comparison with the predictions of eddy-resolving ocean circulation models. The surface of the ocean also rises and falls (relative to the geoid) due to thermal expansion, so the altimeter can be used to monitor variations in the upper ocean's heat content; the maximum range of surface elevation, several decimetres, occurs in the Tropics (figure 5). To monitor these changes of eddy motion and heat content it will be necessary for the satellite to cover the same ground track to within 1 km every week or so. Less precise repetition would introduce contamination from spatial variation of the geoid, which is an order of magnitude bigger (Wunsch (ed.) 1982). A more ambitious goal is to use the satellite altimeter to measure the steady hydrostatic pressure distribution on the geoid, and

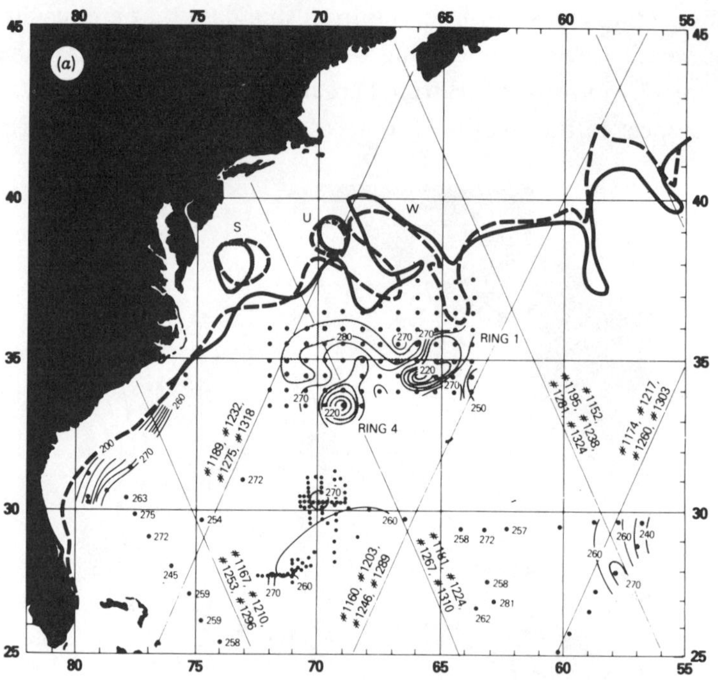



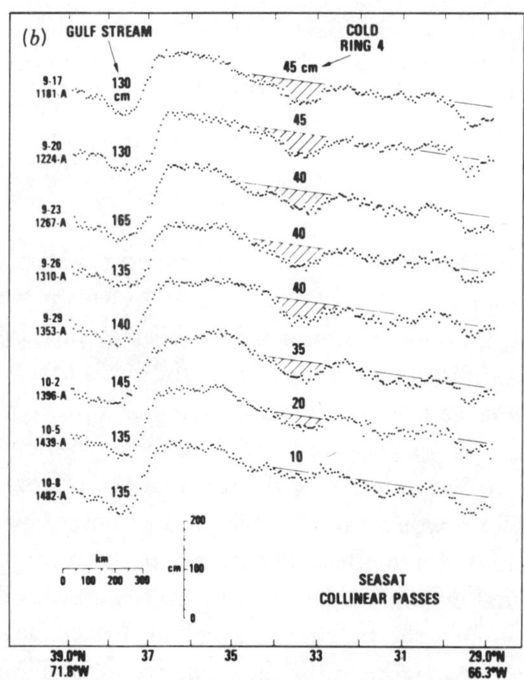



FIGURE 4. Observations of transient dimpling of the sea surface due to oceanic weather systems monitored by the Seasat altimeter during successive passes at 3-day intervals. (a) Seasat ground tracks in the vicinity of the Gulf Stream, 14–27 September 1978; (b) successive altimetric measurements along the set of tracks shown in (a) passing through ring 4, showing the evolution of more than 21 days of the Gulf Stream and a large, strong ring. (From Cheney & Marsh (1981); reproduced with permission.)

hence to calculate the surface geostrophic velocity field. Wunsch (ed.) (1982) has discussed a dedicated satellite mission for that purpose, and the supporting measurements needed to measure the geoid to the required accuracy (a few centimetres). These are the novel products uppermost in the minds of oceanographers planning WCRP-O experiments. Others are under development, including methods of subsurface profiling (Schwiesow 1980; Leonard 1980), but experience with GARP suggests that planning must proceed on the basis of methods that have been first tested on a satellite a decade beforehand.

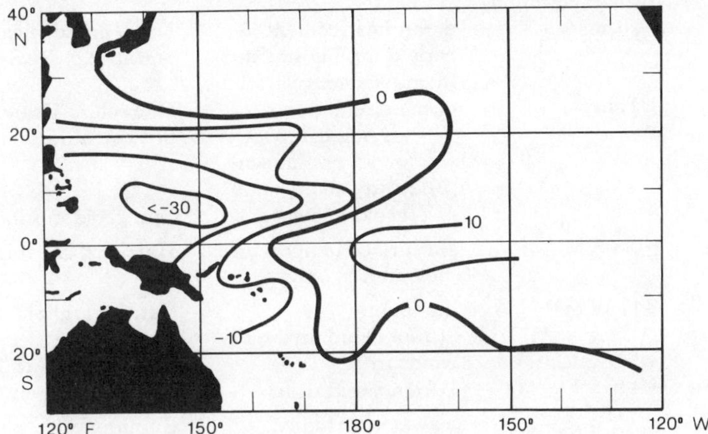


Figure 5. Inter-annual variation (October 1975 minus October 1976) of sea level (centimetres) due to changes in upper-ocean heat content in the western Pacific Ocean. This map is based on tide gauge measurements; satellite altimeter measurements will permit similar mapping around the World Ocean. (From Wyrtki (personal communication 1982); reproduced with permission.)

### Tracking and communication

Although not strictly within the terms of reference of this paper, it would be a serious omission to neglect the supporting role of satellites in WCRP-O experiments. Precise navigation has always been a vital matter for oceanographers, and it has become even more important now that we try to take account of synoptic-scale 'weather' systems in the ocean. Conventional satellite navigation is the standard method of fixing the positions of ships and buoys. The forthcoming Global Positioning System (GPS), which will offer much higher accuracy and effectively continuous fixes (as against every hour or so at present), will greatly simplify ship surveys of intense currents on the high seas. It also opens the way to precision measurement of upper ocean velocity structure by shipborne doppler acoustic profiler, a major tool for the future. GPS may also play a key role in monitoring the orbits of satellites carrying radar altimeters.

Interrogation of drifting buoys (position and atmospheric pressure) by the Argos system was a crucial ingredient of FGGE. It is now possible to hang thermistor chains under drifting buoys to measure the temperature profile, which is also routinely extracted and transmitted to shore by the Argos system. Automatic satellite communication of meteorological and expendable-bathythermograph data from ships of opportunity is now a practical proposition and may supersede some aspects of the existing real-time system, IGOSS, which is based on manual h.f. telemetry via the meteorological global trunk system. Ships of opportunity and drifting buoys offer cost-effective platforms for monitoring the ocean, and will be used intensively during the WCRP-O experiments.

# J. D. WOODS

*The principal satellite measurements for WCRP ocean monitoring*

Table 3 lists the principal measurements required for WCRP-O experiments, with an indication of the expected accuracy.

TABLE 3. SATELLITE MEASUREMENTS OF THE OCEAN FOR WCRP EXPERIMENTS

| name | estimated error | product for WCRP | comments and references |
|---|---|---|---|
| radar scatterometer (SCAT) | (wind velocity) $\pm 1.7$ m s$^{-1}$, $\pm 17°$ | wind stress<br>wind stress curl | Guymer, this symposium<br>Guymer *et al.* (1983) |
| radar altimeter (ALT)<br>class *b* | (surface elevation)<br>$\pm 10$ cm | ocean heat content<br>eddy dimpling statistics<br>intense current variability | large, non-dedicated satellite, in low orbit (e.g. ERS-1) |
| class *a* | $\pm 2$ cm† | ocean circulation<br>  (i) varying component<br>    (geoid not known)<br>  (ii) steady component<br>    (geoid known) | Wunsch & Grant (1982)<br>dedicated mission (e.g. TOPEX)<br><br>from geoid mission |
| infrared radiometer (IR) | $\pm 0.3$ K† | sea-surface temperature<br>(cloud-free only) | Harries *et al.*, this symposium |
| visible radiometer (VR) | $\pm 17$ W m$^{-2}$<br><br>(wind velocity)<br>$\pm 4$ m s$^{-1}$, $\pm 20°$ | insolation<br>(from cloud analysis)<br>wind stress<br>(from cloud drift) | Gautier (1982)<br><br>Krishnamurti & Krishnamurti (1980)<br>Wylie & Hinton (1982) |
| colour scanner (CS) | ? | seawater turbidity<br>depth distribution<br>of solar heating | Robinson, this symposium |
| microwave scanner (MS) | ?<br>$\pm 1.2$ K (*a*)<br>$\pm 0.5$ K (*b*)<br>? | sea ice cover<br>sea-surface temperature<br>(through cloud)<br>precipitation | Untersteiner (1983)<br>(*a*) Hofer *et al.* (1981)<br>(*b*) Bernstein (1982)<br>Atlas & Thiele (eds) (1981) |

† Design target.

TABLE 4. OCEAN-OBSERVING SATELLITES PLANNED FOR THE NEXT DECADE

| launch | satellite | SCAT | ALT*a* | ALT*b* | IR | VR | CS | MR | SAR | sponsor | status (7 Aug. 1982) |
|---|---|---|---|---|---|---|---|---|---|---|---|
| 1984 | DMSP | — | — | — | — | — | — | × | — | U.S.A.F. | approved |
| 1984 | Geosat | — | — | × | — | — | — | — | — | U.S.N. | approved |
| 1986 | NOSS | × | — | — | × | × | × | × | — | U.S.N./N.O.A.A./ N.A.S.A. | cancelled |
| 1986 | MOS-1 | — | — | — | × | × | × | × | — | Japan | approved |
| 1988 | ERS-1 | × | — | × | × | — | — | × | — | E.S.A. | approved |
| 1987 | NOAA-H | — | — | — | × | — | — | — | — | N.O.A.A. | approved |
|  |  | — | — | — | — | — | × | — | — | N.A.S.A. | proposed |
| 1987? | SPOT-2 | — | — | × | — | × | — | — | — | C.N.E.S. | proposed |
| 1988? | TOPEX | (×) | × | — | — | — | — | — | — | N.A.S.A. | proposed |
| 1988? | ROSS | × | — | × | — | — | — | × | — | U.S.N. | proposed |
|  |  | (NOAA-D bus) |  |  |  |  |  |  |  | N.O.A.A. | proposed |
|  |  | × | — | — | — | — | — | — | — | N.A.S.A. | proposed |
| 1988? | ERS-1 | — | — | — | — | — | — | — | × | Japan | proposed |
| 1989? | Gravsat | (supports TOPEX by determining geoid) |  |  |  |  |  |  |  | N.A.S.A. | proposed |
| 1989 | ERS-2 | × | — | × | — | — | — | — | × | E.S.A. | proposed |
| 1990 | Radarsat | — | — | — | — | — | — | — | × | Canada | proposed |
|  |  | × | — | — | — | — | — | — | — | N.A.S.A. | proposed |
| 1990? | MOS-2 | × | — | × | — | — | × | × | — | Japan | tentative |

*Ocean-related satellite missions planned for the* 1980s

The information in table 4 comes from Goody (ed.) (1982). The lead time for planning a new satellite mission is about 10 years, so oceanographers already have a reasonable idea of what, all being well, will be available for WCRP oceanographic experiments. There seems a reasonable prospect of achieving global monitoring of the ocean by a number of satellites from different agencies during the period 1987–93. That period will therefore be designated an 'intensive observing period' for the WCRP-O experiments. Measurements *in situ* would be centred on the intensive period, but also start earlier and end later. Some ship and buoy programmes have already begun.

## OCEANOGRAPHIC EXPERIMENTS IN THE WCRP

The three streams of the WCRP pose different problems for the oceanographer. In this section, I summarize briefly the oceanographic experiments that are being planned for each, emphasizing the role to be played by satellite observations. An essential background task for all climate research is the collection of a new global data base that will yield greatly improved information about the climatology (i.e. the regular seasonal cycle and statistics of the inter-annual fluctuations) in the variables listed in table 2. Satellite observations will make a major contribution to the new data set. The new methods adopted in collecting it will provide experience relevant to future operational monitoring of the oceans for climate prediction. The WCRP Pilot Ocean Monitoring Study (POMS) seeks to encourage new cost-effective methods, including remote sensing by satellites.

*WCRP stream* 1: *extended-range weather forecasting*

In short-term weather forecasting it is assumed that the boundary conditions of the atmosphere remain constant. No attempt is made to take account of changes in the global pattern of surface fluxes of energy and water entering the atmosphere from the ocean during the period of the forecast, which is limited to about 10 days by error growth in the atmospheric general circulation model (AGCM) used to make the forecast. It is nevertheless desirable to make use of accurate initial conditions for the g.p.f.o. In practice that means knowing the sea-surface temperature (s.s.t.), either from the seasonal mean distribution available in standard climatologies (see, for example, Robinson 1976; Robinson *et al.* 1979), or from maps that have been recently updated by using ship, buoy and satellite data collected during the preceding week or month. Such updating is worth while if there are sufficient data available to resolve the inter-annual thermal anomalies.

The possibility of extended-range weather forecasting, essentially by the same kind of AGCMs as are used today for short-term forecasting, rests on the tendency of certain atmospheric circulation systems to persist for longer than the average limits of predictability. The hope is that some kind of forecast might be achievable on periods up to a month or even a season ahead. The g.p.f.o. changes significantly during the period of the forecast. The change must be allowed for either by assuming it proceeds at the climatological mean seasonal rate (starting from either the climatological mean distribution or from an updated initial condition) or, more ambitiously, by calculating the changing g.p.f.o. during the period of the forecast from information derived solely from the AGCM. The success of Denman & Miyake (1973) in using a one-dimensional model of the ocean mixed layer to calculate s.s.t. changes from Ocean

Weather Ship (O.W.S.) meteorological observations, has encouraged National Meteorological Centres around the World to add ocean mixed-layer subroutines to their AGCMs. The assumption is that on periods of 1–3 months the s.s.t. (and therefore the g.p.f.o.) is insensitive to inter-annual changes in the ocean circulation and oceanic planetary waves, and, more critically,

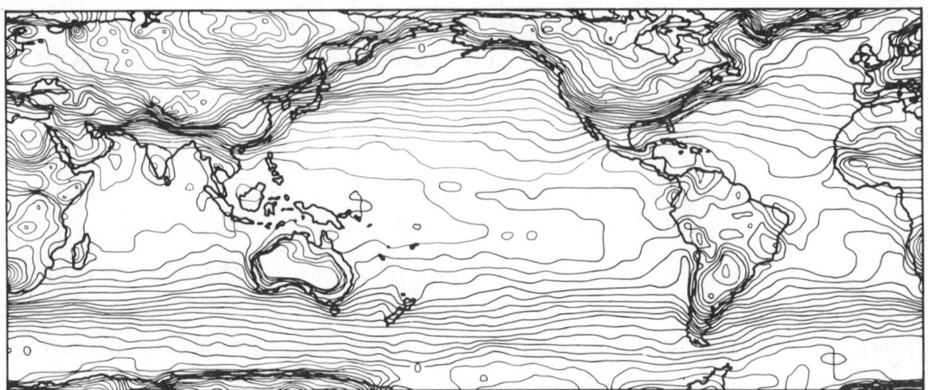

FIGURE 6. The first map of the surface temperature of the world based on satellite infrared observations made during January 1979. Contour interval is 2 K. (From Goody (1982); reproduced with permission.)

that the mixed layer subroutines can be made accurate enough to justify their use, i.e. they must predict the development of inter-annual anomalies of s.s.t. during the period of the forecast. That is a tough specification to meet: s.s.t. anomalies must be calculated to better than 1 K over 3 months. Recent studies of the accuracy of existing mixed-layer models suggest they cannot meet this specification; Thompson (1976) showed they have errors of more than 1 K at O.W.S.

Sensitivity studies by Hofmann (1982) showed that some, but not all, of the discrepancy could be attributed to random errors in the meteorological data used to drive the mixed layer model; the main consequence of which is that the mixed-layer subroutines cannot be tuned accurately to O.W.S. data. When used with an AGCM they will give consistent (consistently wrong?) results. To test the performance of mixed layer models in AGCMs it wil be necessary to compare the predicted s.s.t. anomaly development with the results of global monitoring by all available means, but especially by satellite infrared (figure 6). As we saw in table 2, the inter-annual deviations are small, only a kelvin or so over most of the World Ocean, so the specification for satellite monitoring of s.s.t. is around the $\frac{1}{3}$ K that Harries et al. (this symposium) believe may be possible by 1990.

What happens if the results of global tests with mixed-layer models in AGCMs prove discouraging? Extended-range weather forecasting will then have to proceed less ambitiously, using the climatological values for the seasonal progression of s.s.t., starting from the observed initial conditions. The emphasis will then fall on the collection of the best possible empirical description of the climatological mean seasonal development of the s.s.t. to the same accuracy.

### *WCRP stream 2: short-term climate prediction*

Thermal anomalies occur spontaneously throughout the planetary climate system as the result of short-term fluctuations in the weather. The atmosphere has only a short memory for such anomalies (a few weeks at most), but other elements of the climate system, which can respond to this 'random' forcing by the atmosphere, have longer memories. Attention has been focused on the ocean because it not only has a longer thermal memory than the atmosphere, but it is also mobile, so thermal anomalies created in the ocean by atmospheric forcing might not stay where they are created (as do anomalies on land, for example deserts or forests), but they may move, creating feedback to the atmosphere that is both delayed and displaced. The creation of an oceanic thermal anomaly at one location followed by its emergence to the atmosphere elsewhere, where it may lead to significant climate response, is called an 'oceanic teleconnection'. If the detailed physics of these transient thermal anomalies in the ocean were understood, and if it were possible to calculate their residence time in the ocean and the subsequent response of the atmosphere when they emerge, then short-term climate prediction might become a possibility. Attention has been focused on a number of candidates.

1. The first, and best known, is that winters in Europe are influenced by the advection of warm water eastwards across the North Atlantic by the Gulf Stream. The idea can be traced back to a paper by Sabine (1846) who suggested that the mild winter of 1821/2 was due to the 'warm water of the Gulf Stream spreading itself beyond its normal bounds . . . to the coast of Europe, instead of terminating as it usually does about the meridian of the Azores'.

2. The second is that thermal anomalies created in mid-ocean, notably in the Pacific, as a Markov response to random forcing by the atmosphere relax slowly (over a period of months) modifying the climate. The generation of such anomalies has been studied by Hasselmann (1981), while Namias (1976) and Davis (1976, 1978) have produced statistical evidence that the atmospheric circulation in winter does show some correlation with the sea-surface temperature the previous autumn.

3. The third possibility is that anomalous growth of sea ice persists for several months, reducing the area of ocean available to create relatively warm, moist maritime air masses that are such a large factor in the climate at locations many hundreds of kilometres away from the ocean. Lemke *et al.* (1980) have examined the statistical evidence.

4. A fourth possibility, which is attracting great attention at present, is the transmission of thermal anomalies by planetary waves inside the ocean, just as anomalies are propagated around the globe by planetary waves in the atmosphere (Hoskins 1983). Wyrtki (1975) proposed that anomalies in the seasonal cycle of upper-ocean heat content in the western Pacific are transmitted eastwards at a predictable speed by an equatorial Kelvin wave, arriving on the American coast some three months later. Modelling studies (O'Brien *et al.* 1981) have simulated the phenomenon and shown how the Kelvin wave then travels poleward along the American coasts from where Rossby waves propagate the signal back out into the open ocean across a broad front, creating a massive winter anomaly in the sea-surface temperature, of the kind observed to occur every few years, the 'El Niño' years (figure 7). The anomalous thermal structure of the ocean created by these waves is important in its own right because of the crippling effect it has on the world's greatest fishery (Glantz & Thompson 1981), but also because it leads to a significant change in the rate of energy release to the atmosphere. Statistical analysis by Horel & Wallace (1981), Wallace & Gutzler (1981) and Shukla & Wallace (1983) have shown that the atmosphere responds vigorously to such tropical energy flux anomalies, demonstrating globally what Walker (1923) and Bjerknes (1969) had previously shown in selected locations, the famous 'southern oscillation'. Recent theoretical studies by Webster (1981) and Hoskins (1983) have revealed the sensitive manner in which the atmosphere responds to the El Niño. The response occurs in winter when the Westerlies are close enough to the Equator to overlap

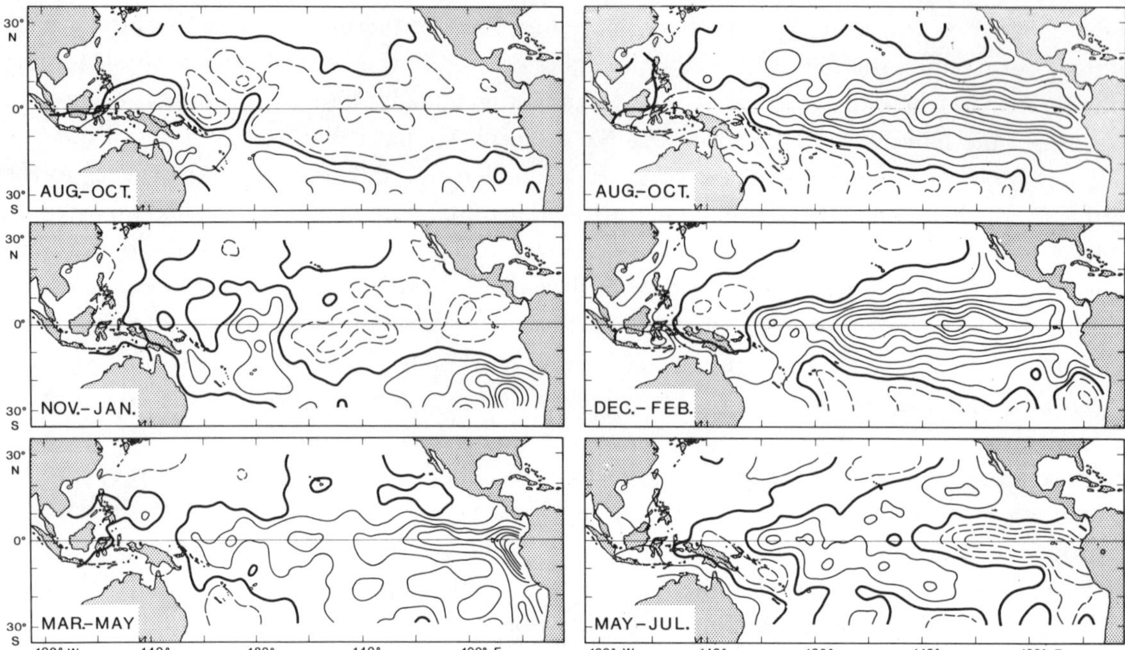

FIGURE 7. A sequence of sea-surface temperature anomaly maps showing the development and decay of an El Niño event. (Based on the ensemble analysis of Rasmussen & Carpenter (1982); reproduced with permission.)

s.s.t. anomalies that have been radiated beyond the Equator by the coastal Kelvin waves and Rossby waves. Thus a combination of planetary waves, first in the ocean, then in the atmosphere, collaborate at one season every few years to radiate climatic disturbances around the globe. The last major event occurred in 1972. Recent measurements off the Pacific coast of America suggest that an El Niño event is also under way this year (1982/3). The Northern Hemisphere winter has exhibited strong anomalies, with spring flowers blooming and birds nesting in northern Germany in December. Formal statistical analysis of this new event will be greatly aided by the advances in understanding made during the last decade. The prospects for climate forecasting based on coupled ocean–atmosphere teleconnections seem promising, but a great international investment in observations and modelling is now needed to set the subject on a sound scientific basis.

The relative importance of these four candidates is now being actively studied by modellers. The method is to use an atmospheric general circulation model to calculate the response of the global climate to simulated anomalies of sea-surface temperature or sea-ice extent. The aim is to discuss which, if any, of the observed ocean surface anomalies produce atmospheric climate anomalies that are statistically significant, i.e. that contain a significant fraction of the total climate variance, most of which is *not* correlated with the test ocean anomaly. Early results suggest that mid-latitude s.s.t. anomalies can produce significant climate response, but only if they are several kelvins (Haney 1979), whereas the observed anomalies are rather smaller, only about 1 K (Barnett & Davis 1975; Weare 1977). On the one hand, significant global climate response has been found to sea-ice anomalies by Warshaw & Rapp (1973), Newson (1973) and Herman & Johnson (1978), and potentially useful responses to equatorial anomalies of the El Niño type, described by Rasmussen & Carpenter (1982), have been revealed by modelling studies by Pan & Oort (1983). The global atmospheric circulation models still contain worrying flaws (for example, in the distribution of the annual heat budget (Dobson *et al.* 1982)), so predictions based on them must be treated with caution. Nevertheless, the results so obtained suggest that the best prospects for short-term climate prediction lie in tropical sea-surface anomalies associated with the El Niño phenomenon. The WCRP stream 2 activity has therefore been focused onto that fourth candidate with a major international project called TOGA (Tropical Oceans and Global Atmosphere).

TABLE 5. REQUIREMENTS FOR TROPICAL OCEAN MONITORING FOR RESEARCH INTO INTER-ANNUAL VARIABILITY OF THE GLOBAL CLIMATE

*(a) coverage*

| | |
|---|---|
| duration | at least 10 years |
| regional limits | the tropical ocean (half the surface area of the World Ocean) |

*(b) sampling density*

| | |
|---|---|
| temporal resolution | 1 month |
| spatial resolution | 2° latitude; 10° longitude |

*(c) measurement accuracy*

| | | | feasibility | |
|---|---|---|---|---|
| | | | | satellite |
| variable | accuracy | *in situ* | 1985 | 1990 |
| surface temperature | 0.3 K | yes† | no | yes |
| surface salinity | 0.1 kg m$^{-3}$ | yes† | no | no |
| sea level | 2 cm | yes† | no | yes |
| total energy flux | 15 W m$^{-2}$ | no | no | no |
| precipitation | 3 cm month$^{-1}$ | no | no | yes |
| surface wind speed | 2 m s$^{-1}$ | yes† | yes† | yes |
| surface wind direction | 20° | yes† | yes† | yes |

† Measurement accurate, but sampling inadequate.

The key factor in all four candidates is the development of anomalies in the global pattern of fluxes of energy and water from the ocean to the atmosphere, associated with anomalous heat content of the upper ocean, or anomalous ice cover. There is an urgent need to improve the documentation of the regular seasonal cycle and inter-annual variability of the g.p.f.o. It is needed both as the raw material for further statistical studies designed to reveal new aspects of global air–sea interaction on a period of a few years, and as time series to be used in testing models of short-term climate change. The specification for such measurements depends on region and season and whether or not an El Niño event is occurring. The values given in table 5 can be treated as a rough guide. They are based on Gill (ed.) (1983), who suggested, as a criterion, values that are 20% of the Rasmussen & Carpenter (1982) El Niño ensemble analysis, which is being widely used in modelling studies. Although most of the required measurements could in principle be made *in situ*, it would be unrealistically expensive to do so with the specified sampling density, which must be sustained over a period of at least a decade if the measurements are to capture the important, but relatively rare, El Niño events. Monitoring by satellite is already making significant contributions, especially with regard to solar heating of the ocean and the surface wind stress derived from cloud drift (figure 8), but the accuracies achieved in other variables (temperature, precipitation) are not yet good enough. The situation could change dramatically by 1990 as the result of improvements in algorithms used to extract the required variables and a new generation of satellites. Most of the requirements should then be achieved. But the most important measurement of all, the energy flux at the surface, will still remain beyond the reach of measurements both *in situ* and by satellite. At present there seems no alternative to filling the gap by an expensive programme of aircraft measurements of the kind made during GATE and JASIN (Nicholls *et al.* 1983).

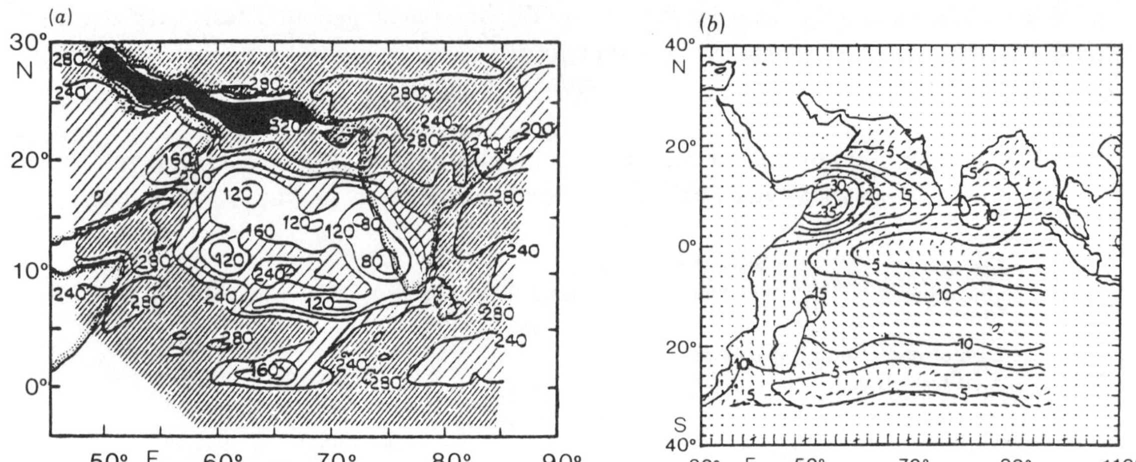


Figure 8. Maps of (*a*) solar energy flux into the sea (watts per square metre) from Gautier (1982), and (*b*) wind stress (centinewtons per square metre) from the cloud drift analysis of Wylie & Hinton (1982). These distributions for the Indian Ocean during the period 10–19 June 1979 illustrate the potential of existing meteorological satellites (in this case GOES-1) to yield oceanic products useful for climate research.

### *WCRP stream 3: longer-term climate prediction*

The third stream of the WCRP is concerned with decadal timescales. The major preoccupation is climate change due to pollution of the atmosphere by $CO_2$ and other radiatively active gases, or changes land or river use, or a change in the solar constant. How would the climate respond to such changes in the boundary conditions of the planetary climate system? The response is best studied by considering the slow changes that will occur in the global circulations of energy and water. In the atmosphere these lead to changes in mean temperature and precipitation and in the regional distributions of temperature and rainfall over the continents, with consequences for agriculture, energy consumption, transport and other factors that influence the habilitability and economy of our planet. The decadal scale is important because major decisions are made by government and industry that are based on assumptions about the pattern of life decades ahead. Climate change is one factor in that pattern; it would be attractive to make reliable forecasts of decadal climate change, rather than assuming as at present that the present climate will persist. The first stage is to predict how the boundary conditions will change over coming decades. So far nothing can be done about predicting future changes in the solar constant, but fairly realistic predictions can be made about the rate at which man will pollute the planet (Bach *et al.* (eds) 1979). The second task is to calculate how the global circulations of energy and water will be altered by a given scenario for future pollution.

Most attention has been focused on $CO_2$ pollution, which is observed to be increasing in the atmosphere (Kellogg 1981) and which is expected to reach double the present concentration in the next century. The $CO_2$ acts as a steadily thickening blanket that allows the Earth to reach a higher temperature before regaining thermal equilibrium with solar heating. The thermal capacity of the planetary climate system introduces a lag between the increase in $CO_2$ and attainment of the new climate equilibrium. The question for climatologists are:

(*a*)  How much will the concentration of atmospheric $CO_2$ change?

(*b*)  How long will the global climate change lag behind the $CO_2$ change?

(*c*)  What will the new equilibrium temperature régime be like?

(*d*)  What will the new water régime be like?

(Water and energy circulation are linked through latent heat.)

The ocean is the key element in solving these problems, for two quite distinct reasons: (1) its ability to remove some of the extra $CO_2$ from the atmosphere, and (2) its thermal lag to increased heating caused by the extra $CO_2$ that remains in the atmosphere. A little more will be said about each of these below; the reader is referred to extended discussion in Bretherton (1982) and Woods (1983) for further information.

1. The rate of increase of atmospheric $CO_2$ depends both on the anthropogenic sources and on the rate at which some of the excess is accommodated in the ocean. So far only about half of the $CO_2$ pollution since the start of the Industrial Revolution remains in the air: the rest has been accommodated by the ocean, where it does not affect the climate. The processes of oceanic accommodation of $CO_2$ involves a complex chain of physical, chemical, biological and sedimentological processes that need not detain us here (see, for example, Bolin *et al.* (eds) 1979). But the effect of those processes depends significantly on physical processes in the upper boundary layer of the ocean, and on the circulation of the ocean. Models used by geochemists to calculate changes in the oceanic carbon cycle do not include a detailed description of ocean circulation based on the laws of ocean dynamics (Bolin (ed.) 1981), and are therefore not entirely satisfactory.

2. Although it is not certain exactly when it will occur, there seems little doubt that eventually the $CO_2$ left in the atmosphere will reach double its present concentration. When that occurs, the net surface flux of infrared radiation at the sea surface will differ from the present flux on average by $1.2$ W m$^{-2}$ due to the $CO_2$ alone, plus $2.3$ W m$^{-2}$ because the $CO_2$ will have directly warmed the air, and (when the new global thermal equilibrium is reached) by another $12$ W m$^{-2}$ due to increased water vapour in the atmosphere, making a total of $15.5$ W m$^{-2}$ (values from Ramanathan 1981). In equilibrium, the ocean loses heat by infrared radiation, evaporation and conduction at the same rate as it gains heat from the Sun and atmosphere. The extra $CO_2$ disturbs this balance, which is only regained when the sea-surface temperature has risen sufficiently. On these timescales, the atmospheric temperature and precipitation are slave to the sea-surface temperature. The thermal aspect of the $CO_2$ problem is therefore to calculate how rapidly the sea-surface temperature rises in response to the modified net infrared balance at the surface. The key question is how much of the oceanic water column is warmed. The lag in climate response to $CO_2$ pollution will be a few years if only the top $50$ m 'mixed layer' is involved; it will be several decades if the whole of the 'warm-water sphere' has to heat up; and it will be centuries if the 'cold-water sphere' must heat up too. Existing ocean models have been used to predict climate response to doubled atmospheric $CO_2$ (Bryan *et al.* 1982). But it is generally agreed that more has to be learnt about the global circulation of the ocean and the processes of water mass formation before we can have confidence in such calculations.

The availability of reliable, quantitative models of the circulation of the ocean and the transformation of water mass temperature and salinity are a *sine qua non* for prediction of climate change over a period of decades in response to pollution by $CO_2$ and other changes in the

[ 111 ]

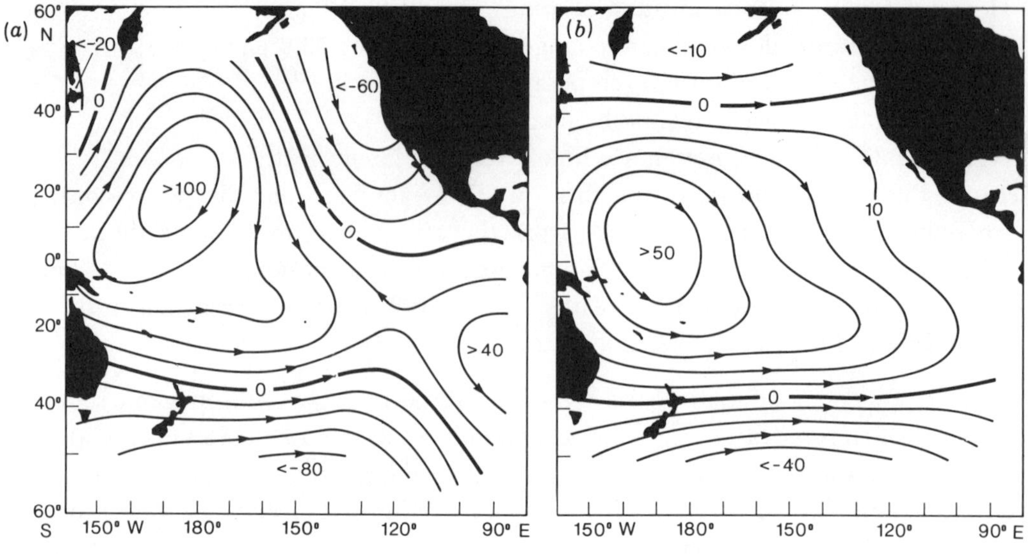

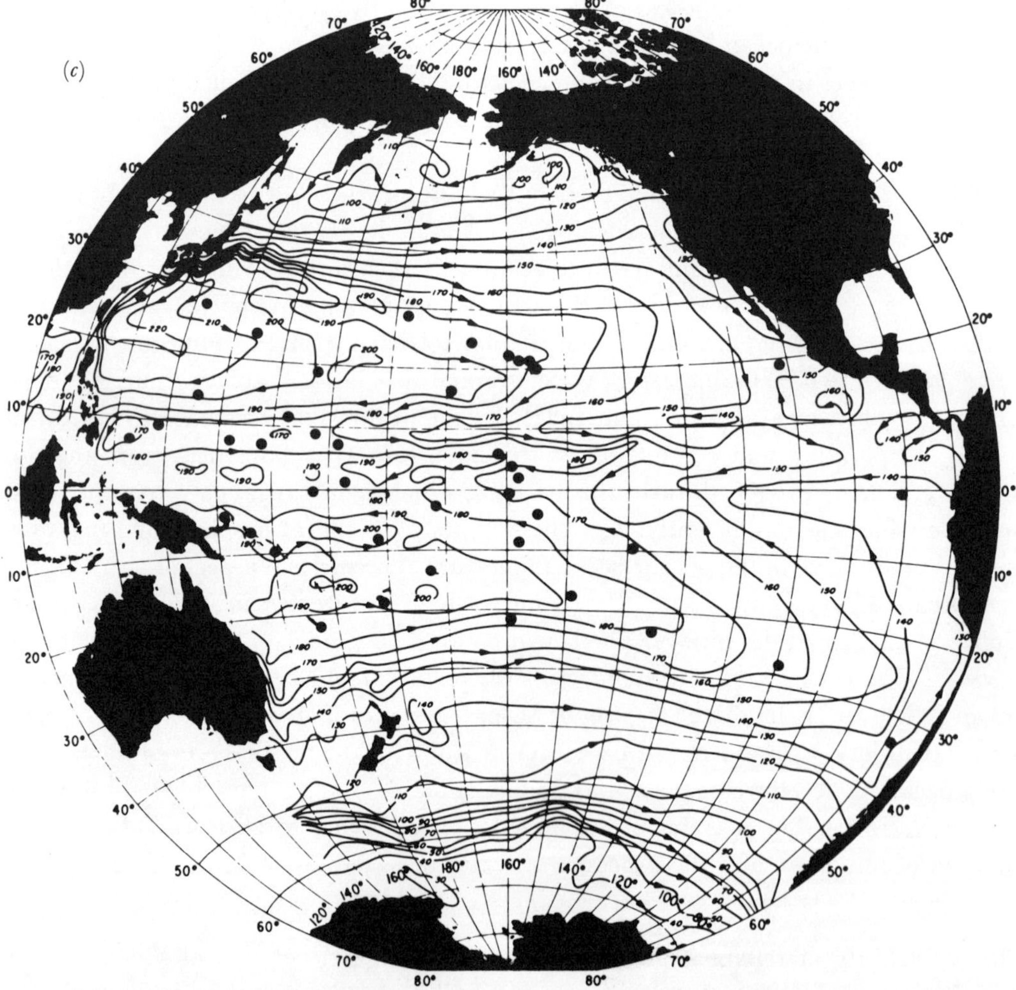

FIGURE 9. A comparison of two estimates of the distribution of sea-surface elevation in the Pacific Ocean: (*a*) from the Seasat altimeter, (*b*) from the classical dynamical method, as used by Wyrtki (1979), after spatial filtering to permit comparison. (From Wunsch (1983); reproduced with permission.) The unfiltered Wyrtki map (*c*) is shown for comparison (reproduced with permission). One of the central aims of the World Ocean Circulation Experiment (WOCE) is to collect a satellite altimeter data set that will permit the production of surface elevation maps with spatial resolution similar to that of Wyrtki's, but not dependent on the assumption inherent in the classical dynamical method that there is a level of no motion at a depth of 1 km.

boundary conditions of the planetary climate system. Existing models of the global ocean, such as that of Bryan & Lewis (1979), have succeeded in describing the principal features of the observed three-dimensional distributions of temperature, salinity and currents, and they have yielded estimates of the meridional heat flux similar to values derived by other methods (Bryden 1982). Nevertheless they are based on simplifications of unresolved processes that give cause for worry (see Wunsch 1983). For example, the parametrization of energetic eddies and waves with scales of about 100 km, the oceanic equivalent of weather systems in the atmosphere, has been shown to be important in a series of studies by Holland (1977). Also, Worthington (1981) has expressed the general feeling of unease about our ability to parametrize water mass formation in such global models. And Garrett (1979) has argued that the representation of mixing across density surfaces in the interior of the ocean is not all that it should be. The general feeling in the oceanographic community is that a new generation of ocean models will be needed to tackle the decadal climate problem. The expectation is that the general rise in computer power should make it possible by the end of the century to do what the meteorologist does with regard to the weather systems in atmospheric climate models; namely to resolve them. But the key processes of water mass transformation, both in the upper boundary layer and in the deep interior of the ocean, will have to be parametrized in ocean climate models for the foreseeable future, as clouds are in atmospheric climate models.

As work proceeds on new climate models of the World Ocean, there will develop an increasingly urgent need for a description of the ocean circulation and scalar distributions with which the model results can be compared. The construction of such an empirical description of the World Ocean is not easy. Oceanographers cannot follow the approach of meteorologists in the Global Atmospheric Research Programme (GARP) who designed the 1979 Global Weather Experiment to serve a similar purpose. The meteorologists could base their description on measurements of directly relevant variables (wind velocity, temperature, pressure, etc.), according to a sampling schedule that resolved the weather systems. The oceanographer must base his description on measurements that are less directly relevant (e.g. chemical tracers) and which under-sample the energetic weather systems. Consequently the empirical description of the ocean circulation that emerges is itself heavily dependent on a diagnostic model. The potential for uncertainties in the description of the ocean is much greater than in the GARP description of the atmosphere.

The description of the World Ocean that we have at present is based largely on hydrographic data collected over the past 100 years, but especially during a small number of major expeditions (notably the German *Meteor* expedition of 1925–7 and the International Geophysical Year expedition of 1958). There have recently been significant advances in the methodology of diagnostic models used to estimate the circulation from those data (Wunsch & Grant 1982). New types of data have also been added to the inventory, notably a wide range of chemical tracers surveyed by the Geosecs expeditions of the 1970s (Broecker 1981), satellite-tracked drifting floats used extensively for the first time in 1979 (Richardson 1983), deep-drifting, acoustically tracked floats used widely in the exploration of the ocean 'weather' in the 1970s (Rhines 1977), and the new range of satellite measurements listed in table 3.

The planners of the WCRP have now decided that the major oceanographic activity in stream 3 should be to seek a new description of the World Ocean circulation and water mass transformation in readiness for the coming generation of climate-related ocean models. Plans are now being prepared for a World Ocean Circulation Experiment (WOCE) intended to produce that description by the end of the century. The aim will be to bring together the new observing techniques and the new diagnostic modelling techniques in a coordinated experiment that will for the first time study the World Ocean as a whole.

The commitment to a global approach, echoing the GARP Global Weather Experiment, puts the satellite observations at the centre of the design. They alone offer the necessary global coverage and without them there would be no WOCE. The following systems are seen as demanding highest priority:

(1) the *altimeter*, to measure the surface pressure distribution and therefore the surface geostrophic current, for (i) statistics of the ocean weather systems, (ii) the varying component of large currents or gyres, and (iii) the steady large-scale circulation;

(2) the *scatterometer*, to measure surface wind stress and hence the Ekman transport and Ekman pumping;

(3) the *infrared radiometer*, to measure sea-surface temperature.

The first two of these are vital for the first goal of WOCE, namely to determine the ocean circulation. The third is added to them as the barest minimum contribution (from satellites) to the second goal of WOCE, namely to determine water mass conversion. In practice all the other systems listed in table 3 will contribute to the second goal, too. The experiment will also involve a major programme of measurements *in situ*. The preliminary list drawn up by the steering group for WOCE includes the following items:

(1) a global ship programme of coast–coast sections of top-to-bottom hydrographic and light (i.e. requiring small samples) chemical profiles;

(2) a sparser global sampling programme of 'heavy' chemicals (i.e. those requiring larger samples);

(3) large numbers (perhaps thousands?) of deep drifting neutrally buoyant floats, with pop-up capability to permit satellite tracking and data communication;

(4) basin-scale acoustic tomographic arrays;

(5) a global network of tide gauges or pressure gauges (to support the altimetric programme);

(6) current-meter arrays at special regions (e.g. overflow regions between ocean basins).

It would be premature to present detailed specifications for the WOCE measurements, whether by satellite or *in situ*. But the experiment is being designed with careful regard to the planned ocean-observing satellites listed in table 4. In particular it is hoped that there can be an intensive observing period lasting for about 5 years starting about 1988. The value of the satellite measurements will be greatly enhanced if they coincide (e.g. cross-calibration of altimeter measurements) and the measurements *in situ* may prove crucial for extracting the full information from them, as the happy coincidence of JASIN and Seasat showed (Guymer *et al.* 1983). These points will be clarified by meetings of the WOCE steering group over the coming years. Meanwhile, it is clear that WOCE is a very ambitious project offering to make a major contribution to solving a key issue in climate research and demanding a major commitment of resources over a period of about 5 years.

## Conclusion

Satellite monitoring of the ocean will be a central ingredient of all three streams of the World Climate Research Programme, and it will be an essential for the major experiments WOCE and TOGA.

[ 114 ]

## REFERENCES

Atlas, D. & Thiele, O. W. (eds) 1981 *Precipitation measurements from space*. Workshop report, Goddard Laboratory for Atmospheric Sciences, N.A.S.A., Greenbelt, U.S.A. (358 pages.)

Bach, W., Pankrath, J. & Kellogg, W. (eds) 1979 *Man's impact on climate*. (325 pages.) Amsterdam: Elsevier.

Barnett, T. P. & Davis, R. E. 1975 In *Long term climatic fluctuations* (*WMO Technical Note* no. 421), pp. 439–450.

Bernstein, R. L. 1982 *J. geophys. Res.* **87**, 7865–7872.

Bjerknes, J. 1969 *Mon. Weath. Rev.* **97**, 163–173.

Bolin, B. (ed.) 1981 *Carbon cycle modelling* (*SCOPE* 16). Chichester: John Wiley.

Bolin, B., Degens, E. T., Kempe, S. & Ketner, P. (eds) 1979 *The global carbon cycle* (*SCOPE* 13). Chichester: John Wiley.

Bretherton, F. P. 1982 *Prog. Oceanogr.* **11**, 93–129.

Broecker, W. S. 1981 In *Evolution of physical oceanography* (ed. B. Warren & C. Wunsch), pp. 434–460. Cambridge, Mass.: M.I.T. Press.

Bryan, K., Komro, F. G., Manabe, S. & Spelman, M. J. 1982 *Science, Wash.* **215**, 56–58.

Bryan, K. & Lewis, L. J. 1979 *J. geophys. Res.* **34**, 2503–2517.

Bryden, H. K. 1982 In *Time series of ocean measurements* (ed. D. Ellett) (*WCP Report Series* no. 21), pp. 3–12. Geneva: World Meteorological Organization.

Cheney, R. E. & Marsh, J. G. 1981 *J. geophys. Res.* **86**, 473–483.

Cracknell, A. P. (ed.) 1981 *Remote sensing in meteorology, oceanography and hydrology*. (542 pages.) Chichester: John Wiley.

Dantzler, H. L. 1976 *Deep Sea Res.* **23**, 783–794.

Davis, R. 1976 *J. phys. Oceanogr.* **6**, 249–266.

Davis, R. 1978 *J. phys. Oceanogr.* **8**, 233–246.

Denman, K. L. & Miyake, M. 1973 *J. phys. Oceanogr.* **3**, 185–196.

Dickson, R. R., Gould, W. J., Gurbutt, P. A. & Killworth, P. D. 1982 *Nature, Lond.* **295**, 193–198.

Dobson, F. E., Bretherton, F. P., Burridge, D. M., Crease, J., Kraus, E. B. & Vonder Haar, T. H. 1982 *The CAGE experiment: a feasibility assessment report* (*WCP Report Series* no. 22). Geneva: World Meteorological Organization.

Eagleson, P. S. 1982 *Land surface processes in atmospheric general circulation models*. (560 pages.) Cambridge University Press.

Garrett, C. 1979 *Dyn. Atmos. Ocean.* **3**, 239–266.

Gautier, C. 1982 *Trop. Ocean-Atmos. Newsl.* **12**, 5–6.

Gill, A. E. (ed.) 1983 *The Tropical Ocean and Global Atmosphere project* (*WCP Report Series*). Geneva: World Meteorological Organization.

Glantz, M. H. & Thompson, J. D. (eds) 1981 *Resource management and environmental uncertainty. Lessons from coastal upwelling fisheries*. (491 pages.) New York: Wiley-Interscience.

Goody, R. (ed.) 1982 *Global habitability*. Greenbelt, U.S.A.: N.A.S.A.

Gower, J. F. R. (ed.) 1982 *Oceanography from space*. New York: Plenum.

Guymer, T. H., Businger, J. A., Katsaros, K. B., Shaw, W. J., Taylor, P. K., Large, W. G. & Payne, R. E. 1983 *Phil. Trans. R. Soc. Lond.* A **308**, 253–273.

Haney, R. L. 1979 *Rev. Geophys. Space Sci.* **17**, 1494–1507.

Hasselmann, K. 1981 In *Climate variation and variability* (ed. A. Berger), pp. 481–500. Dordrecht: D. Reidel.

Herman, G. & Johnson, W. T. 1978 *Mon. Weath. Rev.* **106**, 1649–1664.

Hofer. P., Njoku, E. G. & Waters, J. W. 1981 *Science, Wash.* **212**, 1385–1387.

Hofmann, K. 1982 Diploma thesis, University of Kiel.

Holland, W. R. 1977 In *The sea* (ed. E. D. Goldberg, I. N. McCave, J. J. O'Brien & J. H. Steele), vol. 6, pp. 3–46. New York: Wiley.

Horel, J. D. & Wallace, J. M. 1981 *Mon. Weath. Rev.* **109**, 813–829.

Hoskins, B. J. 1983 *Q. Jl R. met. Soc.* **109**, 1–21.

Houghton, J. T. & Morel, P. 1983 In *The global climate* (ed. J. T. Houghton). Cambridge University Press. (In the press.)

Kellogg, W. W. 1981 In *Man's impact on climate* (ed. W. Bach, J. Pankrath & W. Kellogg), pp. 313–320. Amsterdam: Elsevier.

Keshavamurty, R. N. 1982 *J. atmos. Sci.* **39**, 1241–1259.

Krishnamurti, T. N. & Krishnamurti, R. 1980 *Deep Sea Res.* **26**, 29–61.

Lemke, P., Trinkl, E. W. & Hasselmann, K. 1980 *J. phys. Oceanogr.* **10**, 2100–2120.

Leonard, D. A. 1980 In *Remote measurement of underwater parameters* (ed. H. Dolezalek) (Royal Norwegian Council for Scientific and Industrial Research, Space Activity Division, Report no. SAD-91-T), pp. 13–14.

Mintz, Y. 1981 In *Precipitation measurements from space* (ed. D. Atlas & O. Thiele), Greenbelt, U.S.A.: N.A.S.A. Goddard Laboratory.

Namias, J. 1976 *Mon. Weath. Rev.* **104**, 1107–1121.

Newson, R. L. 1973 *Nature, Lond.* **241**, 39–40.

Nicholls, S., Brummer, B., Fiedler, F., Grant, A. L. M., Hauf, T., Jenkins, G. J., Readings, C. J. & Shaw, W. J. 1983 *Phil. Trans. R. Soc. Lond.* A **308**, 291–309.

O'Brien, J. J., Busalacchi, A. & Kindle, J. 1981 In *Resource management and environmental uncertainty. Lessons from coastal upwelling fisheries* (ed. M. H. Glantz & J. D. Thompson), pp. 159–212. New York: Wiley.

Pan, Y. H. & Oort, A. H. 1983 *Mon. Weath. Rev.* (In the press.)

Ramanathan, V. 1981 *J. atmos. Sci.* **38**, 918–930.

Rasmussen, E. M. & Carpenter, T. H. 1982 *Mon. Weath. Rev.* **110**, 354–384.

Richardson, P. 1983 *J. geophys. Res.* (In the press.)

Rhines, P. 1977 In *The sea*, vol. 6 (ed. E. D. Goldberg, I. N. McCave, J. J. O'Brien & J. H. Steele), pp. 189–318. New York and London. Wiley–Interscience.

Robinson, M. K. 1976 *Atlas of North Pacific Ocean monthly mean temperatures and mean salinities of the surface layer.* (U.S. Naval Oceanographic Office, Reference Publication no. 2.) (262 pages.)

Robinson, M. K., Bauer, R. A. & Schroeder, E. H. 1979 *Atlas of North Atlantic - Indian Ocean monthly mean temperatures and mean salinities of the surface layer.* (Washington: U.S. Naval Oceanographic Office, Reference Publication no. 18.)

Sabine, E. 1846 *Lond. Edinb. Dubl. phil. Mag.*, April.

Saunders, P. M. 1976 *J. mar. Res.* **34**, 155–160.

Schwiesow, R. L. 1980 In *Remote measurement of underwater parameters* (ed. H. Dolezalek) (Royal Norwegian Council for Scientific and Industrial Research, Space Activity Division, Report no. SAD-91-T), pp. 10–11.

Shukla, J. & Wallace, J. M. 1983 *Mon. Weath. Rev.* (In the press.)

Stewart, R. W. 1982 In *Proc. 33rd Congress, International Astronomical Federation, Paris.*

Tabata, S. 1978 *J. phys. Oceanogr.* **8**, 970–986.

Tabata, S. & Gower, J. F. R. 1980 *J. geophys. Res.* **85**, 6636–6648.

Thompson, R. O. R. Y. 1976 *J. phys. Oceanogr.* **6**, 496–503.

Untersteiner, N. 1983 In *The global climate* (ed. J. T. Houghton). Cambridge University Press. (In the press.)

Walker, G. T. 1923 *Mem. Indian met. Dep.* **24**, 75–131.

Wallace, J. M. & Gutzler, D. S. 1981 *Mon. Weath. Rev.* **109**, 784–812.

Warshaw, M. & Rapp, R. R. 1973 *J. appl. Met.* **12**, 43–49.

Weare, B. C. 1977 *Q. Jl R. met. Soc.* **103**, 467–478.

Webster, P. J. 1981 *J. atmos. Sci.* **38**, 554–571.

Woods, J. D. 1983 In *The global climate* (ed. J. T. Houghton). Cambridge University Press. (In the press.)

Worthington, L. V. 1981 In *Evolution of physical oceanography* (ed. B. Warren & C. Wunsch), pp. 42–69. Cambridge, Mass.: M.I.T. Press.

Wunsch, C. (ed.) 1982 *The TOPEX report.* Greenbelt, U.S.A.: N.A.S.A.

Wunsch, C. 1983 In *The global climate* (ed. J. T. Houghton). Cambridge University Press. (In the press.)

Wunsch, C. & Grant, B. 1982 *Prog. Oceanogr.* **11** (1), 1–59.

Wylie, D. P. & Hinton, B. B. 1982 *J. phys. Oceanogr.* **12**, 186–189.

Wyrtki, K. 1975 *J. phys. Oceanogr.* **8**, 530-532.

Wyrtki, K. 1979 *Eos, Wash.* **60**, 25–27.

## *Discussion*

P. K. TAYLOR (*Institute of Oceanographic Sciences, Wormley, U.K.*). The accuracy of the measurements of surface winds *in situ* quoted by Professor Woods, $\pm 5$ m s$^{-1}$, sounds surprisingly poor. What is the source of this estimate, and should greater effort and resources be devoted to increasing the accuracy, and temporal and spatial distribution, of observations *in situ*?

J. D. WOODS. The estimate of $\pm 5$ m s$^{-1}$ was based on evidence from a combination of sources. First, my observations at sea: I have noted that sporadic malfunction of anemometers on ships (for example, at particular headings to the wind) can introduce such errors. Second, visual estimates of sea state are unlikely to be better than one point on the Beaufort scale, which rises in steps of 5 m s$^{-1}$ in the upper part of the range, which is particularly important for global climatology. Third, comparisons of wind fields derived independently from the altimeter and scatterometer on Seasat, and from merchant ship observations exhibit differences as large as 5 m s$^{-1}$. Of course, some of the difference can be attributed to inadequate sampling by ships; only part is due to measurement error. I agree with Dr Taylor that *in situ* measurements by anemometers on purpose-built buoys can yield data with errors much smaller than

$\pm 5$ m s⁻¹, and that careful analysis of research ship anemometer records can also yield excellent results, as he and his colleagues showed in the Royal Society JASIN expedition (*Phil. Trans. R. Soc. Lond.* A**308**, 221–449). However, such high-quality data are seldom available, and do not have the coverage needed to produce wind fields used in global climate research. Such wind fields depend on the analysis of either patchy and less accurate merchant ship data, or satellite data. Experience with Seasat encourages us to believe that the latter will not only solve the sampling problem, but that the data will also have superior accuracy ($\pm 2$ m s⁻¹). The European ERS-1 mission, due to be launched in 1988, will be particularly important for the WCRP. Meanwhile, efforts to improve the coverage and quality of *in situ* wind measurements are being focused onto regions like the Tropical Pacific, where there is an urgent need to support WCRP projects, such as TOGA, that have already begun.

*Phil. Trans. R. Soc. Lond.* A **309**, 361–370 (1983)

*Printed in Great Britain*



# Detection of large-scale ocean circulation and tides

By D. E. Cartwright

*Institute of Oceanographic Sciences, Bidston Observatory, Birkenhead, Merseyside L43 7RA, U.K.*

As emphasized recently by Munk & Wunsch, the traditional methods of monitoring the ocean circulation give data too hopelessly aliased in space and time to permit a proper assessment of basin-wide dynamics and heat flux on climatic timescales. The prospect of nearly continuous recording of ocean-surface topography by satellite altimetry with suitable supporting measurements might make such assessments possible. The associated identification of the geocentric oceanic tidal signal in the data would be an additional bonus. The few weeks of altimetry recorded by Seasat gave a glimpse of the possibilities, but also clarified the areas where better precision and knowledge are needed. Further experience will be gained from currently projected multi-purpose satellites carrying altimeters, but serious knowledge of ocean circulation will result only from missions that are entirely dedicated to the precise measurement of ocean topography.

## 1. Introduction

Measuring and trying to understand the dynamics of the general circulation of the oceans has always been the central problem of physical oceanography. The more that detailed measurements are made, the more complicated the dynamics appears to be, with mesoscale eddies dominating the picture of what was once thought of as steady basin-wide gyre systems. The small scales of this relatively fine structure, of order 100 days and 100 km, now make the conventional methods of monitoring ocean circulation, by means of ships, tracked floats, and arrays of moored instruments, seem rather inadequate, except for localized experiments. This situation is emphasized, colourfully, by Munk & Wunsch (1982), who predict a new generation of ocean-monitoring experiments in the last decade of this century, employing the combined techniques of acoustic tomography and radar altimetry from satellites. The present paper is concerned only with the second of these techniques.

Estimating ocean circulation by means of radar altimetry involves measuring the elevation of the smoothed ocean surface relative to a fixed equipotential surface to 0.1 m precision or better. The anomaly sought is usually less than the tidal variation of the surface, so the tidal elevation must first be removed from the direct measurements. This is particularly important for the solar tides, which tend to have long-period aliases in the altimetry signal. Many accounts of the subject ignore the tides, assuming that they can be removed by means of a computer model. The best tidal models available (Parke & Hendershott 1980; Parke 1982; Schwiderski 1980) are probably capable of giving 0.1 m precision in most oceanic regions, but this is not yet proven and there is a need to evaluate the tide directly from the altimetric measurements, especially in the extensive areas where they have not been measured previously by other means. Extracting information about the tides can therefore be regarded as an important by-product of the process of extracting the ocean circulation.

## 2. Definitions and theory

The fundamentals of satellite altimetry have been described by many authors, for example Wunsch & Gaposchkin (1980), whose notation I shall largely use. Here, I outline only those concepts essential as a basis for the subsequent discussion.

Let $N(\theta, \lambda)$ be the height of the gravitational equipotential surface known as the 'geoid' above a standard ellipsoid of reference, where $\theta, \lambda$ denotes geographical latitude and longitude coordinates. In older books on geodesy and even in fairly recent papers on satellite theory, the term geoid is used synonymously with the sea surface, but such an approximation is tolerable only when precision of no better than 1 or 2 m is assumed. In the present context, the sea surface differs essentially from the geoid by the quantity $\zeta(\theta, \lambda, t)$ which appears as an important variable in ocean dynamic theory. The *mean* value of $\zeta$ over the oceans is supposed to be zero by definition. If $S$ is the instantaneous height of the satellite above the wave-smoothed sea surface as measured by the altimeter's radar pulse (with atmospheric and ionospheric corrections), and $R$ is its height above the reference ellipsoid as computed from gravitational theory and tracking data, then the desired sea surface height anomaly is given by

$$\zeta(\theta, \lambda, t) = R - S - N. \tag{1}$$

### (a) Circulation

Most of the currents in the ocean are in geostrophic balance with the local pressure gradients. Exceptions are in the Ekman boundary layer beneath strong winds and in the neighbourhood of strong jets. Allowances can be made for non-geostrophic forces in altimetry calculations, given supplementary data such as the wind-stress; indeed, monitoring of the wind-stress is considered to be an essential adjunct to future altimetric exercises intended for circulation studies. I shall assume geostrophy here, to demonstrate the essential principles. Tidal accelerations are of course assumed to be eliminated.

In a field of spatially variable density $\rho(x, y, z)$, where $x, y, z$ are a right-handed set of local Cartesian coordinates with $z$ vertically upwards, elementary principles show that the velocity $v$ in the (arbitrary) horizontal direction of the $y$ axis is related to the density gradient in the $x$ direction by the 'thermal wind equation'

$$\frac{\partial(\rho v)}{\partial z} = \frac{g}{f}\frac{\partial \rho}{\partial x}, \tag{2}$$

where $f = 2 \sin \theta \times$ (Earth rotation frequency) is the Coriolis requency and $g$ is gravitational acceleration. The problem of deducing the velocity field $v(x, y, z)$ by integrating (2) with respect to $z$ from a supposed equipotential 'level of no motion' and the known hydrographic data for $\rho$ has exercised oceanographers for decades. Various deep levels of no motion have been suggested and disputed, and some deny the reality of the concept altogether. Most realistically, Wunsch (1978, 1981) determines the unknown velocity field at an arbitrary reference level by solving an inverse problem posed by various continuity equations and horizontal boundary conditions. However, this method has to rely on the effective consistency of many hydrographic sections taken at different times and places.

Specifying the surface topography $\zeta(x, y)$ by altimetry enables the geostrophic velocity field at all depths normal to a satellite track to be integrated unambiguously from the surface downwards. If $z = 0$ is taken at the level of the geoid, then the pressure at a depth $d$ below the geoid is

$$p(-d) = \rho(0)g\zeta + p_0 + g\int_{-d}^{0} \rho(z)\,\mathrm{d}z, \tag{3}$$

where $\rho(z)$ has been written for $\rho(x, y, z)$ and $p_0$ is the atmospheric pressure at the surface. The geostrophic assumption then gives

$$v(x, y, z) = \{f\rho(0)\}^{-1}\left\{g\rho(0)\frac{\partial\zeta}{\partial x} + \frac{\partial p_0}{\partial x} + g\int_{-d}^{0}\frac{\partial\rho(z)}{\partial x}\,\mathrm{d}z\right\}. \tag{4}$$

If the mean density field is not known, then putting $d = 0$ in (4) at least gives the surface velocity. Finally, remembering that $v$ is the component of horizontal velocity normal to (and to the left of) the sub-orbital track, the full vector velocity is determined only at the crossing points of such tracks, and then only if they do not intersect at too acute an angle.

### (b) Tides

The tides are not geostrophic, because they are governed by both geostrophic and inertial forces of comparable magnitude. However, the variations in time of the topography, $\zeta(\theta, \lambda)$ itself, or more appropriately of its spatial gradients, contain practically all the required dynamic information if properly analysed. Vertical structure of the tidal current is unobtainable, because baroclinic motions do not deform the surface to any measurable extent; only the zero-order baroclinic mode with current uniform in depth is relevant, and this is derivable from the spatial gradients of the tidal topography. For optimum extraction of the tidal signal, it is desirable to minimize the noise caused by atmospheric loading by combining the topography from (1) with the surface pressure $p_0$, obtaining the 'sub-surface pressure' at the geoid:

$$g\rho(0)\zeta' = g\rho(0)\zeta + p_0. \tag{5}$$

I shall refer to $\zeta'(\theta, \lambda)$ as defined by (5) as the 'corrected topography'.

The initial object of tidal analysis of the corrected topography is to evaluate the spatial distribution of the constant admittance functions.

$$Z_2^m(\omega; \theta, \lambda), \quad m = 0, 1, 2,$$

of the ocean tide to the spherical harmonics of the tide-generating potential of order (species) $m$ and degree 2 with respect to frequency $\omega$. Alternatively, it is convenient to think of the decomposition of $\zeta'(\theta, \lambda)$ into time-harmonics of frequency $\omega_i$ and amplitude

$$H_i(\theta, \lambda) = |Z_2^m(\omega_i; \theta, \lambda)|\Gamma_i$$

with phase-lag

$$G_i(\theta, \lambda) = m\pi - \arg\{Z_2^m(\omega_i; \theta, \lambda)\},$$

where $\Gamma_i$ are the amplitudes of discrete harmonics of the luni-solar tide-generating potential, tabulated by Cartwright & Edden (1973). (In practice, of course, $H_i$ and $G_i$ have to be regarded as slowly variable quantities depending on the position of the Moon's node). Possible methods of mapping $Z$ or the discrete harmonics $(H_i, G_i)$ will be considered later. For the present there are two further fundamental points to be noted.

Firstly, the tidal parts of $\zeta'(\theta, \lambda)$ are strictly components of the *geocentric* tide, which differs from the usual definition of the ocean tide by the addition of the vertical component of the Earth tide. The latter is supposed to be calculable by inverting an integral formula for the loading effect of the relative ocean tide, but there are possible differences from the theoretical formulae (for example, a loading phase lag) that make it worth while to evaluate the Earth tide directly by analysis of the corrected topography and comparison with conventional tide measurements.

The second noteworthy point is that, because in tidal analysis we are only concerned with the temporal variations of $\zeta'$, its spatial variations may in principle be eliminated as an arbitrary constant part. This constant part may also include unremoved parts of $N(\theta, \lambda)$, which enters $\zeta'$ through (1), and so tidal analysis does not depend so critically on precise knowledge of the form of the geoid as does the estimation of the mean circulation. However, use of this property depends on frequent repetitions of the same sub-orbital track. Similarly, errors in the atmospheric loading $p_0$, though undesirable, will in the long run be eliminated by their lack of correlation with the tide-generating potential (ignoring a minute and correctable effect due to the solar atmospheric tide).

### 3. LIMITATIONS IN PRECISION

Both circulation and tidal studies require altimetry to measure $\zeta'$ continuously to an accuracy of about 0.1 m. More precisely, because the spatial gradients are the most relevant aspect of the topography, $\mathrm{grad}(\zeta')$ should be measurable to about 0.1 m in 1000 km for gyre-scale circulation and to 0.03 m in 100 km for mesoscale eddies. Both scales are equivalent to currents at mid-latitude of order 0.003 m s⁻¹, from the first term of (4). Lower accuracy is of course tolerable for grosser features of the circulation and for tides of greater slope-amplitude.

From (1) and (5) it is clear that any error in specifying $S$, $N$, $R$ or $p_0$ will affect the precision of the corrected topography $\zeta'$ or its gradients, on which all dynamical calculations depend. Errors in hydrographic data are not so serious because the surface current and tides are independent of $\rho$, but there are limitations in regions like the southeast Pacific, where the density field is less well known, or in regions of rapid thermocline variation. The errors to which $S$, $N$, $R$ and $p_0$ are subject are different in their magnitudes and origin; they are reviewed individually below.

*Altimeter height, S.* Instrumentally, altimeter technology was shown by the Seasat calibration experiment (Kolenkiewicz & Martin 1982) to be precise to a few centimetres, but several corrections have to be made to convert the pulse delay to a true path length through some 800 km of space and atmosphere. There is a small correction, due to the electron density in the ionosphere, that can be supplied adequately by operating at two frequencies, and another expressible in terms of $p_0$ itself. The latter is no worse than the correction for atmospheric loading in (4) or (5), and this may be obtained to sufficient accuracy over the more important oceans from meteorological agencies. (Ideally, the spacecraft should carry a pressure-sounding instrument.) Water vapour in the atmosphere accounts for a substantial variable delay in travel time, and must be measured directly from the spacecraft by radiometer. Sea waves introduce a bias that is not yet entirely understood, but the wave-height measuring function of the altimeter (Tucker, this symposium) enables one to avoid at least severe wave states.

*Geoid height, N.* The geoid topography is of course the largest of the geophysical signals in altimetry, with spatial variations up to 100 m amplitude. It is known to about 0.1 m precision in a few sea areas such as the North Sea and part of the northwest Atlantic from dense gravity measurements. Elsewhere, the best geoid models derived from satellite orbit perturbations (e.g.

GEM-L2 from Lageos; see also Lerch *et al.* (1981)) are claimed to be correct to better than 0.1 m in their longest-wavelength components (over 6000 km). Short wavelengths, on the other hand, say less than 1000 km, can also be accurately determined by averaging the altimetry itself over many passes of past satellites, on the assumption that such steep variations in the ocean circulation are quasi-random. There is, however, a worrying range of medium-scale geoid variations for which present knowledge is inadequate. Possible dedicated satellite–satellite tracking missions such as Gravsat, dedicated to precise geopotential determination, may materialize. Otherwise, some classes of ocean monitoring by altimetry can only proceed by inference. Wunsch (1981) has proposed a model of the mean dynamic topography in the North Atlantic, derived from hydrography by inverse methods, which when subtracted from the mean altimetric geoid would provide a tolerable 'interim' gravitational geoid as a basis for dynamic monitoring. The same method could in principle be extended to other ocean areas where sufficient hydrography is known.

*Orbit altitude, R.* This is potentially the largest of all sources of error; it depends a great deal on the nature and global distribution of the systems used to track the satellite. Laser systems are very precise but are restricted to clear skies. Doppler systems (e.g. Tranet) give only 1–2 m accuracy but they have usually been better distributed and operate in all sky conditions. (Tranet-2 is a new-generation doppler system planned to give an accuracy of about 0.1 m.) However, even with optimum use of islands for tracking stations, large areas of oceanic orbit must remain untracked, and here the orbit altitude must be computed from the gravitational field of the Earth with non-trivial allowances for atmospheric drag and solar radiation. In fact the gravity field at 800 km altitude is not known well enough for the 0.1 m precision, but specially tailored approximations may be derived to suit an individual orbit, achieving altitudes precise to about 0.7 m (Lerch *et al.* 1982). A continuous 'Global Positioning System', (G.P.S.), based on satellite–satellite and ground–satellite tracking, is said to be insufficiently precise for this work at present.

If we accept the limitations of tracking and gravity models, much empirical refinement of the altitude can be achieved by using the property that the radial error is dominated by a once-per-revolution harmonic with a few much smaller harmonics due to errors in the gravity model (Lerch *et al.* 1982). Over a limited area this can alternatively be simulated by a 'bias-and-tilt' correction, adjusted to minimize the differences in the tidally corrected altimetry at the network of points where ascending and descending orbits cross. Marsh *et al.* (1982) used the last method to determine precise mean ocean topography over a considerable area of the northeast Pacific, with residual r.m.s. 'errors' in the range 0.05–0.20 m (some of which could be due to real surface motion). There is, however, a danger of removing a genuine tilt in the sea surface due to a steady circulation.

Finally, there is a limitation to resolution of the tidal signal in the topography, and hence in its removal from the circulation signal, due to synchronism with solar tides. It is understood that for the present purposes the orbit must be adjusted so that its Earth-track is repeated every $P$ days, where $P$ is typically in the range 3–10. Since the satellite's node regresses very slowly for the high orbital inclinations needed to cover a large range of latitude, this condition makes for a period of repeated crossing of a given Earth point close to an integral number of sidereal days, thus sampling the sidereal tides $K_1$ and $K_2$ at a very long or infinite period. Most commonly, the inclination is forced to be slightly retrograde in order to make the orbit 'Sun synchronous', for special reasons of power conservation. Sun synchronism shifts the 'frozen

[ 123 ]

tide' from $K_2$ to $S_2$, which is a larger tidal constituent, while $K_1$ and $K_2$ are aliased into $1/y$ and $2/y$ frequencies respectively, to be confused with seasonal terms of interest in the circulation. There are similar problems in resolving the diurnal from the semidiurnal tides in a Sun-synchronous orbit. How to deal most effectively with this problem is outside the scope of this review, but synchronous tides may be avoided by lowering the limiting latitude to 60–70°, thus increasing the rate of nodal regression. Such an orbit is planned for the TOPEX mission, which is dedicated to the altimetric monitoring of ocean circulation (TOPEX Group 1981).

### 4. Results from recent missions

The only spacecraft to carry altimeters have so far been Skylab, Geos-3 and Seasat. Skylab was a pioneering experiment; although it gave encouraging profiles of the geoid to 5 m accuracy, its tracking data were too coarse to enable detection of any dynamic signals in the ocean surface. Geos-3 was more precise, both instrumentally and in its tracking support, and its large volume of data recorded in some sea areas has enabled accurate mean altimetric geoids to be constructed by minimizing crossing-point differences (Marsh *et al.* 1980).

The combined circumstances, of orbital error being confined to large wavelengths and excellent geoidal definition in the northwest Atlantic, enabled Leitao *et al.* (1979) to give the first demonstration of the 1 m steep rise of $\zeta$ across the Gulf Stream and a 0.5 m hump corresponding to a warm eddy to its north. These and other diagnoses were confirmed by AXBT data and infrared imagery from N.O.A.A. satellites. Much work in the same area was done with greater precision from the Seasat altimetry, notably by Cheney & March (1981), whose diagram of eight repeated passes over a slowly changing cold ring south of the Gulf Stream is now famous.

Cheney & Marsh (personal communication) have also experimented with smoothing the mean topography derived from global Seasat altimetry and the GEM-L2 geoid to obtain a circulation pattern that may be considered accurate at wavelengths greater than about 6000 km. Bruce Douglas of N.O.A.A. has done the same type of experiment over a 3 day arc. Draft copies of both maps appear to reproduce the major known gyre patterns, with some anomalies.

As well as its greater overall precision, the short-lived career of Seasat included an important period of 25 days when its orbit was arranged to repeat its Earth-track precisely with a period of 3 days 13 min. The results from the repeat orbits showed that this condition is essential if one is to distinguish temporal variations in $\zeta$ from spatial variations in $N$ in areas where the geoid $N$ is not accurately known. The most outstanding results of this sort are the mapping of the mesoscale variability of the ocean from the variation of $\mathrm{grad}(\zeta')$ about its 25 day mean value at each point. (This measure can be roughly converted into kinetic energy at crossing-points.) The procedure was apparently first suggested and carried out by Menard (1982) for the areas bordering the Gulf Stream and the Kuroshio. His map of the latter region is reproduced in figure 1. A similar map covering the world's oceans has been produced by Cheney *et al.* (1982). As well as a surprisingly large variability in the vicinity of the Agulhas and Falkland currents and over much of the Southern Ocean, their map also shows interesting large zones of quiescence in the eastern parts of the Atlantic and Pacific oceans.

All the investigations mentioned above were preceded by removal of the tidal elevation from $\zeta'$ by means of one of the computer models mentioned in §1, and paid no further attention to any possible residual tidal signal. A few researchers, however, have examined the tidal content

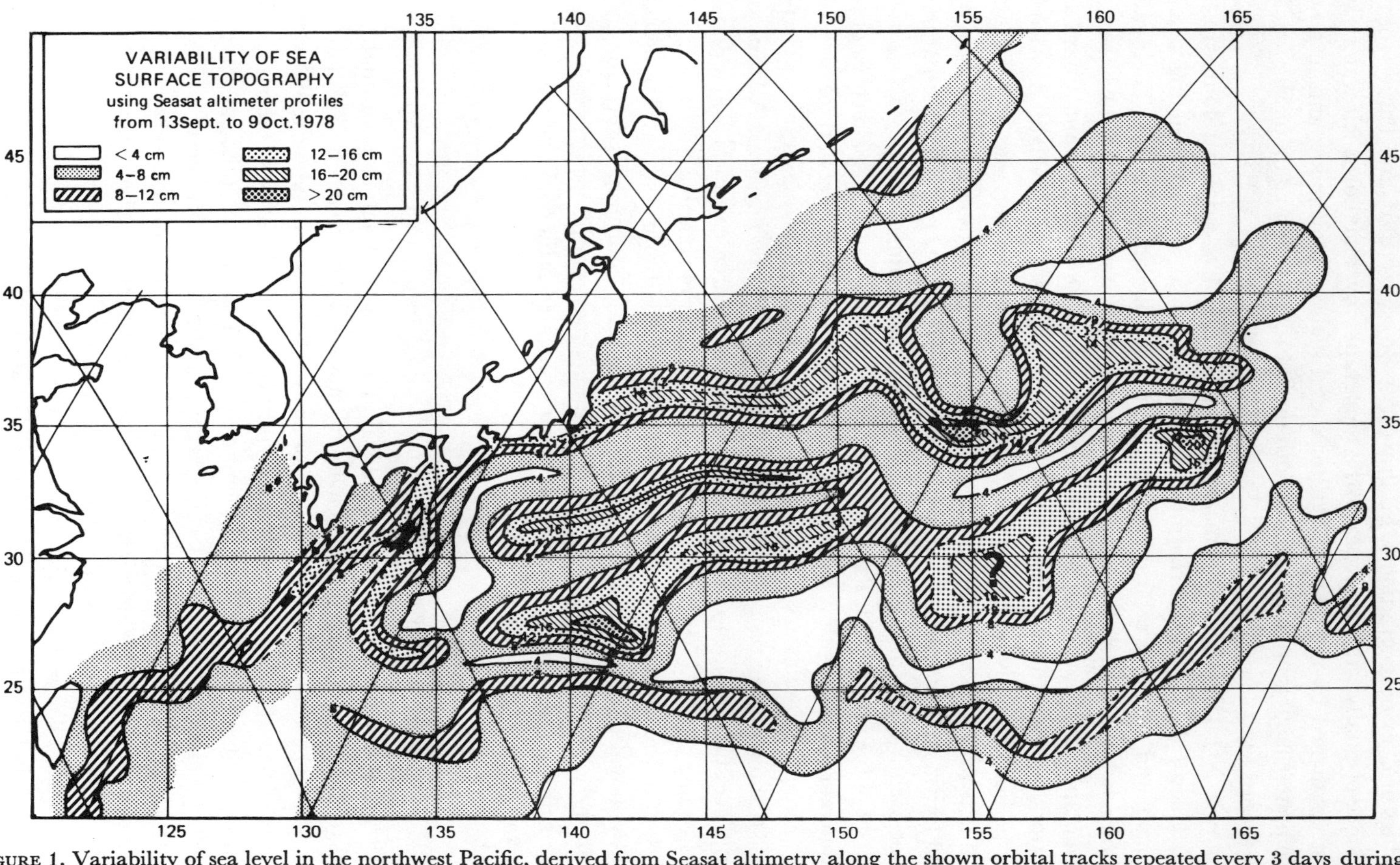



FIGURE 1. Variability of sea level in the northwest Pacific, derived from Seasat altimetry along the shown orbital tracks repeated every 3 days during 13 September to 9 October 1978. Shaded areas denote intervals of r.m.s. variation from less than 4 cm (white) to more than 20 cm. (From Menard (1982), by permission.)

of raw values of $\zeta'$ or of its gradients. Le Provost & Brossier (1981) and Parke (1981) showed clear evidence of tidally varying slopes in the topography over shallow shelf waters, where the signal is strengthened by the compressed horizontal scale of movement, and in any case the oceanic computer models are inadequate. Both these authors relied again on the repeated orbit period of Seasat.

Diamante & Nee (1981) fitted a mean sea surface and five tidal harmonics to the (non-repeating) Geos-3 data over the Bermuda calibration area, after subtracting a fairly good geoid estimate and adjusting orbital biases and tilts in an iterative solution. Their results compare reasonably well with deep-sea tidal measurements in the area, unlike those of other authors who appear to have been unable to obtain adequate tidal constants from Geos data.

Again working with the 25 days repeated orbit of Seasat, Cartwright & Alcock (1981) devised a method of tidally analysing the spatial differences in $S$ between all adjacent crossing points in a network, without geoidal correction, but after corrections for $p_0$ and the Body Tide of the solid Earth. All harmonic constituents were lumped together into a three-parameter tidal solution for each $S$-difference, these differences being constrained to sum to zero round each mesh element of the orbital grid. Results for the $M_2$ constituent over a sizable area of the northeast Atlantic agreed well with a tidal map based on an extensive network of direct measurements, surprisingly for such a short span of satellite data. The same authors (Cartwright & Alcock 1983) show similar results for an even larger area of the northeast Atlantic (figure 2), and a less successful attempt to map the tides in the noisy area east of Newfoundland.

Using the same data set, Mazzega (1982) has fitted a set of complex spherical harmonics up to order (3, 3) to the $M_2$ tide in the Indian Ocean, with an approximate averaging assumption for partial removal of other tidal harmonics. His resulting tidal map is surprisingly realistic. Here, as in the work of Cartwright & Alcock, simplifying assumptions necessary to produce a plausible result from the very short span of Seasat's repeated orbit could be removed or refined if applied to a year or more of good data.

## 5. Summary and future prospect

There is no doubt from the above results that precision in satellite altimetry and in its associated technology has passed the elementary stage of rough geoid determination and has reached a point where information about major current features and tides can be extracted. The performance of Seasat is generally admitted to have surpassed all expectations. Nevertheless, the precisions of its orbit $R$ and of the known geoid $N$ were not sufficient for even a brief determination of ocean-wide circulation, except perhaps at extremely long wavelengths. Orbital precision for Seasat was limited by its relatively low altitude of 800 km and by the bulky shape of its antennae, designed to perform 'synthetic aperture radar' imagery, thus introducing indeterminate drag forces and solar radiation pressures.

The precision in all aspects of altimetry required for a serious monitoring of ocean circulation (perhaps somewhat less precision is needed for tidal studies), is really very demanding of modern space technology. It requires something a little better than even 'Seasat', and this can only be achieved by a satellite system completely dedicated to this one objective. It must have a greater altitude to avoid atmospheric drag, a reasonably symmetrical shape, a globally distributed set of tracking stations, a radiometer for water vapour assessment, a scatterometer for wind-stress, and preferably a sounder for surface pressure. Its orbit must produce a repeated

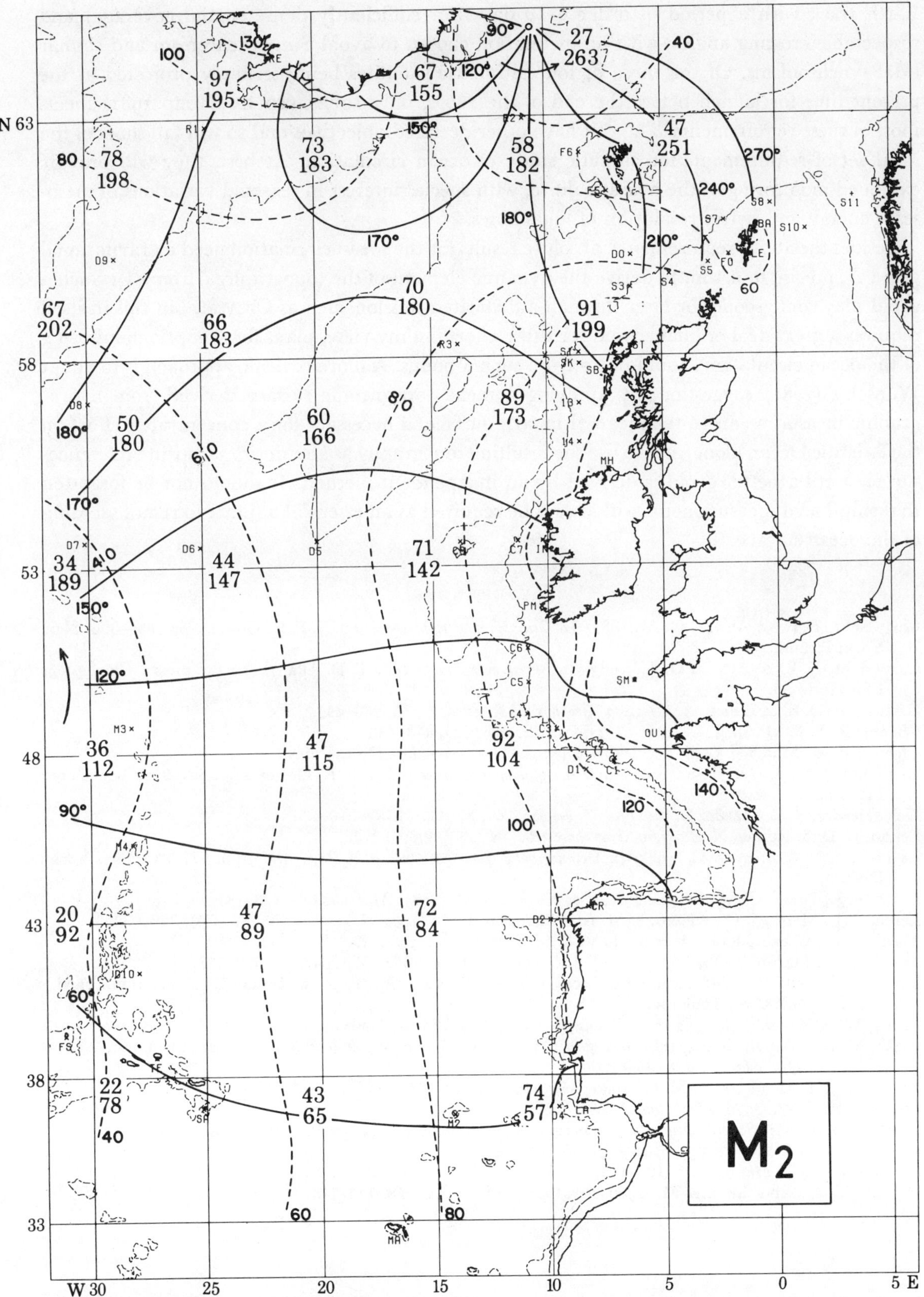


FIGURE 2. Tidal constants for $M_2$ in the northeast Atlantic from bottom pressure measurements (contours) and from analysis of differences in Seasat altimetry between orbital crossing points. Pairs of large figures give amplitude in centimetres (upper) and phase lag in degrees (lower) at crossing points, resulting from altimetric analysis. (From Cartwright & Alcock (1983).)

Earth track with a period of order 6–10 days (for sufficiently dense spatial coverage) and respectable crossing angles, with an inclination chosen to avoid Sun-synchronism and similar tidal synchronisms. Of the three or four altimetric satellites being seriously projected at the present time for launch before the end of the 1980s, only TOPEX (TOPEX Group 1981) meets most of these requirements. The rest have other declared objectives and so will fall short of the ideal set of requirements for climatic study of ocean circulation. At best, they will provide extended monitoring of the Seasat calibre, with special interest in seasonal variations of meso-scale energy and better resolution of lunar tides.

Besides the above requirements, absolute results for the mean circulation need a gravitational geoid $N$, precise to 0.1 m, to derive the dynamic element of the topography $\zeta$ from (1). Such a geoid may come sooner or later through a dedicated mission such as Gravsat, but this in itself requires a great deal of finance and scientific effort. In my view, plans for synoptic monitoring of the ocean circulation should not rely on such a bonus. A more realistic approach is to adopt Wunsch's (1981) suggestion, by applying a mean topographic surface derived from hydro-graphic measurements to the mean altimetric surface, if necessary for a concentrated effort on the Atlantic Ocean alone, and to use the resulting topography as an interim substitute reference-surface until a better geoid is derived by an independent method. It should not be forgotten that ship-based measurements will always be required as an essential adjunct to remote sensing of the ocean surface.

### REFERENCES


Cartwright, D. E. & Alcock, G. A. 1981 In *Oceanography from space* (ed. J. F. R. Gower), pp. 885–895. New York: Plenum.

Cartwright, D. E. & Alcock, G. A. 1983 In *Seasat over Europe* (ed. T. D. Allan). (In the press.) Chichester: Ellis Horwood.

Cartwright, D. E. & Edden, A. C. 1973 *Geophys. Jl R. astr. Soc.* **33**, 253–264.

Cheney, R. E. & Marsh, J. G. 1981 *J. geophys. Res.* **86** (C1), 473–483.

Cheney, R. E., Marsh, J. G. & Beckley, B. D. 1982 *J. geophys. Res.* (In the press.)

Diamante, J. M. & Nee, T.-S. 1981 In *Oceanography from space* (ed. J. F. R. Gower), pp. 907–918. New York: Plenum.

Kolenkiewicz, R. & Martin, C. F. 1982 *J. geophys. Res.* **87** (C5), 3189–3197.

Leitao, C. D. & Huang, N. E. 1979 *J. geophys. Res.* **84** (B8), 3969–3973.

Le Provost, C. & Brossier, C. 1981 In *Oceanography from space* (ed. J. F. R. Gower), pp. 927–932. New York: Plenum.

Lerch, F. J., Putney, B. H., Wagner, C. A. & Klosko, S. M. 1981 *Mar. Geod.* **5**, 145–187.

Lerch, F. J., Marsh, J. G., Klosko, S. M. & Williamson, R. G. 1982 *J. geophys. Res.* **87** (C5), 3281–3296.

Marsh, J. G., Cheney, R. E., Martin, T. V. & McCarthy, J. J. 1982 *Eos, Wash.* **63**, 178–179.

Marsh, J. G., Martin, T. V., McCarthy, J. J. & Chovitz, P. S. 1980 *Mar. Geod.* **3**, 359–378.

Mazzega, P. 1982 The M2 oceanic tide recovered from Seasat altimetry in the Indian Ocean. (Unpublished.) C.N.E.S./G.R.G.S., Toulouse.

Munk, W. & Wunsch, C. 1982 *Phil. Trans. R. Soc. Lond.* A **307**, 439–464.

Parke, M. E. 1981 In *Oceanography from space* (ed. J. F. R. Gower), pp. 919–925. New York: Plenum.

Parke, M. E. 1982 *Mar. Geod.* **6**, 35–81.

Parke, M. E. & Hendershott, M. C. 1980 *Mar. Geod.* **3**, 379–408.

Schwiderski, E. W. 1980 *Mar. Geod.* **3**, 161–255.

TOPEX Group 1981 *Satellite altimetric measurements of the ocean.* Pasadena: Jet Propulsion Laboratory.

Wunsch, C. 1978 *Rev. Geophys. Space Phys.* **16**, 583–620.

Wunsch, C. 1981 *Mar. Geod.* **5**, 103–119.

Wunsch, C. & Gaposchkin, E. M. 1980 *Rev. Geophys. Space Phys.* **18**, 725–745.

*Phil. Trans. R. Soc. Lond.* A **309**, 371–380 (1983)
*Printed in Great Britain*

**371**

# Observation of ocean waves

By M. J. Tucker

*Institute of Oceanographic Sciences, Crossway, Taunton, Somerset TA1 2DW, U.K.*

[Plate 1]

The European Space Agency plans to launch an Earth Resources Satellite (ERS-1) in a few years' time with a view to establishing an operational system of such satellites. It will carry two microwave devices giving information on waves: a precision altimeter and a synthetic aperture radar (SAR). Careful assessment of the potential performances of these instruments is therefore being carried out.

In the precision altimeter, the leading edge of the returned radar pulse is smeared by the rough sea surface. The degree of smearing is highly correlated with the significant waveheight and appears to be independent of other parameters. The main fundamental limitation to operational use comes from the sampling variability, which necessitates long averaging times. The SAR gives pictures of the sea surface that often show wave patterns, but their precise interpretation is an exceedingly complex problem, which is still not properly understood.

Even with three satellites in orbit, coverage in U.K. latitudes would be only once a day along the sides of a diamond-shaped grid with a side of approximately 500 km. An initial assessment indicates that this coverage is probably enough to be very useful in the open ocean, but that this limitation and the size of the altimeter 'footprint' become increasingly serious as a coast is approached.

## 1. Introduction

At present we can measure waves satisfactorily only by using sensors at the sea surface, and it is impracticable to put out large numbers of these to cover a wide area. Remote sensing provides the promise of covering large areas with one instrument.

There are at present only two types of wave sensor that are serious candidates for use in satellites, and both were tested on Seasat during 1978. They are the precision radar altimeter and the synthetic aperture radar. Both work at microwave frequencies. Optical sensors, even if practicable in principle, would not be acceptable operationally because of the loss of data due to cloud cover, and radio wavelengths would not be practicable because of the need to use large directional aerials to achieve adequate signal:noise ratios, and possibly also because of the interference they would cause.

The orbits of Earth-observing satellites are designed to be at approximately constant altitude, so that they can be described by two main parameters: this altitude and the inclination to the Earth's axis. The altitude of the sort of satellite with which we are concerned is fixed fairly tightly between two constraints: it must be outside the drag of the atmosphere, but it must not be too high because the sensors are close to their signal:noise limits and the higher the orbit the worse these get. At higher altitudes the attainable spatial resolution is also less. Thus the range of practicable altitudes is about 650–800 km. A satellite at 668 km does precisely $14\frac{2}{3}$ orbits per day, and one at 770 km does $14\frac{1}{3}$ orbits per day. In each case, therefore, they will cover the same track after 3 days. Such an orbit is termed 'Sun-synchronous', and a typical track of the

[ 129 ]

372	M. J. TUCKER

former is shown in figure 6. The speed of such a satellite in its orbit is approximately 7.5 km s$^{-1}$, and the speed over the ground approximately 6.8 km s$^{-1}$. Seasat had an altitude of approximately 800 km. It was not Sun-synchronous during the first part of its life, but was moved to a 3-day Sun-synchronous orbit 24 days before it failed.

## 2. The radar altimeter

The more straightforward of the two wave sensors is the precision radar altimeter, and a detailed description of the one mounted on Seasat is given by Townsend (1980). The carrier frequency was 13.5 GHz, the effective pulse length was just over 3 ns in time or approximately 10 cm in space, and the pulse repetition frequency was 1020 Hz. The echo from a rough sea shows a sharp initial rise due to the echo from immediately below the satellite, followed by a decaying tail from the rest of the patch of sea illuminated by the pulse (figure 1). The steepness of the rise depends on the waveheight. The precise mechanism is still being researched, but qualitatively one can think of echoes coming from crests and troughs of waves over an area a few kilometres in diameter immediately below the satellite. The important point is that the mechanism is sufficiently clear for us to be certain that the steepness of the rise of the echo pulse depends primarily on waveheight and not significantly on wavelength or any other wave parameter.

The steepness of the rise is measured by sampling the echo at three points. The system causes the centre one to follow the mid-point of the rise, and the separation of the other two is switched automatically to give optimum sampling as the waveheight varies. A number of corrections have to be applied, for example, to take account of the tilt of the beam from vertical.

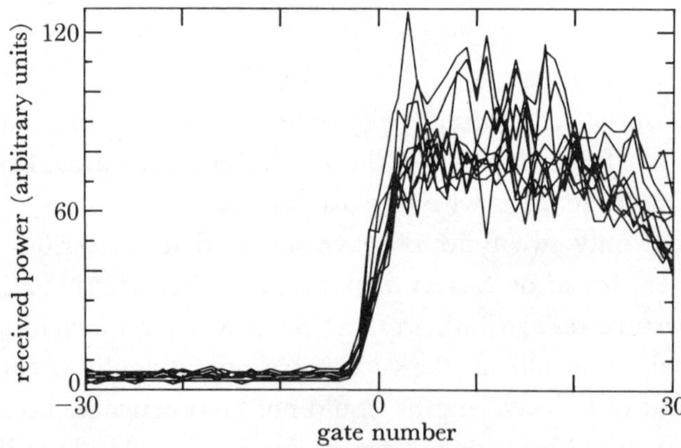

Figure 1. Ten of the smoothed returns from the radar altimeter on Seasat. Each return results from the averaging of 50 raw pulses and all the ten shown were taken within a flight time of 1 s. Each gate on the horizontal axis has a width of 3.124 ns.

As in nearly all wave-measuring devices, one of the main limits to accuracy is the inherent sampling variability. In the present case this appears as a random variability in the radar echo strength. When an echo is being received from a random surface such as that of the sea, the echo strength in any range gate is randomly chosen from a probability distribution whose mean (or 'expected value') is the information we require. The departure from the expected value is random on two timescales: these can be seen in figure 1 as the 'wiggles' within a single return, and the changes between successive returns.

[ 130 ]

Within the return from a single transmitted pulse, the amplitude randomizes after a time equal to the pulse length. Thus the random components of the returns within our three gates are uncorrelated. If the whole system were stationary, the returns from successive transmissions would be identical. However, the satellite is moving and looks at the sea from a different position for each transmission: the sea surface also moves, but this has less effect. The precise time it takes for this movement to produce an independent new 'look' at the sea surface depends on various factors including the waveheight, but the systems are designed so that with moderate or rough seas, each transmission produces an effectively independent look. In these circumstances, when smoothing over $N$ echoes that have been through a square-law detector (as in Seasat), the proportional standard deviation $\sigma$ of the smoothed values about their mean is simply $1/\sqrt{N}$. When $N = 50$, $\sigma \approx 14\%$: the scatter in figure 1 appears to be approximately this value. Estimates of waveheight derived from the differences between three gates on the rise are obviously going to have considerably more proportional scatter. Thus further smoothing is necessary if meaningful estimates of waveheight are to be obtained.

The relevant part of the Seasat specification was that when computed waveheights are averaged over 1 s, their standard deviation should be less than 10% of the mean. Webb (1981, amplified by personal communication) found that in practice he had to take 21 s averages to reduce the variability to 2%. This corresponds to a track length of approximately 140 km. The significance of this will be discussed later.

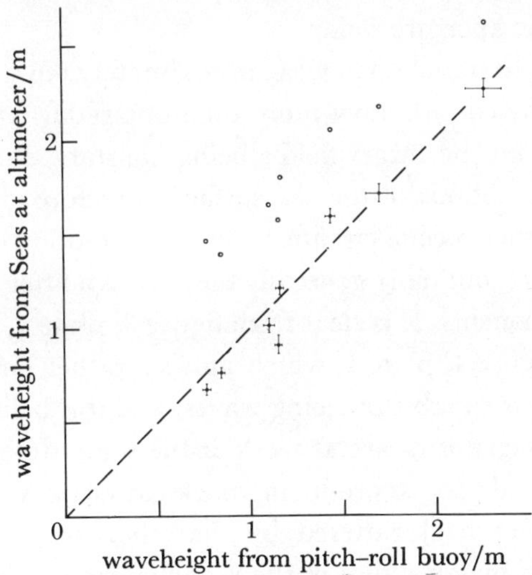


FIGURE 2. Comparison of estimates of significant waveheight from the Seasat altimeter with those from surface buoys. The solid circles with confidence bars show altimeter data that have been averaged and corrected by the J.P.L. algorithms. The open circles represent averaged but otherwise uncorrected data from the altimeter (from Webb 1981).

Webb (1981) compared the altimeter measurements of significant waveheights with measurements taken by surface buoys during the JASIN experiment and after all corrections had been made, the agreement was excellent considering that the measurements were not made at exactly the same place (figure 2). Chelton *et al.* (1981) report comparisons with 87 buoy measurements and get indications that the altimeter gave waveheights approximately 50 cm higher that they should have been for significant waveheights greater than 2 m. The reason for this discrepancy does not appear to be known. However, with more research we can expect that such an altimeter can achieve a 10% accuracy in waveheight measurement.

### 3. Synthetic aperture radar

The second wave sensor is the synthetic aperture radar, or SAR. A good description of the principles is given by Tomiyasu (1978), and the characteristics of the one on Seasat are described by Jordan (1980).

A satellite-borne SAR consists basically of a coherent radar looking at right angles to the flight path. In Seasat, the swath examined covered from 240 to 340 km to the right of the point beneath the spacecraft, the radar wavelength used was 23.5 cm, and the along-track resolution obtained (at a range of approximately 880 km) was 25 m. To obtain this resolution, a real aerial would have to be approximately 4 km long. The resolution in range is governed by the pulse length and in Seasat this was also 25 m on the sea surface.

The beam width of an ordinary radar is governed by the aperture of its aerial. Simply speaking, at a given point within the beam the phases of the signals coming from each part of the aperture have to be effectively the same so that they add constructively, whereas outside the beam they are different and add destructively. If we had a small aerial moved progressively across a large aperture, transmitting a signal each time it moved its own length, and measured the phase (relative to the transmission) of the received signal at a given target point, then when the large aperture had been filled one could add the received signals vectorially and get the same result as from a single transmission from the large aperture. With a little thought it will be seen that the same is true for the echo received by the radar from a fixed point target. This is the basic concept of synthetic aperture radar.

The time taken for Seasat to travel 4 km was approximately 0.6 s. This is long enough for the sea surface to have moved by considerably more than one radar wavelength. The principle of the SAR depends, of course, on the target field's being constant during the period of aperture synthesis, and the complex motions of the sea surface therefore degrade the along-track, or azimuthal, resolution. In rather special circumstances these motion effects can actually increase the visibility of long low waves, but more generally they act as a filter removing short-wavelength azimuthally travelling components. It is clear from figure 3, plate 1, which shows waves refracting round an island, and figure 4, plate 1, which shows a rather larger-scale image of waves on the deep ocean, that the SAR is capable of seeing waves, and it is believed that at least three and possibly four modulation mechanisms are at work at the same time. In all cases it is generally assumed that the basic physical process producing backscattering is 'Bragg resonance', in which the incident radar waves are backscattered by that short-wave component of the surface roughness whose wavelength matches that of the radar waves on the sea surface and therefore gives a coherent return ($\lambda_{\text{water}} = \lambda_{\text{radar}}/2 \sin \theta$, where $\theta$ is the angle of incidence). One of the easiest modulation mechanisms to understand in concept is the change in tilt of the sea surface towards or away from the satellite. Another is the way that the short Bragg resonant waves are

---

### Description of plate 1

FIGURE 3. Waves refracting round the Island of Foula, west of the Shetland Isles. Seasat SAR; orbit 1149; image 30 km × 24 km; $y$ axis true north; SAR range direction is 56° east of north. (Image made by Space Department, Royal Aircraft Establishment.)

FIGURE 4. An image of waves from Seasat orbit 0762 at 60° 11′ N, 6° 41′ W; image 12.8 km × 9.4 km; $y$ axis is SAR incremental range direction, and is 56° east of north. (Image made by Marconi Research Centre, Chelmsford.)

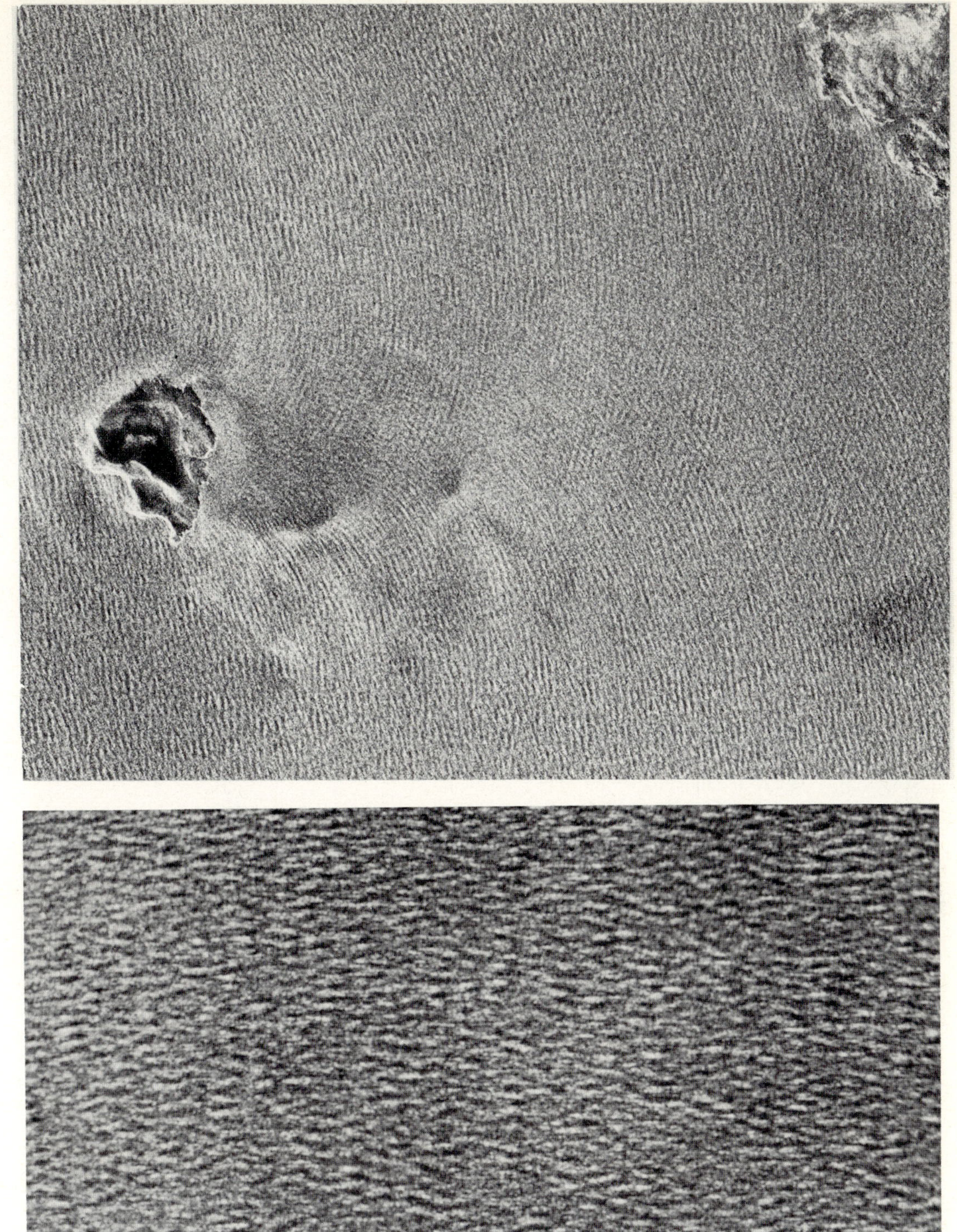

FIGURES 3 AND 4. For description see opposite.

stretched and compressed by the much longer waves that are being imaged. There is some indication that the variation of wind speed over the wave profile plays a part. Finally, as has been mentioned, in some circumstances the interaction of the surface motion due to long low waves with the aperture synthesis process can produce an imaging mechanism: however, this is a subtle process not readily explained in a short paper.

Figure 5 is the directional spectrum of the image in figure 4 (calculated digitally via Fourier transforms), and shows that apart from the longest waves (i.e. the lowest wavenumber waves), the spectrum of the image is symmetrical about the range axis and has a fairly constant width in the azimuthal direction. In this particular case we have wind speed and waveheight (but not a directional spectrum) measured nearby on the sea surface, and we can deduce that the true

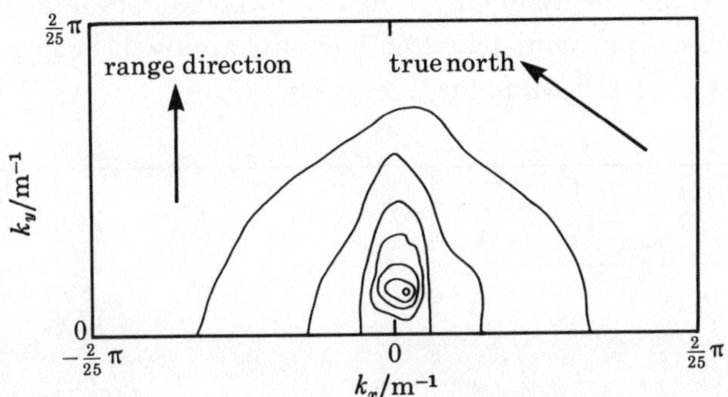


FIGURE 5. The two-dimensional spectrum of the image in figure 4 obtained digitally via Fourier transforms. Contours are at 2:1 intervals of spectral density. Note that apart from the low-wavenumber swell peak, contours are approximately symmetrical about the $k_x = 0$ axis, and that in the middle ranges of wave number the 3 dB azimuthal width is approximately constant. The omnidirectional background is due to speckle, which is not discussed in this paper.

wave spectrum is likely to be much broader in its range of directions at medium and higher wavenumbers than the image spectrum. Thus there appears to be a filter operating that removes the components with higher azimuthal wavenumbers, and this is due to the interaction of the sea surface motions with the aperture synthesis process. In the case under discussion the oceanographic situation is fairly simple and we can calculate the approximate characteristics of this filter (Tucker 1983), and its bandwidth agrees with that of the measured spectrum at medium wavenumbers (assuming that the azimuthal bandwidth of the real waves is much broader).

However, the way in which sea waves are imaged by a SAR is a complex matter, which is still far from being properly understood. Thus we are at present unable to extract waveheight information from the images to any useful precision, and are even unsure to what extent we can extract useful information on wavelengths and directions. This is an area of very active research at the present time.

Of course, the wave-measuring role should not be considered as the only justification for flying a SAR over the seas, as will be clear from Dr Raney's contribution to this symposium. There is also a fascinating report by Fu & Holt (1982) showing examples of all the identifiable oceanographic phenomena that were seen on Seasat SAR images.

## 4. Operational considerations

In this context there are two types of requirement for wave data: real-time data for use in the execution of marine operations, and statistical or 'climate' data for use in the design of structures and for the planning of marine operations. In fact, the first application really requires forecast data. A number of wave forecasting systems are in operation, including one run by the U.K. Meteorological Office using, as input, data from their meteorological forecasting model (Golding 1980). However, it is helpful to have real-time data to supplement and check the predictions of these models.

A single satellite gives rather sparse coverage in space and time, as will be seen from figure 6. Consecutive 'up' tracks on the Earth's surface are separated by approximately 2700 km in latitude at the Equator, and by 1350 km at 60° N. Twelve hours later the same area will be crossed by 'down' tracks. Approximately 24 h later the area will be crossed by tracks 900 km to the west at the equator, or 450 km to the west at 60° N.

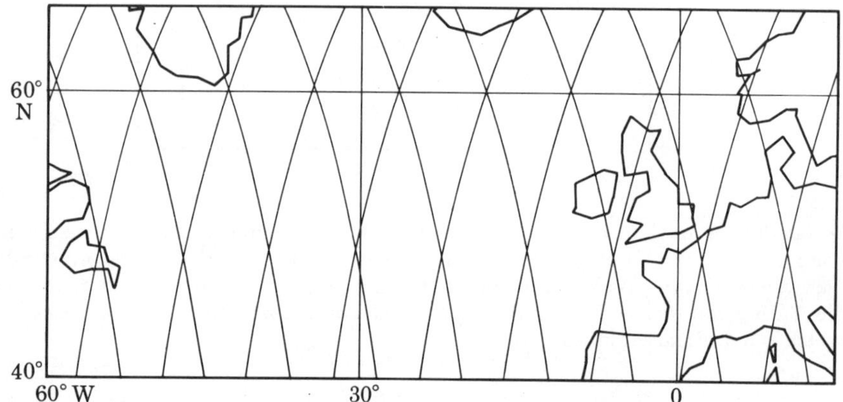


FIGURE 6. The tracks of a single satellite over the North Atlantic during a period of 3 days.
Height 668 km, orbit inclination 98°.

A system of three satellites arranged to cover the same tracks at daily intervals would give better coverage, but at U.K. latitudes one would still only get daily readings along the sides of a roughly diamond-shaped grid of about 500 km sides.

This raises the two related questions: 'How local are short-term wave conditions?' and 'How local is the long-term wave climate?' Near coasts, the conditions vary on a scale roughly equal to the distance from the nearest coast (this is a gross oversimplification but good enough as an order of magnitude for present purposes), so in this case the answer to both questions is 'very local'. However, the situation is not so clear over the open ocean far from land. Challenor (1983) has made a preliminary study of this problem by using eight successive passes of the Seasat altimeter over the same track over a period of 24 days (all that was available owing to the short life of Seasat). The track chosen was a 'down' track passing from approximately NE to SW and close to O.W.S. *Lima* (57° N, 20° W). Data on significant waveheight ($H_s$) from the altimeter were plotted and are shown in figure 7. Variation along the tracks (each approximately 2000 km long), is remarkably slow, even in the vicinity of storms, and the few bumps, which are probably due to fronts, are comparatively insignificant. The average of these eight tracks is shown in figure 8 and shows only a slow and modest drop towards the southern end of the track studied. Challenor compares these space scales with the time variation of waveheight at O.W.S. *Lima*, using the group velocity of the waves, and gets general agreement. Some idea of the sort of climatic information that could be available is given by Chelton *et al.* (1981), who have analysed all the data from Seasat.

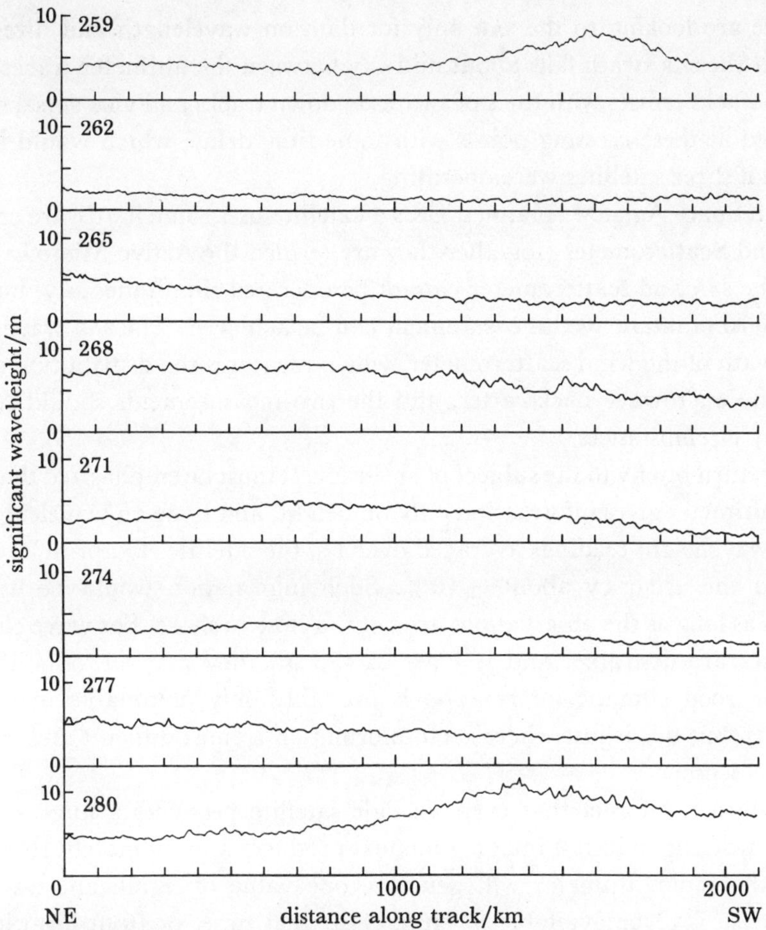

FIGURE 7. Significant waveheight from the Seasat altimeter during eight passes over the same track at 3-day intervals. The tracks passed over O.W.S. *Lima* (57° N, 20° W) and went approximately from NE to SW (from Challenor 1983).

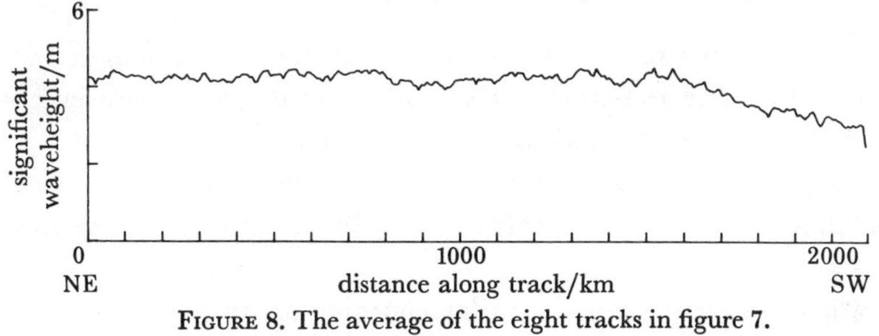

FIGURE 8. The average of the eight tracks in figure 7.

This evidence is, of course, very limited, but as far as it goes it agrees with what one might expect. Thus the coverage of the deep oceans away from the coast by the altimeter may well be adequate for the determination of waveheight climate (given a long enough duration of

[ 135 ]

observation), and with three satellites the frequency–spacing combination will give a reasonable, though not ideal, short-term coverage of the same areas for operational use. If detailed wind maps could be made available by a satellite scatterometer system, for example, then these would supplement the pattern by showing where there were high winds producing locally rough seas.

At present we are looking to the SAR only for data on wavelength and direction, and it is unfortunate that the SAR swath falls about midway between the altimeter tracks. However, the altimeter on up tracks crosses with the SAR swath on down tracks and vice versa, so that the data can be correlated at these crossing points with some time delay, which would in fact be comparatively short if three satellites were operating.

The European Space Agency's planned ERS-1 satellite uses some hardware common to both the SAR and Wind Scatterometer (together they are termed the Active Microwave Instrument, or AMI). Thus the SAR and scatterometer cannot be operated simultaneously, but a reasonably satisfactory pattern of interleaved measurement can be achieved. The SAR wave measurements will be in the swath of the wind scatterometer, which measures the distribution of surface wind by measuring the microwave backscatter, and the two measurements should supplement one another in many circumstances.

We can now return briefly to the subject of accuracy. It must be emphasized that as established at present, the altimeter gives information only on height, and none on wavelength or direction of travel. With waveheight readings averaged over 1 s, the satellite 'footprint' is approximately 10 km long and the accuracy about $\pm 10\%$. Such information would be useful for many operational uses as long as the observations are not too close to shore. For wave climate purposes higher accuracies are desirable, and it looks as though they may be obtainable by longer averaging. Thus good climatic information is probably only obtainable in the open ocean away from coasts, but this is just where such information is most difficult and expensive to get by using surface sensors.

Finally, one must not forget that even a single satellite produces a huge amount of data. When the SAR is working in its full imaging mode it produces approximately $10^8$ bits of data per second! Even the radar altimeter will generate one value of significant waveheight every second. If this data is to be available in either near real time, or from historical archives, a major system has to be set up to process it and bank it.

## 5. CONCLUSIONS

A precise radar altimeter on a satellite can measure waveheight to useful accuracy for operational purposes over oceans and to within perhaps 10 km of the coast. However, the time and spatial coverage is rather sparse for applications near the coast.

The altimeter can possibly give an accuracy adequate for wave climate work but with a poorer spatial resolution, which means that it can only be used much further from coasts than the 10 km quoted above.

The degree to which synthetic aperture radar can give useful information on wavelength and direction is not yet established. This poses a complicated problem requiring further research. The mechanisms of modulation are not yet well enough understood to extract waveheight information from it.

REFERENCES

Challenor, P. G. 1983 In *Seaset over Europe* (*Proceedings of a conference in London*, 14–16 April 1982) (ed. T. D. Allan). Chichester: Horwood. (In the press.)

Chelton, D. B., Hussey, K. J. & Parke, M. E. 1981 *Nature, Lond.* **294**, 529–532.

Fu, L. L. & Holt, B. 1982 *Seasat views oceans and sea ice with synthetic aperture radar.* N.A.S.A. Jet Propulsion Laboratory Publication no. 81-120.

Golding, B. W. 1980 In *Power from sea waves* (ed. B. Count), pp. 115–134. Academic Press.

Jordan, R. L. 1980 *IEEE Jl ocean. Engng* **OE-5**, 154–164.

Townsend, W. F. 1980 *IEEE Jl ocean. Engng* **OE-5**, 80–92.

Tomiyasu, K. 1978 *Proc. IEEE* **66**, 563–583.

Tucker, M. J. 1983 In *Seasat over Europe* (*Proceedings of a conference in London*, 14–16 April 1982) (ed. T. D. Allan). Chichester: Horwood. (In the press.)

Webb, D. J. 1981 *J. geophys. Res.* **86**, 6394–6398.

## Discussion

E. D. R. SHEARMAN (*Department of Electronic and Electrical Engineering, University of Birmingham, U.K.*). In discussing the observation of gravity waves by a SAR, Mr Tucker mentioned various mechanisms by which gravity waves modulate the capillary waves, which are the primary scatterers detected by a microwave radar. One mechanism, which he did not mention, is roughness modulation, the variation of the amplitude of the capillary waves from one part of a gravity wave to another, a phenomenon that one can observe visually on waves in a tank or at sea. Would Mr Tucker comment on the contribution that this mechanism makes to the overall process of gravity-wave detection by SAR?

As a comment, rather than a question, I should like to point out another technique for area surveillance of gravity waves, namely high-frequency (h.f.) radar, of which the ground-wave version should permit continuous observation of wave development over an area extending some 100–200 km out from the sea coast. In the 6–8 years between now and the launch of the next generation of radar satellites for ocean surveillance, there is time for a network of h.f. radars to be installed and to be generating invaluable sea-truth for the satellites. Of the two techniques, satellites give world coverage but on a time and space sampled scheme, whereas h.f. radar gives continuous coverage over a limited area, the two thus being truly complementary. (Sky-wave radar gives poorer quality data over a considerably greater area.)

I suggest that it behoves those planning the satellite experiments to ensure good coordination of the two types of measurements.

M. J. TUCKER. The mechanisms of modulation are discussed briefly in the published paper. Detailed analyses of the 'hydrodynamic modulation' effect referred to by Professor Shearman, and of the mechanisms of modulation by tilt and by the interaction of sea-surface velocities with the aperture synthesis process, are given in the following review paper: W. R. Alpers, D. B. Ross & C. L. Rufenach (*J. geophys. Res.* **86** (C7), 6481–6498 (1981)).

However, these three mechanisms do not adequately explain recent experimental results from radars mounted on towers in the sea, and it seems likely that one or more other mechanisms play a part. Table D1 is instructive: it shows the attenuation times and distances of water waves due to viscosity alone. There are additional mechanisms for dissipating energy from steep waves, so the decay times and distances quoted are maximum possible values. (Typical values of the relevant parameters are used in these computations.)

### Table D1

| wavelength/m | phase velocity | group velocity | radian frequency | amplitude decay to $1/e$ | |
|---|---|---|---|---|---|
| | | | | $\overline{s}$ | $\overline{m}$ |
| 1.0 | 1.249 | 0.6252 | 7.85 | 8332.0 | $1.07 \times 10^4$ |
| 0.25 | 0.626 | 0.3160 | 15.74 | 520.7 | $3.40 \times 10^3$ |
| 0.10 | 0.401 | 0.2119 | 25.19 | 83.3 | $3.64 \times 10^1$ |
| 0.04 | 0.272 | 0.1780 | 42.70 | 13.3 | 4.88 |
| 0.02 | 0.233 | 0.2146 | 73.07 | 3.33 | 1.48 |
| 0.01 | 0.248 | 0.3086 | 155.6 | 0.83 | 0.53 |

These figures show that for the longest wavelength radars (for example, L-band), the energy of the Bragg resonant ripples persists for many cycles of the predominant sea waves, and travels slowly compared with the phase velocity of these. Thus the compression of the sea surface in the crests of the long waves compresses the ripple energy there, and the reverse happens in the troughs. However, for the shortest wavelength radars (for example, Ku band) the energy of the Bragg resonant ripples decays quickly and an approximate local equilibrium is achieved between generation and decay as they ride over the long-wave profile. In this case it is plausible, for example, that variation in surface wind speed over the wave profile can modulate the backscatter.

It is perhaps worth commenting on the common use of the term 'capillary wave' in this context, because I believe it to be misleading. By using a typical value for the surface tension of clean water, the contribution of this to the restoring force equals that of gravity for waves with a wavelength $\lambda$ of approximately 1.7 cm. For longer waves, the effect of surface tension rapidly diminishes, and is 5 % for $\lambda \approx 5.2$ cm. For the Seasat SAR, the Bragg resonant wavelength was approximately 34 cm, and for the proposed SAR on ERS-1 it is approximately 7.2 cm. Thus these are both firmly in the gravity-wave part of the wave spectrum. Even for X-band, the Bragg resonant waves will usually be on the gravity-wave side of the dividing line, depending on the angle of incidence. The term 'ripple' would be better to distinguish these short waves from longer ones.

D. E. CARTWRIGHT (*I.O.S. Bidston, Birkenhead, U.K.*). I wonder if too many people are assuming that ERS-1 will necessarily have a 3 day repeat? I am not in touch with the latest decisions by E.S.A., but those concerned with using the altimeter data for currents and tides would certainly prefer an exact repeat period in the region of 6–10 days, to produce a grid of smaller mesh in a somewhat longer time. This would make negligible difference in the spacing between Earth-tracks beneath consecutive revolutions, which is most relevant to Mr Tucker's problem of defining the global wave field. For example, altering the 'Seasat' orbit from a repeat period of 3 days to one of 10 days would only entail the difference between $14\frac{1}{3}$ and $14\frac{3}{10}$ revolutions per day.

*Phil. Trans. R. Soc. Lond.* A **309**, 381–395 (1983)
*Printed in Great Britain*

381

# Observations of sea-surface temperature for climate research

By J. E. Harries, D. T. Llewellyn-Jones, P. J. Minnett,
R. W. Saunders† and A. M. Zavody
*Rutherford Appleton Laboratory, Chilton, Didcot, Oxfordshire OX11 0QX, U.K.*

The measurement of global sea-surface temperature (s.s.t.) from space, with high absolute accuracy, is one of the important requirements of the World Climate Research Programme (W.C.R.P.). This paper considers the definition of measurement aims based on considerations of specific types of scientific problem, and gives as examples discussion of two particular problems, first the possible influence of Pacific s.s.t. on the lower stratosphere, and second the role of s.s.t. in the cloud–climate feedback process. Following this, a brief review is presented on current status in satellite measurements of s.s.t. with both infrared and microwave techniques, and the paper concludes with a description of a future s.s.t.-measuring instrument, the Along-Track Scanning Radiometer (ATSR).

## 1. Introduction

The oceans are a major store of heat in the Earth's climatic system, though the details of how this heat energy is transported around the globe are only beginning to be understood (Hastenrath 1982). In their effect upon the atmosphere, and therefore on the climate as it is sensed by man, the oceans act as a vast thermal reservoir and so our present insubstantial understanding of what processes govern the interactions of the oceans and the atmosphere represents a major gap in our ability to understand, and eventually to predict, the climate.

The influence of the oceans on the atmosphere is determined by a number of parameters, but two in particular – direct radiation and release of latent heat through evaporation – are directly related to the temperature of the sea surface, the layer of ocean directly in contact with the air. Thus, while remote sensing of sea-surface temperature (s.s.t.) from space can give us information only from the skin depth, which may be only a few micrometres (infrared) or few millimetres (microwave) deep, and which may be at a rather different temperature from the bulk of the surface water (Paulson & Simpson 1981), this information is of great value to climate research.

Thus satellite remote sensing of s.s.t. has a major role to play in climate research. The accuracy of the determination of s.s.t. must, however, be very high indeed. A target of $\pm 0.2$ K absolute accuracy has been quoted by scientists involved in the planning of the World Climate Research Programme (World Climate Research Programme 1981). This measurement accuracy is needed on a global scale, and must be made in the presence of not only instrumental noise and systematic errors, but through the partially absorbing and emitting atmosphere which, even in the most transparent 'windows', can bias the observed temperature by between 1 and 20 K depending on atmospheric conditions. This is because the main absorber in the 10–12 μm atmospheric window, which is widely used for s.s.t. measurements, is water vapour, so that the atmospheric bias is particularly great in tropical regions (where most of the oceanic heat storage occurs), and is very variable around an orbit of the Earth. As a further indication of the importance of atmospheric effects, we should note in passing that the National Oceanic

† Present address: E.S.O.C./M.D.M.D., Robert-Bosch-Strasse 5, 6100 Darmstadt, F.R.G.

[ 139 ]

and Atmospheric Administration has ceased issuing s.s.t. data from spaceborne sensors in tropical regions owing to the influence of aerosols in the stratosphere after the El Chichon volcanic eruption in March 1982. Despite these very serious difficulties in the science of measurements, the potential rewards to geophysics from having accurate s.s.t. measurements globally are so great that major efforts to overcome all problems are justified.

The paper will begin with a short discussion of the accuracy requirements for the measurements, on the basis of the type of scientific problems to be addressed. This will be followed by a discussion of two particular scientific problems that demand high-accuracy measurements of s.s.t. for their solution: the first concerns a possible 'tele-connection' between the Pacific Ocean surface temperature and the humidity of the stratosphere; the second concerns the role of s.s.t. in determining the sign and magnitude of cloud–climate feedback. Subsequently, a brief review of current status in measurements of s.s.t. from space, by both infrared and microwave passive remote sensing, will be given. The paper will conclude with a description of a future instrument, the Along-Track Scanning Radiometer, which will introduce new techniques with the aim of improving accuracies of s.s.t. measurements to a level where they will be of direct value to climate research.

## 2. Definition of required accuracies

To establish the specification of future measurement systems, the W.C.R.P. (World Climate Research Programme 1981) has considered a variety of types of geophysical problem in which s.s.t. might be an important factor, defined in terms of the temporal and spatial scales of the processes involved, and have suggested what measurement accuracies are required in each case. These definitions are given below.

### (a) Large-scale processes

An example is inter-annual anomalies, scale 20° longitude by 10° latitude, lasting for several months. Such anomalies can probably cause changes in the atmospheric global circulation, or might be indicators of past atmospheric fluctuations. Typical peak amplitudes are *ca.* 2 K; an absolute accuracy of *ca.* 0.2 K is therefore required. S.s.ts can be averaged in time and space (e.g. for 30 days over 300 km), although smaller scales are important in terms of aliasing (e.g. eddies). Very important information about large-scale processes is obtained if measurement accuracy $\Delta T \leqslant 0.5$ K; much less information if $\Delta T \approx 1$ K; virtually no information if $\Delta T > 1$ K.

### (b) Mesoscale processes

Examples are meanders of boundary currents, or eddy-shedding processes. Such mesoscale processes may be early indices of climatic change. S.s.t. is required in terms of a time series of patterns, which may be related to underlying dynamics. Data are required on scale of several kilometres and several days.

### (c) Small-scale processes

Examples are oceanic and continental shelf fronts. Small-scale processes may be significant for climate modelling, and need to be parametrized. Small-scale processes are also important in the understanding of the sampling problem implicit in larger-scale measurement. A high spatial resolution (not more than 1 km) is required. S.s.t. data can be usefully combined with visible imagery radiation in such studies.

With these various points in mind it is possible to define measurement specifications for the three classes of problem, and these are summarized in table 1.

TABLE 1. ACCURACIES OF MEASUREMENTS OF SEA-SURFACE TEMPERATURE

*(a)  large-scale processes*

| | |
|---|---|
| absolute temperature accuracy | $\pm 0.2$ K |
| spatial averaging interval | 200–300 km |
| temporal averaging interval | 20–40 days |
| type of data product | isotherm contours in map coordinates |

*(b)  mesoscale processes*

| | |
|---|---|
| absolute temperature accuracy | $\pm 1.0$ K |
| spatial averaging interval | 5–10 km |
| temporal averaging interval | 3.5 days |
| type of data product | isotherm contours in map coordinates |

*(c)  small-scale processes*

| | |
|---|---|
| absolute temperature accuracy | $\pm 2.0$ K |
| spatial averaging interval | 1.0 km |
| temporal averaging interval | instantaneous |
| horizontal gradient accuracy | 0.5 K/1.0 km |
| type of data product | gridded images located to $\pm 10$ km |

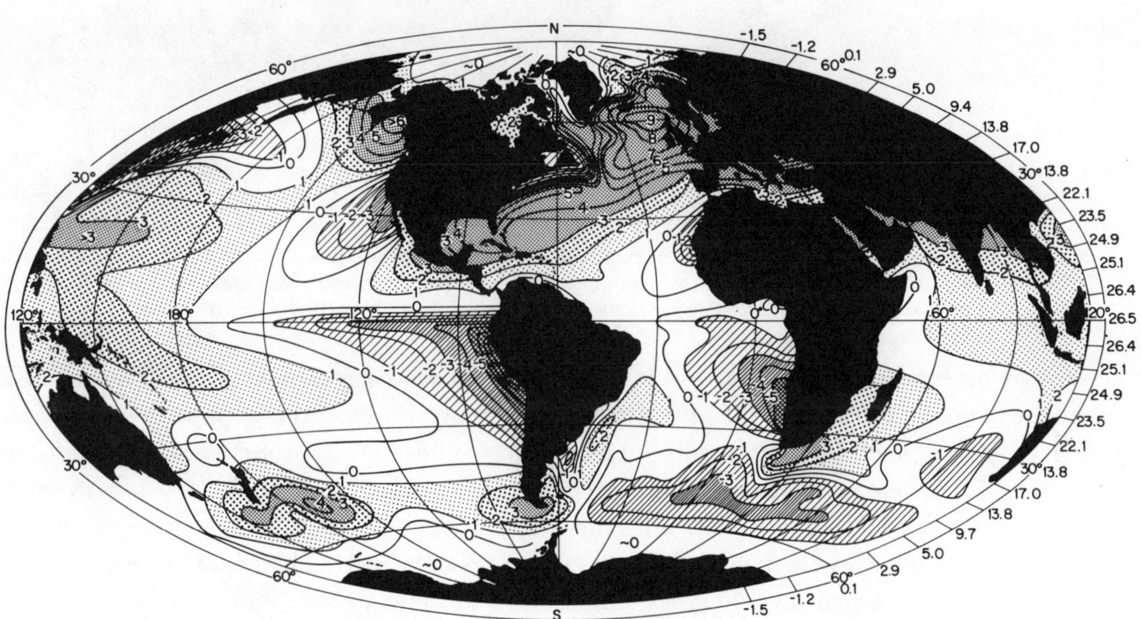

FIGURE 1. The deviations from zonal average sea-surface temperature (s.s.t.), adapted from Bjerknes (1969). Numbers at right are the zonal means, calculated excluding land surfaces. Contours are departures from zonal means, in kelvins; hatched areas are cold anomalies; stippled areas are hot anomalies.

## 3. EXAMPLES OF RESEARCH USE OF S.S.T. DATA

### *(a)  A possible tele-connection between the Pacific Ocean and the stratosphere*

Figure 1 (taken from Bjerknes (1969)) shows the deviation of s.s.t. from the long-term zonal average in a given latitude band (i.e. the zonal mean anomalies in s.s.t.) over ocean regions. This figure contains a number of details, but the point that we wish to stress here is that the Pacific Ocean exhibits a warm-to-cold gradient from west to east of about 7 K along the

Equator. This represents a vast thermal heat engine. As Bjerknes (1969) points out, this intense gradient of s.s.t. anomaly drives an equatorial surface easterly circulation. This circulation forms a closed cell – the Walker circulation – with a rising branch over the western Pacific and descending branch over the colder eastern waters.

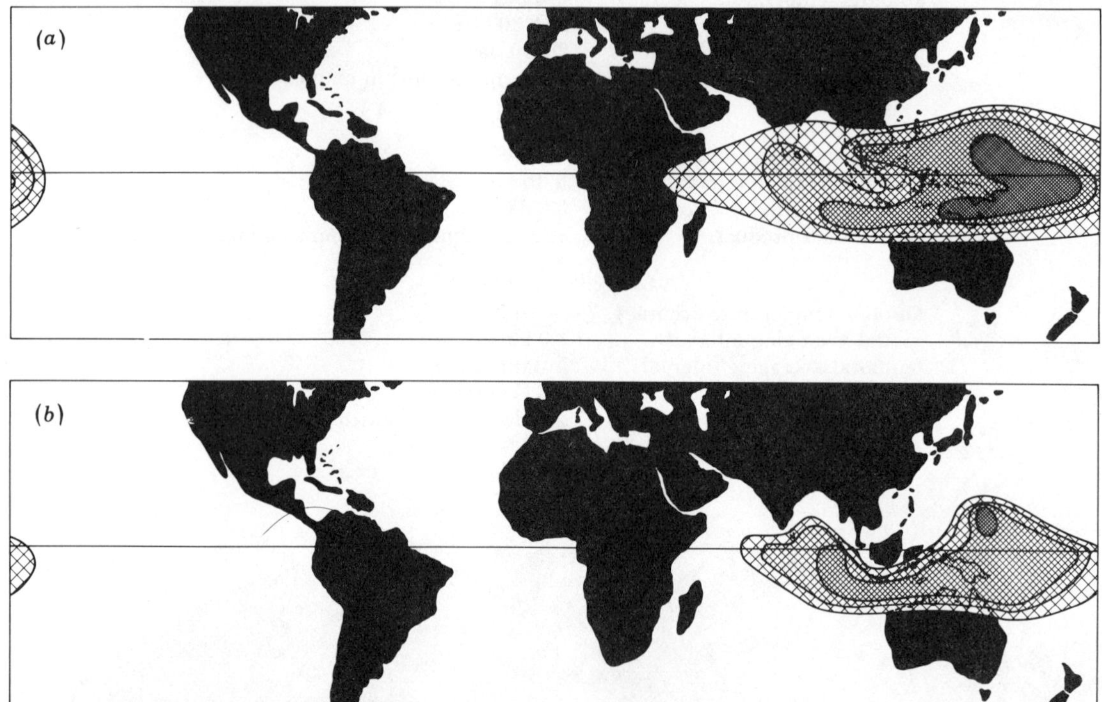

FIGURE 2. Contours of frequency of 100 mbar level temperature occurring below $-82.4\,^{\circ}\mathrm{C}$. Data are averages for 1970–80, and monthly means for (a) January and (b) March are shown. Contours are for 10%, 30%, 50% and 90%, and are darker for higher percentages. (Taken from Newell & Gould-Stewart (1981).)

It is to be expected that the rising branch, and indeed the return flow (westerly at high levels), will not be confined to the upper troposphere, but will affect the tropopause region and the lower stratosphere. A recent study by Newell & Gould-Stewart (1981) has shown that the local tropopause temperatures (or, more precisely, the temperatures at the 100 mbar† level were actually considered) are colder in the western Pacific – southeast Asia sector of the equatorial zone than anywhere else on Earth, at this pressure level. Their study considered the fraction of time for which the 100 mbar temperature was lower than $-82.4\,^{\circ}\mathrm{C}$ (which corresponds to the frost-point at 100 mbar for a volume mixing ratio of $3.5 \times 10^{-6}$, which is representative of mean observed water vapour mixing ratios in the lower stratosphere (see, for example, Harries 1976)). In other words, they studied those parts of the atmosphere that could physically be responsible for freeze-drying stratospheric air as suggested in the Brewer–Dobson theory (Brewer 1949). Newell & Gould-Stewart found that in data taken over the period 1970–80 there was a large area centred at about 160° W and on the Equator, where in January 90% of the 100 mbar temperatures were lower than $-82.4\,^{\circ}\mathrm{C}$. This area shrank seasonally to a minimum in about July when in a restricted area over the Indian subcontinent and southeast

† 1 mbar = 100 Pa.

Asia the frequency of temperatures this low was only 10 %, and zero everywhere else. Then the area and frequency increased again through the autumn, to a maximum in January. Figure 2 illustrates some of Newell & Gould-Stewart's results, for January and March.

In returning to the discussion of the Walker circulation it is therefore tempting to consider the physical connection that might exist between the Pacific Ocean 'heat engine' and the dryness of the lower stratosphere. The basic circulation system is shown in figure 3*a* (taken from Julian & Chervin (1978)), which depicts a schematic cross section across the equatorial Pacific with warm waters in the western ocean. The rising branch of the Walker circulation is particularly vigorous, partly because of the direct heating by the ocean but also because of the latent heat of evaporation due to the large amount of moisture picked up by the surface easterly part of the circulation. This powerful ascending motion leads to extreme cooling at tropopause levels, presumably once the air has been dried by precipitation and cloud formation during the ascent. Again, satellite cloud photographs quoted by Bjerknes (1969) seem to confirm this process.

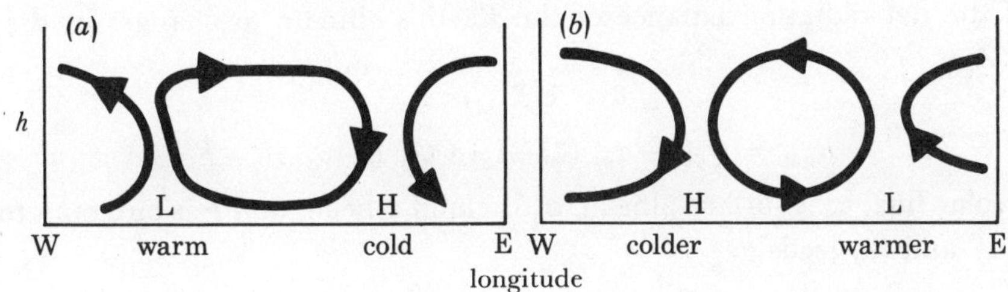


Figure 3. Schematic cross section, looking north, of the equatorial Pacific. Main features of atmospheric circulation are shown by arrows. (*a*) 'Normal' Walker circulation; (*b*) disturbed circulation during the 'El Niño' event.

Thus the suggestion of a 'tele-connection' between the Pacific and the stratosphere relies on the following series of qualitatively understood steps (a more quantitative assessment remains to be carried out):

(i)   warm W Pacific, cold E Pacific, along the Equator;

(ii)  direct Walker circulation (tropical easterlies at surface);

(iii) strong ascending branch in W Pacific (direct heating and moisture leading to wet adiabatic ascent);

(iv)  very low tropopause temperatures;

(v)   control of humidity of stratosphere by 'freeze-drying'.

If this conjecture is true, there are two very interesting corollaries: first, this region must be the entrance for most stratospheric air, because dryness seems to be a universal feature of the stratosphere, once well clear of local tropopause effects (Harries 1976); and second, we have seen a marked annual cycle in the 100 mbar tropopause temperatures as reported by Newell & Gould-Stewart, and it is of interest to further consider what connection this might have with the annual variability in stratospheric humidity observed in northern mid-latitudes by Mastenbrook (1980), Harries (1973) and others.

Before we leave this subject, it is interesting to consider the 'El Niño'. This is an oceanographic event that affects fishing off the western coast of South America, about which much has been written (for instance by Julian & Chervin (1978), Volkov (1981) and Weare (1982)). During El Niño the eastern tropical Pacific becomes much warmer than usual, and this disrupts

[ 143 ]

fish feeding cycles. The phenomenon is associated with a number of other processes, but it is difficult to unravel which are precursors to others because of the obvious feedback nature of atmosphere–ocean interactions. However, El Niño is clearly of some potential relevance to the mechanism proposed above, and so we shall consider it briefly here.

The phenomenon (Weare 1982) is associated with a warming (*ca.* 3 K) in the eastern Pacific and a slight cooling (*ca.* 0.5 K) in the west. The surface easterlies reverse (see figure 3*b*), the 'head' of water that had been built up by the previous prevailing easterlies is released, and an equatorially trapped wave propagates eastwards across the equatorial ocean, with some reflexions and poleward propagation occurring as this wave intersects with the South American continent. It is believed that these events trigger the Southern Oscillation. The main question in the present discussion is, if course, the potential effect of the complex of phenomena known as El Niño on the stratospheric water vapour cycle, since the Walker circulation injection process described above is disturbed.

### (b) Cloud–climate feedback

Schneider (1972) suggested the use of a parameter, $\delta$, which would be a measure of the sensitivity of the net radiation balance of the Earth's climate to changes in the fraction of cloud cover, $A_c$.

$$\delta = \partial R_{net}/\partial A_c, \tag{1}$$

where
$$R_{net} = Q_0(1-\alpha_s)(1-A_c) + Q_0(1-\alpha_c)A_c - F. \tag{2}$$

In (2), $Q_0$ is solar flux, $\alpha_s$ is surface albedo, $\alpha_c$ is cloud albedo, and $F$ is outgoing thermal flux. Combining (1) and (2) leads to

$$\delta = -Q_0(\alpha_c - \alpha_s) - \partial F/\partial A_c. \tag{3}$$

From a consideration of (1) and (2) we can see that if $\delta < 0$ the cloud albedo effect dominates, and if $\delta > 0$ the cloud greenhouse effect dominates.

We should also note that the outgoing flux, $F$, is a very strong function of, among other things, the s.s.t. Also, the cloud parameters $A_c$, $\alpha_c$ and $T_c$ (cloud-top temperature) are important.

Theoretical estimates of $\delta$ (Hartmann & Short 1980) range from $-35$ to $-100$ W m⁻². When we remember that typical net fluxes of heat absorbed by the oceans are of the order of 50 W m⁻², we can see that the theory is not in a position to establish a satisfactory understanding of the significance of cloud–climate feedback, and that an accurate empirical determination based on direct measurements is very important indeed. To obtain useful accuracy G. Molnar (personal communication 1981) has shown that errors in measurements of s.s.t. and other parameters need to be reduced to the following levels:

$$\left.\begin{array}{l} \Delta T_{s.s.t.} \approx 0.2 \text{ K}; \\ \Delta T_c \approx 0.2 \text{ K}; \\ \Delta A_c \approx 0.03. \end{array}\right\} \tag{4}$$

Such measurement accuracies are very demanding indeed: however, the importance of accurate s.s.t., and other, data in achieving an understanding of this fascinating problem is clear.

### 4. Recent satellite measurements of s.s.t.

Satellite measurements of s.s.t. are, of course, already available. For example, N.O.A.A. normally issues data derived from the Advanced Very-High-Resolution Radiometer (AVHRR) instrument on the Tiros/NOAA polar orbiting satellites (although this service has recently been interrupted in the zone 10°–40° N, by the effects of aerosols from the El Chichon volcanic eruption earlier in 1982). Also, the Jet Propulsion Laboratory (J.P.L.) has published a global map of s.s.t. for January 1979, based on data from the High-Resolution Infrared Sounder (HIRS/2) and Microwave Sounding Unit (MSU) instruments. However, the accuracies of such data do not yet meet the requirements laid out in § 2 above. In this section we consider current status in both infrared and microwave remote sensing of s.s.t.

### (a) Infrared results from the Advanced Very-High Resolution Radiometer Mk 2

At the R.A.L., workers have been comparing s.s.t. data from the AVHRR/2 instrument on the NOAA 7 satellite with in-situ s.s.t. data provided by a number of collaborating institutions (see appendix 1): detailed results may be found in Saunders *et al.* (1983), but a brief summary will be given here.

The s.s.t. ($T_s$) is a complicated nonlinear function of the measured radiances; however, it is usually sufficient to use a linear approximation of the form

$$\text{(day)} \quad T_s = C_0(\alpha) + C_1(\alpha)\,T_{11} + C_2(\alpha)\,T_{12}, \tag{5}$$

$$\text{(night)} \quad T_s = C_0'(\alpha) + C_1'(\alpha)\,T_{11} + C_2'(\alpha)\,T_{11} + C_3'(\alpha)\,T_{3.7}, \tag{6}$$

where $T_s$ is surface temperature, $\alpha$ is airmass and $T_{11}$, $T_{12}$ and $T_{3.7}$ are apparent temperature measured in the 11, 12 and 3.7 µm channels. It is sometimes necessary, e.g. in the tropics, to introduce higher-order terms.

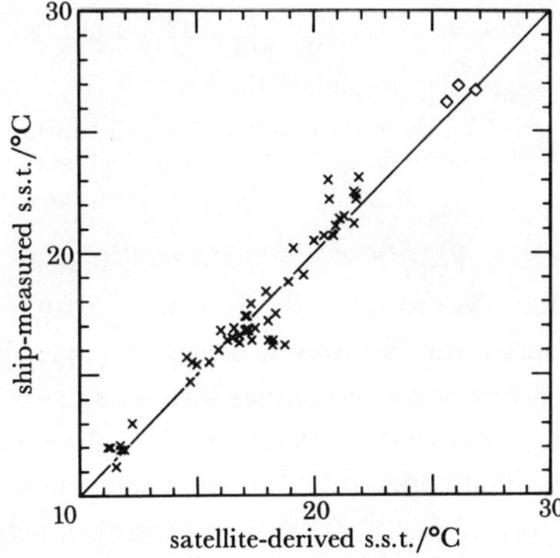


FIGURE 4. S.s.t. measurements from the NOAA 7 AVHRR/2 instrument, compared with ship and buoy data. Latitude range is 5° S to 65° N, and number of cases is 53. Standard deviation of the difference between AVHRR and ship is ± 0.88 K, with no significant systematic difference. ×, North Atlantic points; ◇, tropical points. (After Saunders *et al.* (1983).)

The coefficients $C_n$, $C_n'$ have not been determined from empirical models of atmospheric transmission as has been the normal practice by previous workers, but have been derived directly from spectroscopic parameters. In this analysis the effects of clouds in the field of view have been eliminated by using a statistical histogram technique described by Harris *et al.* (1981). The comparisons with surface data have been made in the region 37° to 65° N, and some additional data in tropical regions.

[ 145 ]

The results obtained for a sample of 53 cases, between approx. 10 and 27 °C, are

$$\Delta T_{\text{r.m.s.}} = \pm 0.88 \text{ K}$$

and

$$\Delta T_{\text{systematic}} = -0.09 \text{ K}.$$

The results are shown in figure 4.

The fact that there is no large systematic bias in these results suggests that the modelling procedures used are basically correct. The r.m.s. scatter in the results requires further investigation, as the sampling techniques used for the ground observations obviously contribute to this figure.

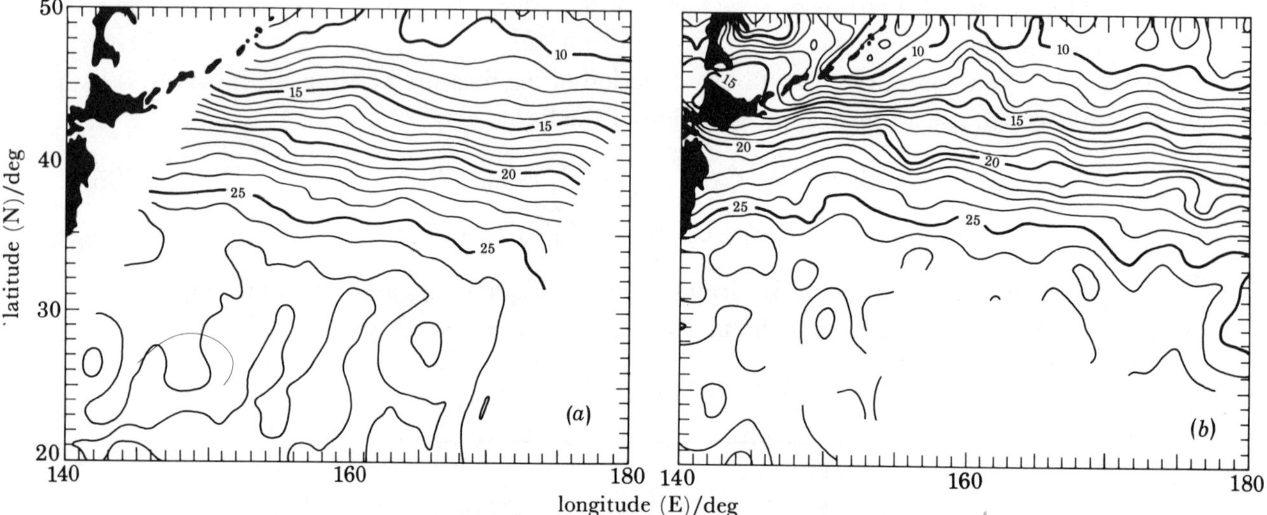


FIGURE 5. S.s.t. mapped data for period 7 July to 6 August 1978, for the northwestern Pacific area: (a) from the Seasat SMMR instrument, and (b) from ship data. Contours are in degrees Celsius. (From Bernstein (1982).)

### (b) Microwave SMMR results

The Scanning Multichannel Microwave Radiometer (SMMR) was a passive radiometer instrument carried on both Seasat and Nimbus 7. Bernstein (1982) has reported a study of s.s.t. based on the Seasat data. Figure 5 shows a comparison of s.s.t. maps generated (a) from SMMR and (b) from ship data for the northwestern Pacific region, showing the generally good correlation between the two maps. The r.m.s. difference between these mapped data is $\pm 0.75$ K, with a bias, or systematic error, of $+0.22$ K (SMMR warmer). These data apply to the period 7 July to 6 August 1978, and several thousand individual measurements are included in both cases. The structure of the detail in the two results is somewhat different, however, with the

TABLE 2. NUMBERS OF DATA POINTS $(N)$ AND THEIR STANDARD DEVIATIONS $(\sigma)$

ABOUT THE MAPPED SURFACES OF FIGURE 5

| dates | SMMR | | ship | |
|---|---|---|---|---|
| | $N$ | $\sigma/\text{K}$ | $N$ | $\sigma/\text{K}$ |
| 7–18 July 1978 | 1732 | 1.96 | 3573 | 2.39 |
| 19–27 July 1978 | 1032 | 1.05 | 2034 | 1.72 |
| 28 July to 6 Aug. 1978 | 1105 | 0.97 | 916 | 1.55 |
| overall, 7 July to 6 Aug. 1978 | 3869 | 1.51 | 6523 | 2.10 |

Seasat data rather more featureless than the ship data. A more quantitative assessment has also been attempted by Bernstein (1982) and is reproduced in table 2.

The reason suggested by Bernstein for the rather greater error in the 7–18 July period (table 1) is due to the fact that more measurements close to land occurred, which were contaminated by the rather intense sidelobes of the microwave device's field of view. Overall the ship data show more scatter, which is perhaps not surprising because many different ships contributed to the average. The final feature to note from table 2 is that the standard deviation of the individual data from the SMMR, about the smoothed mapped data from the SMMR (see figure 5a), is $\pm 1.51$ K, although substantial empirical corrections to the measured radiances based on post-launch validation experiments have been used.

### (c) Summary

Table 3 summarizes these two cases. Since previous analyses by other workers have shown similar results (see, for example, McClain 1981), we may take these new results as fair indications of current status.

TABLE 3. SUMMARY OF S.S.T. MEASUREMENTS

| parameter | infrared (Saunders *et al.* 1983) | microwave (Bernstein 1982) |
|---|---|---|
| $\Delta T_{\text{r.m.s.}}/\text{K}$ | $\pm 0.88$ | $\pm 1.51$ ($\pm 0.75$ mapped) |
| $\Delta T_{\text{systematic}}/\text{K}$ | $-0.09$ (AVHRR/2 colder) | $+0.22$ (SMMR warmer) |
| $\Delta x/\text{km}$ | $50 \times 50$ ($1 \times 1$ pixel) | $150 \times 150$ |
| $N$ | 53 | 3869 |
| range of $T/°C$ | 10–27 | 10–29 |

It should be added that microwave techniques have considerable potential owing to their capacity for penetrating clouds. However, the emission processes are more complex than in the infrared (depending critically upon sea-state for example) and much development is required in this area.

### 5. THE ALONG-TRACK SCANNING RADIOMETER

So far in this paper we have discussed some of the many exciting scientific questions in climate research that require accurate measurements of s.s.t. for their solution. Also, we have shown by the brief review in the previous section that present capabilities of measurement are not yet good enough to provide the accuracy of better than 0.5 K required to resolve many of these questions. In an attempt to improve this situation, a new instrument, the Along-Track Scanning Radiometer (ATSR), is being developed in the U.K. for flight on the first European Space Agency remote sensing satellite, ERS-1, in 1987.

The ATSR will for the first time enable us to view in quick succession the same patch of ocean surface at two different angles through the atmosphere with one instrument. This allows us to eliminate the variable part of the signal in the two measurements, to leave the constant term, which is of course the radiance corresponding to the s.s.t. In this way we avoid the need to rely entirely on multispectral corrections for the atmosphere, which is the technique adopted in present-day infrared sounders. This is particularly important in conditions of high atmospheric correction (e.g. in equatorial regions), and when layers of attenuating material that are optically 'grey' (i.e. spectrally featureless), such as thin cirrus, atmospheric aerosols and hazes, are in the field of view. (This would be of particular value when large amounts of volcanic aerosols are present.) The multispectral approach does, of course, have advantages, and because of this, ATSR will carry three co-registered channels at 3.9, 11 and 12 μm, each capable of two 'looks' at the same scene.

The instantaneous field of view is equivalent to a 1 km × 1 km pixel at the surface in the nadir direction, which is imaged onto the photoconductor detectors by an $f/2.3$ off-axis paraboloid mirror. The aperture is 10 cm, and the detectors are cooled to *ca.* 80 K by a closed-cycle Stirling refrigerator.

TABLE 4. PREDICTED ATSR PERFORMANCE (STANDARD DEVIATIONS IN KELVINS)

operational night and day mode (11 and 12 μm channels only)

| radiometer bias | visibility | 11 and 12 μm | 11 μm | 12 μm |
| --- | --- | --- | --- | --- |
| none | clear | 0.21 | 0.41 | 0.65 |
|  | mixed | 0.36 | 0.51 | 0.78 |
| 1 % change in emissivity of calibration black body between 11 and 12 μm | mixed | −2.2 | 0.7 | — |

clear-air night-only mode (3.7 μm channel available)

| radiometer bias | visibility | 3.7 μm | 3.7 and 11 μm |
| --- | --- | --- | --- |
| none | clear | 0.15 | 0.10 |
|  | mixed | 1.8 | 0.75 |

The ATSR is being built by a consortium consisting of scientists from the Rutherford Appleton Laboratory, Mullard Space Science Laboratory of University College London, Oxford University, and the Meteorological Office. In addition, a small microwave radiometer is being added, to increase the accuracy of the atmospheric water vapour determination needed for the highest accuracy in s.s.t. measurements, by the French Centre de Recherches en Physique de l'Environment Terrestre et Planetaire. A number of other European laboratories are contributing to the ATSR programme in areas such as surface measurements and data investigations.

It is important to point out that the idea of using a two-look technique for measurements from space is not entirely novel. For example, Fleming (1980) discusses using a tomographic technique for increasing the vertical spatial resolution of atmospheric temperature soundings; Chedin *et al.* (1982) discuss the use of polar orbiter (Tiros) and geostationary satellite (Meteosat) data to determine s.s.t. to an accuracy of *ca.* 1 K; and Zandlo *et al.* (1982) report a similar study with the use of Tiros and GOES measurements.

The predicted performance of the ATSR has received considerable attention (Zavody 1981). Calculations have been carried out by using a sophisticated line-by-line computer model of the atmospheric spectrum, to simulate ATSR results. These calculations (see also Minnett *et al.* 1982) have considered a wide range of model atmospheres (polar to tropical), molecular absorption and re-emission, and aerosol scattering. A noise equivalent temperature of 0.5 K was assumed for the radiometer. A summary of the findings is given in table 4: the numbers quoted are standard deviations in kelvins, and the number of cases used (different atmospheres) was 59. The results demonstrate a number of points.

First, if the radiometer has zero bias (systematic error), standard deviations of $\pm 0.21$ K in a clear atmosphere and $\pm 0.36$ K in a mixed or partly cloudy (up to 90 % covered) scene are possible. If, however, a 1 % difference in emissivity of the calibration targets exists at 11 and 12 µm, table 4 shows that a multichannel approach leads to large errors (2.2 K) so that a conventional radiometer (e.g. AVHRR) would be useless, whereas once such an error is detected

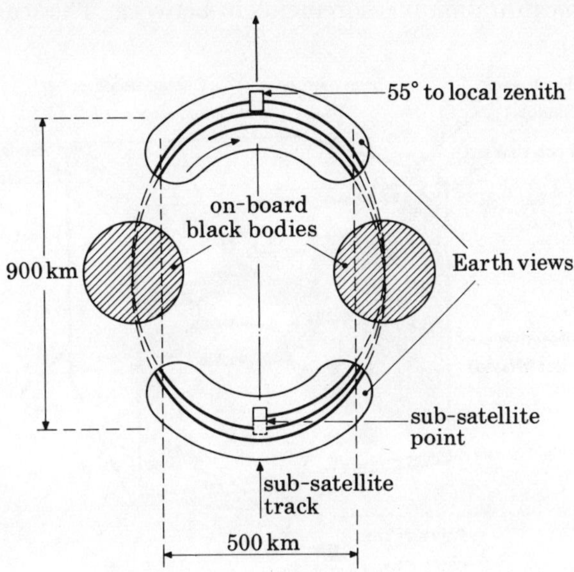


FIGURE 6. The field-of-view scan pattern for the ATSR. A conical scan directs the view in turn to the sub-satellite direction, calibration target 1, a forward direction (about 55° to local zenith) and calibration target 2.

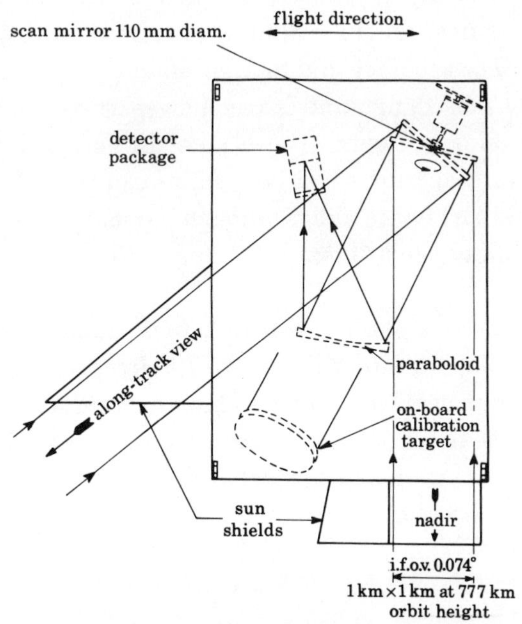


FIGURE 7. The internal optical scheme of the ATSR. Radiation is received by an asymmetrically mounted plane mirror, which performs a conical scan at *ca.* 6 Hz. An $f/2.3$ off-axis paraboloid collects radiation from the scanning mirror and directs it onto the cooled detector assembly. The positions of calibration targets and sunshields are indicated. I.f.o.v., instantaneous field of view.

(e.g. by ground truth comparisons) an ATSR type of device can constrain such errors to *ca.* $\pm 0.7$ K by use of only one channel and two looks. At night the absence of scattered solar radiation means that the 3.7 µm channel can be added, to yield clear-air s.s.t. measurement accuracies as low as $\pm 0.10$ K; however, as is shown by the mixed case, scattering by haze is dominant at these shorter wavelengths. Figure 6 illustrates the scan pattern of the ATSR, which illustrates how a conical scan pattern yields data at the sub-satellite point, at a forward angle of about 55°, and two calibration measurements in between. The scan pattern is synchronized

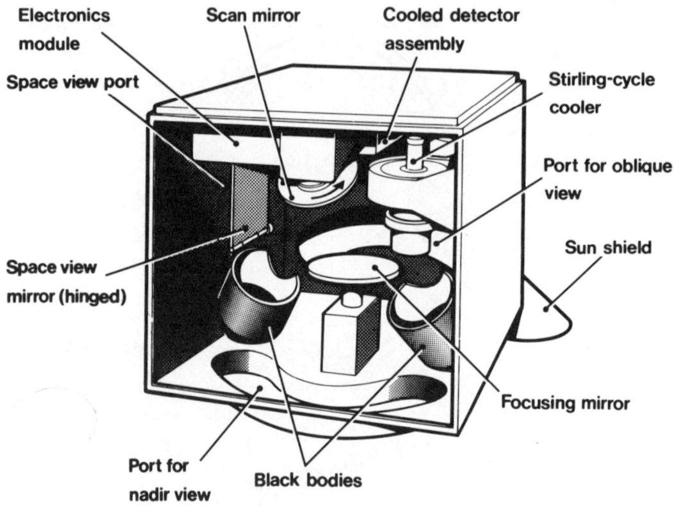


FIGURE 8. A schematic cut-away diagram of the ATSR. The curved apertures for the nadir and forward views are shown, as well as the scanning and the off-axis paraboloid mirrors. The layout of other units, such as the electronics, the calibration black bodies and the Stirling-cycle cooler are shown.

with orbit height and velocity to ensure contiguity between scans. Figure 7 illustrates schematically the internal optical system. Scanning is carried out by an asymmetrically mounted rotating plane mirror. The accuracy and thermal stability of the black-body calibration targets is a crucial part of the ATSR design and is receiving particular attention at M.S.S.L. and the Meteorological Office. Finally, figure 8 shows an artist's cut-away view of the instrument. The curved nadir and forward apertures can be seen, as can the calibration targets. The Stirling-cycle coolers are a vital part of the design, though we shall not discuss them further here. The complete instrument measures 570 mm × 380 mm × 540 mm, weighs 45 kg and consumes a mean power of 50 W.

Much more detail than it is possible to give here is available from the ATSR project team at R.A.L. However, we should add that the ATSR experiment promises to provide measurements of s.s.t. and of cloud parameters $(A_c, T_c)$ with absolute accuracies that perhaps for the first time will be consistent with the needs of climate researchers as quoted in § 2 of this paper. It is therefore a very exciting prospect, though somewhat daunting in view of the very high accuracies and levels of instrumental performance that are being sought.

## 6. CONCLUSION

Measurements of sea-surface temperature at high absolute accuracy are of great importance in climate research. Two examples of scientific problems requiring such data have been given in this paper, though a number of others could have been discussed. The first of these examples referred to the thermal driving of the tropical atmosphere by the Pacific Ocean, with possible implications for the humidity of the stratosphere: if the model proposed is verified, we might at last be approaching a satisfactory understanding of a long-standing problem, the water budget of the stratosphere. Our second example stressed the importance of accurate measurements of s.s.t. in empirically determining the magnitude (and sign) of cloud–radiation feedback processes.

These examples will, it is hoped, have served to stress the importance of accurate s.s.t. data. The remainder of the paper has considered the present and future status of global measurements of s.s.t. from satellites. In this context a number of further points need to be made. First, it is clear that while present instruments perform well, improvement of the order of threefold or fivefold in absolute accuracy are still required. Secondly, the problems in climate research for which we require measurements of s.s.t. need such data on a truly global basis, without gaps, and without problems of temporal or spatial aliasing: present satellite systems do not provide this comprehensive coverage. Thirdly, we must recognize that no one technique has all the answers, and without doubt we shall require – for s.s.t. measurements and probably all others too – a mix of techniques. Particularly, for s.s.t. measurements, scientists should strive to achieve a combination of improved infrared and microwave measuring systems in spacecraft. Finally, as an extension of the last point, we should not forget that surface measurements will always be an important component of any global measurement system, whether for calibration or to study detailed processes inaccessible to satellite remote-sensing systems: the accuracy and reliability of the surface measurements, too, require improvement, because these are often the standards against which we compare our more global data sets.

We gratefully acknowledge the valuable contributions and helpful discussions provided by a number of colleagues in a number of the areas discussed in this paper. In particular, mention should be made of E. J. Williamson, C. G. Rapley, I. Barton and G. Molnar.

APPENDIX 1. COLLABORATING GROUPS SUPPLYING SHIP S.S.T. DATA

Nederlands Centrum voor Oceanografische Gegevens at K.N.M.I., de Bilt, the Netherlands;
   *Tydeman Cumulus* and *Noord Hinder*
Marine Information and Advisory Service, at I.O.S., Wormley, U.K.; *Discovery*
R. D. Pingree, Marine Biological Association, Plymouth, U.K.; *Frederick Russell*
J. Meincke, University of Kiel, F.R.G.; *Meteor*
U. Herrmannsen, University of Kiel, F.R.G.; *Poseidon*
H. M. van Aken, University of Utrecht, the Netherlands; *Tyro*

REFERENCES

Bernstein, R. L. 1982 *J. geophys. Res.* **87**, 7865–7872.
Bjerknes, J. 1969 *Mon. Weath. Rev.* **97**, 163–172.
Brewer, A. 1949 *Q. Jl R. met. Soc.* **75**, 351–363.
Chedin, A., Scott, N. A. & Berroir, A. 1982 *J. appl. Met.* **21**, 613–618.
Fleming, H. 1980 In *Proc. International Radiation Symposium, Fort Collins, Colorado, August 1980*, pp. 87–89.
Gill, A. E. 1980 *Q. Jl R. met. Soc.* **106**, 447–462.
Harries, J. E. 1976 *Rev. Geophys. Space Phys.* **14**, 565–575.
Harries, J. E. 1973 *J. atmos. Sci.* **30**, 1691–1698.

Harris, A. W., Llewellyn-Jones, D. T., Harries, J. E. & Williamson, E. J. 1981 In *Proc. IAMAP 3rd Scientific Assembly, Aug. 1981*, pp. 199–201.
Hartmann, D. L. & Short, D. A. 1980 *J. atmos. Sci.* **29**, 1413–1422.
Hastenrath, S. 1982 *J. phys. Oceanogr.* **12**, 922–927.
Julian, P. R. & Chervin, R. M. 1978 *Mon. Weath. Rev.* **106**, 1433–1457.
Llewellyn-Jones, D. T., Williamson, E. J., Rapley, C. G. *et al.* 1981 The Along-Track Scanning Radiometer: a proposal for an additional package to be flown on ERS-1, to the European Space Agency.
Mastenbrook, H. J. & Daniels, R. E. 1980 In *Atmospheric water vapour* (ed. A. Deepak, T. D. Wilkerson & L. H. Ruhnke), pp. 329–342. Academic Press.
McClain, P. 1981 In *Oceanography from space* (ed. J. F. R. Gower), pp. 83–85. New York: Plenum.
Minnett, P. J., Saunders, R. W., Zavody, A. M. & Llewellyn-Jones, D. T. 1982 In *Proc. Int. Symp. on Problems of Radiative Transfer and Satellite Measurements in Meteorology and Oceanography, Cologne, March 1982*, pp. 58–60.
Newell, R. E. & Gould-Stewart, S. 1981 *J. atmos. Sci.* **38**, 2789–2796.
Paulson, C. A. & Simpson, J. J. 1981 *J. geophys. Res.* **86**, 11044–11054.
Saunders, R. W., Minnett, P. J., Zavody, A. M. & Llewellyn-Jones, D. T. 1983 (In preparation.)
Schneider, S. 1972 *J. atmos. Sci.* **29**, 1413–1422.
Volkov, Y. N. 1981 *Izv. atmos. Ocean. Phys.* **16**, 871–877.
Weare, B. 1982 *J. phys. Oceanogr.* **12**, 17–27.
World Climate Research Programme 1981 *Report of meeting on co-ordination of plans for future satellite observing systems and ocean experiments to be organised within the WCRP, Jan. 1981.* (Available from W.M.O. secretariat, Geneva.)
Zandlo, J. A., Smith, W. L., Menzel, W. P. & Hayden, C. M. 1982 *J. appl. Met.* **21**, 44–50.
Zavody, A. M. 1981 Annex in Llewellyn-Jones *et al.* (1981).

## *Discussion*

P. Wadhams (*Scott Polar Research Institute, University of Cambridge, U.K.*). The atsr discussed by Dr Harries will have a footprint of 1 km² on the ocean surface. When an instrument of this kind passes over sea ice – as it will during the high polar orbit of ERS-1 – it sees a composite surface temperature. A polar icefield is composed mainly (95 % or more) of thick sea ice with a surface temperature close to that of the air, which may be − 30 °C in winter. A small percentage of the surface, however, consists of open or recently refrozen leads with a surface temperature of − 1.8 °C or somewhat less. The massive contrast between these two types of surface temperature allows the composite temperature sensed by the instrument to be a measure of the fraction of the icefield occupied by leads.

J. E. Harries. This is certainly true, though the interpretation of structure within the footprint, or instantaneous field of view, of a remote sensor has to be carried out with great care, taking into account such aspects as the field of view response function and the modulation transfer function of the optical and electronic system: for instance, the response to this structure will depend on its apparent frequency, measured at the instrument output.

P. K. Taylor (*Institute of Oceanographic Sciences, Wormley, U.K.*). The atsr views a given area of ocean through two different atmospheric columns. Is it necessary to assume that each column has similar atmospheric structure and is cloud-free, and to what extent does the oblique viewing angle restrict the ability to detect sea-surface temperature in regions containing clouds?

J. E. Harries. It is necessary to assume that each column has similar atmospheric structure, although on the scales involved atmospheric inhomogeneities are likely to be quite small and their effects on the signals received at the satellite have been shown, in simulations, to be small. Concerning clouds, it is necessary only that the cloud cover within a 50 km × 50 km array of pixels is less than about 90 %; in such cases statistical methods can be used to derive clear-column surface temperature.

The obliqueness of the viewing angle does, for example, change the shadowing or masking effects of clouds, though geometrical corrections can, of course, be made. The most severe problem in this case arises with multiple-layer clouds. While our simulations so far indicate that these problems will not undermine the achievable accuracies with ATSR, they (the problems) are nevertheless important enough to warrant further study.

J. T. Houghton, F.R.S. (*Rutherford Appleton Laboratory, U.K.*). I should like to point out a potentially very important application of the accurate measurement of sea-surface temperature. The questions are often asked: How can climate change be measured? Are measurements available to show how climate changed over the last 100 years and can we observe changes in the future? Regarding temperature change, our current assessments are based on measurements at a few sites where careful records have been kept over a long period. Such sites are limited to a few countries at mid-latitudes in the Northern Hemisphere and therefore represent a very inadequate sample so far as the whole globe is concerned. Consistent measurements of the temperature of the sea surface from satellites would overcome the sampling problem very well and would be very suitable for looking at climatic trends provided that they could be made with sufficient accuracy. An accuracy of say better than 0.2 K would be required. The ATSR, if carefully built and calibrated, should approach this sort of accuracy.

*Phil. Trans. R. Soc. Lond.* A **309**, 397 (1983)
*Printed in Great Britain*

397

# General discussion

E. D. R. SHEARMAN (*Department of Electronic and Electrical Engineering, University of Birmingham, U.K.*). We have had comments from a number of speakers on three planned remote-sensing satellites designed for ocean studies, the European ERS-1, the Canadian Radarsat and the Japanese satellite ERTS-1, all scheduled for 1988–90. If the orbits were coordinated, one suggested revisit interval, namely 3 days, for high-resolution observations could be reduced to a 1-day revisit interval. Could anyone tell us whether an attempt is being made internationally to agree on a single revisit cycle and to coordinate the launches so that the maximum benefit is obtained from the overall effort?

J. T. HOUGHTON, F.R.S. The various space agencies involved are discussing questions of co-ordination. Further, the Joint Scientific Committee of the World Climate Research Programme is organizing meetings to try to ensure the best possible scientific return from the various ocean observation satellites that will be flying at the time mentioned.

*Phil. Trans. R. Soc. Lond.* A **309**, 399–414 (1983)
*Printed in Great Britain*



# A review of Seasat scatterometer data

By T. H. Guymer

*Institute of Oceanographic Sciences, Brook Road, Wormley, Godalming, Surrey GU8 5UB, U.K.*

A microwave scatterometer operating at 14.6 GHz was one of several sensors on Seasat that provided data at the sea surface. The technique by which estimates of the near-surface wind velocity can be obtained is described and studies to assess the accuracy of such data are reviewed. With few exceptions wind speed can be retrieved to $\pm 1.6$ m s$^{-1}$ and direction to $\pm 17°$ for winds between 3 and 16 m s$^{-1}$, and the technique is useful up to at least 25 m s$^{-1}$. The ability to measure wind on a near-global basis with a spatial resolution of 50 km is of great importance for meteorology and oceanography. Some examples of applications of scatterometer data are given, including the horizontal variation of the wind stress curl, the interpretation of synthetic aperture radar imagery and the inference of surface pressure fields. Comments on future satellite scatterometer systems are also made.

## 1. Introduction

One of the most important requirements for a better understanding of ocean circulation and improved weather forecasts is that of an adequate specification of the surface fluxes (Businger & Charnock 1983). In particular, improved knowledge of the surface wind field, which is the principal forcing function for the ocean, is essential to gaining an understanding of the variability of many of the world's surface current systems (O'Brien *et al.* 1982; Robinson 1963). Routine observations over the sea, obtained by conventional methods, lack the necessary spatial resolution, are concentrated in major shipping lanes and can have large errors due to poor calibration and exposure. Microwave remote sensing has developed over the past 10 years to the point where useful information about the wind near the sea surface can be obtained from a spacecraft at several hundred kilometres altitude. A major advance occurred when the first satellite dedicated to oceanography, Seasat-A, was launched by N.A.S.A. in 1978 (Born *et al.* 1981). It returned 100 days of data before a power failure occurred on 10 October, terminating the mission. Such was the volume of data obtained and its relevance to future microwave satellite sensors that extensive analysis still continues and is likely to do so for several years. Several scientific journals have produced issues dedicated to Seasat: *Science, Washington* (29 June 1979), *IEEE Journal of Oceanic Engineering* (April 1980), *Journal of Astronautical Sciences* (October–December 1980) and *Journal of Geophysical Research* (April 1982).

This paper assesses the capabilities of the Seasat-A Satellite Scatterometer (sass) for measuring surface winds over the sea, in relation to its design specifications. Comparisons with surface measurements and with those from other Seasat sensors are used. Finally, examples of the applications of sass data are described.

## 2. The Seasat scatterometer

### (a) Background

Arising from radar research connected with World War II requirements, evidence emerged that backscatter from the ocean showed a dependence on surface winds (Cowan 1946; Kerr 1951; Grant & Yaplee 1957). Airborne measurements by the Naval Research Laboratory using a four-frequency radar (Daley 1973) and by N.A.S.A., operating a 13.3 GHz (Ku-band)

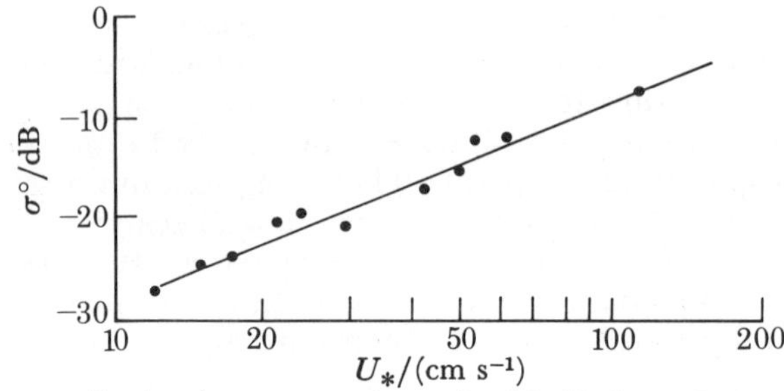

FIGURE 1. Variation of the normalized radar cross section, $\sigma^\circ$, with friction velocity, $U_*$. Upwind, horizontally polarized measurements at an incidence angle of 40° from an airborne scatterometer have been used. (From Jones & Schroeder (1978).)

fan-beam scatterometer (Bradley 1971; Claasen *et al.* 1972) and a 'pencil-beam' 13.9 GHz combined microwave radiometer and scatterometer known as RADSCAT (Jones *et al.* 1977) enabled the dependence of backscatter on radar parameters and near-surface conditions to be quantitatively investigated. This led to the development of the first satellite-borne scatterometer (a mechanically scanned version of RADSCAT) launched on Skylab in 1973. The results all confirmed that the normalized radar cross section, $\sigma^\circ$, depended on the wind speed; Guignard *et al.* (1971) suggested a simple power-law relation. Jones & Schroeder (1978) reanalysed the airborne RADSCAT data in terms of $U_*$, using Cardone's (1970) relation to derive the friction velocity, $U_*$, from wind and stability measurements ($U_*$ is defined by $\rho U_*^2 = |\tau_0|$, where $\rho$ is air density and $\tau_0$ the surface shearing stress). An example of $\sigma^\circ$ as a function of $U_*$ for particular values of incidence and azimuth angle, and horizontal polarization, is shown in figure 1.

The studies showed that higher radar frequencies (*ca.* 2 cm wavelength) were more sensitive to wind speed and that at these frequencies there were significant differences according to polarization. The variation with radar azimuth angle (off-nadir) was such that maxima occur in the upwind and downwind directions with minima close to crosswind, and this allows the possibility of recovering some directional information on the wind. At vertical incidence $\sigma^\circ$ was relatively insensitive to wind speed and the dependence was the reverse of that observed at 30–50°. A good review of these experimental results and of corresponding theoretical developments is contained in Moore & Fung (1979).

The data suggest that, for incidence angles greater than 25°, Bragg or resonant scattering is the dominant mechanism. At Ku-band frequencies the Bragg condition is satisfied for ocean waves of wavelength *ca.* 3 cm, i.e. in the short-gravity wave part of the spectrum. These ripples, which appear in light winds, grow in amplitude with increasing wind speed. However, the response of these short waves as wind speed continues to increase has been a source of some debate. Phillips (1977) has argued that saturation should occur at quite low speeds, the actual value depending on how clean the surface is. In contrast Pierson & Stacy (1973) proposed a spectral form, based on data from a number of sources, which implied a strong wind-speed dependence of the capillary-gravity wave portion of the spectrum and this has been given some credence by Mitsuyasu (1977), who found a clear wind-speed response in measurements from a tower at wind speeds of 8 m s$^{-1}$. Both the aircraft backscatter and ocean wave data suggest that there is less sensitivity to wind for wavelengths greater than 5 cm. An additional complication is the effect of lower frequency waves on the short-gravity waves. Generally, it appears that a two-scale scattering model, in which it is assumed that Bragg scattering from small-scale roughness elements associated with local winds is modified by tilting due to the longer wave, is appropriate. A large degree of empiricism remains, though, in the specification of the relation between $\sigma^\circ$ and wind.

### (b) Wind-velocity determination

The SASS (Grantham et al. 1977) was designed to view the ocean at two azimuths by using four dual-polarized fan-beam antennae aligned so that they pointed $\pm 45^\circ$ and $\pm 135^\circ$ in azimuth relative to the track beneath the satellite. Twelve Doppler filters were used to subdivide the antenna footprint electronically into resolution cells ca. 15 km (across beam) × 70 km (along beam). The incidence angles of the cells ranged from $25^\circ$ to $55^\circ$ for the regions illuminated by both forward and aft beams. Three additional measurements from incidence angles near nadir covered the subtrack, although no directional information can be obtained from them. Various operating modes of the SASS were used, allowing differing combinations of transmitted and received beam polarizations and swath coverage. Engineering aspects of SASS are discussed in detail by Johnson et al. (1980).

Two types of algorithm are needed before wind-velocity estimates are obtained. The first, the sensor algorithm (Bracalente et al. 1980), converts the basic engineering units to the normalized radar cross section, $\sigma^\circ$. Corrections are made for the attenuation of the signal due to atmospheric water along the path by using the brightness temperature measured by the Seasat Scanning Multichannel Microwave Radiometer (SMMR). This can only be achieved for the swath to the right of the satellite track because of the look angle of the SMMR; no correction was made on the left-hand side. Moore et al. (1982) have attempted an evaluation of the attenuation algorithm and conclude that significant improvements in the wind determinations can be made in high precipitation rates.

From the $\sigma^\circ$ data the geophysical algorithm computes the 19.5 m neutral stability wind speed and direction. This is assumed to be the wind speed that would result from a given surface stress in near-neutral stratification and with a dry-adiabatic lapse rate from the surface to 19.5 m and avoids difficulties associated with the effect of air–sea temperature difference on the vertical variation of wind speed. Before the launch of Seasat three algorithms had been proposed (Pierson et al. 1974; Dome et al. 1977; Jones et al. 1978). As a result of comparisons with independent surface data these formulations were subsequently modified. The SASS model function $F(\theta_i, \phi, \epsilon, W)$ is empirical, relating the decadic logarithm of $\sigma^\circ$ to radar parameters and the surface wind vector, and is expressed by

$$F(\theta_i, \phi, \epsilon, W) = \lg \sigma^\circ = G(\theta_i, \phi, \epsilon) + H(\theta_i, \phi, \epsilon) \lg W, \tag{1}$$

where $W$ is the wind speed, $\theta_i$ is the incidence angle, $\phi$ is the azimuth angle between the wind direction and the radar look angle, and $\epsilon$ is the polarization of the beam. The different model

[ 159 ]

functions are characterized by the values of $G$ and $H$ that they use for the various combinations of $\theta_i$, $\phi$ and $\epsilon$. A weighted least-squares method is used to find values for $W$ and $\phi$ that, for a given model function, will produce the best fit to the observations of $\sigma^\circ$. The model function has an approximately $\cos 2\phi$ dependence (Jones *et al.* 1982) and as a result the sum of the

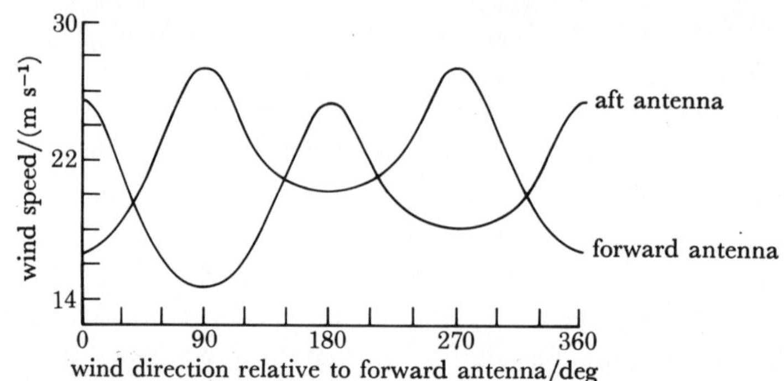



FIGURE 2. The locus of wind-vector solutions for two orthogonal noise-free $\sigma^\circ$ measurements. Directions given by the geophysical algorithm correspond to the intersections of the two curves, which in this case produces four possible directions, only one of which is correct. (From Jones *et al.* (1982).)

squares has several local minima. For a single noise-free $\sigma^\circ$ measurement the locus of solutions is a nearly sinusoidal curve in wind-vector space. For two orthogonal measurements (figure 2) the solutions correspond to the intersection of two curves and vary in number from two to four. These are the so-called directional aliases. They differ widely in direction but are approximately equal in magnitude.

Several methods have been proposed to remove the aliases from Seasat data. Wurtele *et al.* (1982) employ pattern recognition techniques to locate synoptic meteorological features by using SASS data alone and then select a preferred direction on the basis of streamline analysis. For 10 passes giving a total of 3476 comparisons with in-situ data, the correct direction was chosen on 70% of the occasions. A less subjective approach was adopted by Hoffman (1982), in an analysis of a single pass intersecting an intense mid-latitude depression, by direct minimization of an objective function that incorporated SASS and ship reports plus wind fields generated by a forecast model. Additional constraints that could be used are low-level winds derived from satellite cloud tracking (used by Endlich *et al.* 1982), correlation between low-level clouds and convergence in the inferred wind field, and surface pressure data. However, it is not clear that mesoscale features would be properly resolved. It has been suggested (O'Brien *et al.* 1982) that the problem might be alleviated in future scatterometers by the addition of extra antennae as currently planned for the E.S.A. satellite E.R.S.-1, which would reduce the number of aliases. Aliases with only two possible solutions are acceptable because they differ by nearly 180° in direction and so can be readily distinguished.

## 3. ASSESSMENT OF ACCURACY

### (a) *In-situ data*

After the premature end of the Seasat mission, several workshops were held at which winds calculated from the different geophysical algorithms were compared with various surface data sets. The latter included the Gulf of Alaska Seasat experiment (GOASEX) (J.P.L. 1979*a*, *b*), high wind occasions (J.P.L. 1979*c*) and the Joint Air–Sea Interaction Experiment (JASIN) (J.P.L. 1980). Comparisons with GOASEX data revealed biases between antennae indicative of calibration problems and also inadequacies in the model functions, which were subsequently modified to minimize the discrepancies.

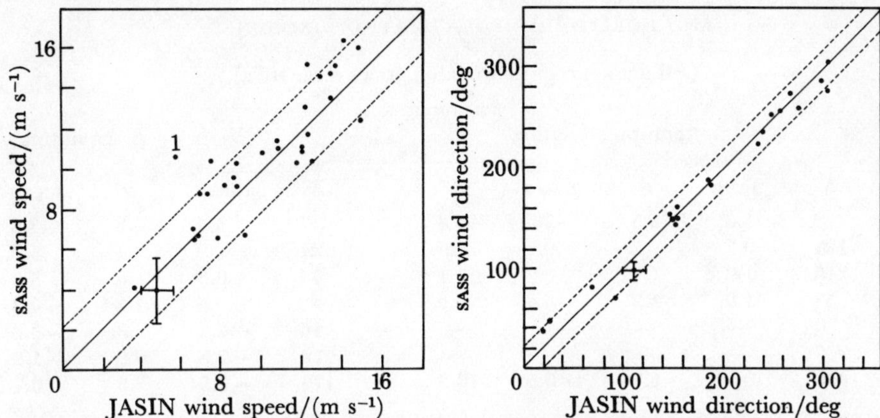

FIGURE 3. Comparison of scatterometer winds (w-7 algorithm) with JASIN winds (60 min means from ships and buoys) for 23 passes. Each point represents the mean of all comparisons for a particular satellite pass, typical standard deviations being indicated (point labelled 1: rev. 557, discussed in the text). Broken lines indicate the design specifications. Directions are with respect to true north. (From Allan & Guymer (1980).)

It was therefore necessary to assess the accuracy of these tuned model functions (now reduced to two, known as CWK and W-7) by using an independent high-quality data set. Fortunately, JASIN, a major international experiment, took place within Seasat's operational lifetime (Pollard *et al.* 1983). A number of ships and buoys sampled surface meteorological variables in a 200 km × 200 km area 300 km off NW Scotland at intervals from 1 min to 6 h. Careful comparisons were made between the sensors (Guymer 1983*a*; Brown *et al.* 1982) to produce an internally consistent data set in which interplatform differences in winds were less than 0.7 m s⁻¹ (0.5 m s⁻¹ for buoys) and less than 5°. The absolute accuracy of the wind speeds is difficult to determine. For convenience one buoy sensor was adopted as a standard to which all other data were corrected. Weller *et al.* (1983) have suggested that this may overestimate speeds by 10%, but such a view is not supported by comparison with aircraft-measured winds suitably corrected for height.

For the comparison two sets of SASS winds were generated for 23 passes by using the CWK and w-7 model functions. The passes were selected so as to give good spatial coverage of the JASIN area over as wide a range of wind speed as possible. Winds varied from 0 to 16 m s⁻¹ with 83% distributed between 6 and 14 m s⁻¹. All measurements were corrected to give the 19.5 m neutral stability wind.

A detailed discussion of the results is given in J.P.L. (1980) and summaries appear in Jones *et al.* (1981*a*, 1982) and Schroeder *et al.* (1982*a*). An example of the comparisons of SASS winds (w-7 algorithm) with 60 min means of autologged ship and buoy data is given in figure 3. Most of the individual comparisons on which the figure is based were within the design specifications and the overall statistics showed that SASS winds were accurate to better than ± 1.7 m s⁻¹ (w-7) and ± 1.5 m s⁻¹ (CWK) in speed and ± 17° in direction for both algorithms.

However, some passes compared less favourably. The most striking was rev. 557 where wind

404        T. H. GUYMER

speeds were up to 15 m s$^{-1}$ higher than measurements *in situ* for a small region near the southern corner of the JASIN triangle. Using JASIN surface and upper-air data and infrared satellite imagery, Guymer *et al.* (1981) were able to infer that a mid-level thunderstorm had affected the region at the time of the anomaly. It was concluded that the sass winds were high because

TABLE 1. Sass minus in-situ winds for storms data set, sass-1 model function, vertical and horizontal polarizations combined

(All data except nadir (after Jones *et al.* 1982).)

*hurricanes*

| in-situ wind speed/ (m s$^{-1}$) | | non-precipitating | | | | | precipitating | | | |
|---|---|---|---|---|---|---|---|---|---|---|
| | | speed | | direction | | | speed | | direction | |
| | $N$ | mean | $\sigma$ | mean | $\sigma$ | $N$ | mean | $\sigma$ | mean | $\sigma$ |
| 0–5 | 8 | 1.6 | 2.5 | −2.2 | 15.3 | 7 | 5.8 | 0.8 | 3.8 | 19.2 |
| 5–10 | 128 | 0.7 | 1.7 | 0.7 | 19.6 | 24 | 1.9 | 2.1 | 5.2 | 28.2 |
| 10–15 | 215 | 1.0 | 1.7 | 0.7 | 14.4 | 45 | 0.5 | 1.5 | 3.9 | 18.3 |
| 15–20 | 53 | 1.0 | 1.2 | −8.4 | 9.0 | 42 | 0.5 | 1.3 | −0.2 | 19.3 |
| 20–25 | 0 | — | — | — | — | 46 | −2.2 | 2.3 | −3.2 | 1.94 |
| >25 | 0 | — | — | — | — | 19 | −7.8 | 6.0 | −1.0 | 24.5 |
| > 5 | 396 | 0.9 | 1.7 | −0.5 | 16.0 | 176 | −0.9 | 3.8 | 0.7 | 21.1 |

*extratropical cyclone*

| | | non-precipitating | | | | | precipitating | | | |
|---|---|---|---|---|---|---|---|---|---|---|
| | | speed | | direction | | | speed | | direction | |
| | $N$ | mean | $\sigma$ | mean | $\sigma$ | $N$ | mean | $\sigma$ | mean | $\sigma$ |
| 0–5 | 0 | — | — | — | — | 0 | — | — | — | — |
| 5–10 | 172 | −0.3 | 2.6 | −10.0 | 18.0 | 28 | −0.1 | 3.0 | −39.0 | 18.0 |
| 10–15 | 102 | −3.2 | 1.7 | −8.0 | 18.0 | 31 | −2.7 | 2.2 | −18.0 | 24.1 |
| 15–20 | 156 | −1.7 | 2.4 | −2.1 | 13.0 | 56 | −2.1 | 1.3 | −5.0 | 13.8 |
| 20–25 | 440 | −1.7 | 2.4 | −2.4 | 17.0 | 285 | 0.6 | 3.7 | −7.2 | 21.0 |
| >25 | 74 | −3.4 | 2.5 | −5.0 | 17.0 | 0 | — | — | — | — |
| > 5 | 944 | −1.7 | 2.6 | −4.5 | 17.3 | 400 | −0.2 | 3.4 | −9.0 | 24.0 |

of the effect of raindrops on the sea surface or because of enhanced backscatter from the rain itself. This pass was excluded from the comparison statistics. On two other passes high standard deviations of the difference between the satellite and JASIN winds were found: both were frontal situations in which real horizontal wind-gradients contributed to the scatter.

Results with both algorithms indicated that agreement with *in situ* data was worst at 20–25° and 50–60° incidence angles, as also reported by Halberstam (1981). After the comparisons with the withheld JASIN data were completed it was decided to produce a new model function, sass-1 (Boggs 1982*a*), tuned so as to minimize the differences. For most wind-speed–incidence angle categories the r.m.s. differences were significantly improved to better than ± 1.3 m s$^{-1}$ and ± 17°. Offiler (1981) extended these comparisons of the sass-1 data set to all passes over JASIN and found similar results.

It should be noted that these figures include uncertainties in surface data that will arise not just from unremoved instrumental errors but also from unrepresentative sampling on the scale for which sass winds are appropriate.

Neither GOASEX nor JASIN provided a validation of the upper wind-speed limit of the sass and, owing to the short lifetime of Seasat, other planned calibration experiments were not held. Fortunately, coverage was obtained over three storms, an intense North Atlantic depression known as the *QEII* storm (see §4*a*) and two Pacific hurricanes, Fico and Ella, which have extended the comparisons. Results are summarized in table 1. A clear tendency for sass to underestimate winds over 20 m s$^{-1}$ is evident in those regions of hurricanes with a high probability of precipitation. In part this is due to an inadequate attenuation correction in the intense rain bands because of the relatively poor SMMR resolution, and nonlinear brightness effects. It is also difficult to obtain meaningful comparisons in regions of large horizontal wind

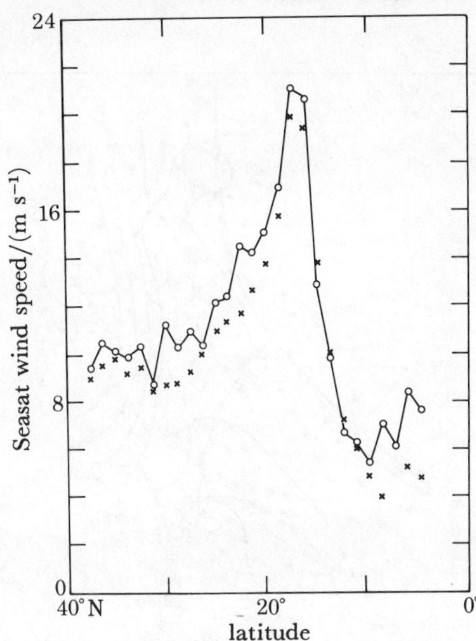



FIGURE 4. A comparison of SASS and SMMR wind speeds for Hurricane Fico, rev. 331. (From Wentz *et al.* (1982), with permission of the American Geographical Union.)

gradients near the hurricane centre. For the *QEII* storm the high wind-speed comparisons were in better agreement even when precipitation occurred. Rainfall in this extra-tropical feature was probably much less intense than in the hurricanes. Interestingly, there are no examples of anomalously high winds as in the JASIN thunderstorm, suggesting that backscatter is dominated by the wind-roughened surface or that the attenuation effects of rain outweigh any increased backscatter from the raindrops.

### (b) Comparisons with other Seasat sensors

Wind speeds can also be obtained from the altimeter (Fedor & Brown 1982) and SMMR (Lipes 1982). Wentz *et al.* (1982) have compared SASS values with both instruments at nadir, and with the SMMR alone for the 100 km off-nadir where the instrument swaths overlap. From an analysis of four passes over the Gulf of Alaska Wentz *et al.* concluded that all three sensors agreed to within 2–3 m s$^{-1}$ for winds between 2 and 10 m s$^{-1}$. However, at higher speeds the altimeter values are much lower. This is due to differences in the model functions rather than in the measured backscatter. I have found similar behaviour in a comparison of altimeter winds with surface data from JASIN and O.W.S. *Lima*.

In the off-nadir comparison Wentz *et al.* found that, in the mean, SMMR winds were 0.03 m s$^{-1}$ greater than those from SASS, with a standard deviation of the difference about the mean of 1.42 m s$^{-1}$. The correlation between the two estimates was 0.95 for speeds between 2 and 20 m s$^{-1}$. Figure 4 shows SASS and SMMR winds from rev. 331, which passed through Hurricane Fico:

[ 163 ]

both sensors show a very similar along-track variation with a marked gradient of wind speed to the south of its centre. For the 27 cells considered the SMMR–SASS difference was $1.1 \pm 1.1$ m s$^{-1}$. It should be noted that this pass was early in Seasat's mission. Allan & Guymer (1982) have suggested that a change in the SMMR wind speed bias occurred on about rev. 900, which accounts for differences between JASIN and GOASEX SMMR comparisons. Nevertheless, these results confirm that SASS winds are accurate to within 2 m s$^{-1}$.

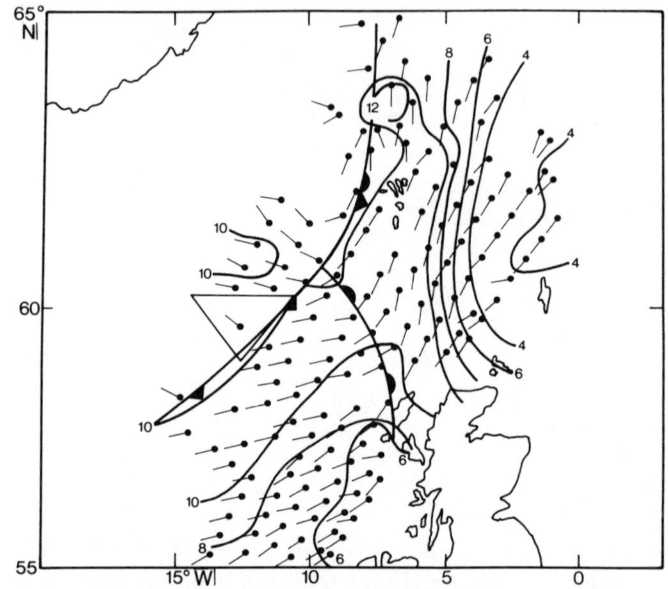



FIGURE 5. Scatterometer wind field for rev. 929, 23h11 G.M.T., 30 August 1978. Speed has been contoured in metres per second. (From Guymer *et al.* (1983).)

## 4. APPLICATIONS OF SEASAT SCATTEROMETER DATA

### (a) *Synoptic mapping of surface wind velocity*

Two consecutive SASS passes (929 and 930) during the JASIN experiment gave coverage of a well defined occluding frontal system on 30/31 August 1978. Guymer *et al.* (1983) analysed the wind field on the first pass after removing the directional ambiguities in a manner similar to that of Wurtele *et al.* (1982) but also allowing routine pressure analyses every 6 h and JASIN surface winds to influence the choice of the preferred direction. The resulting wind field is shown in figure 5 with about 20% of the deduced wind vectors shown for clarity. A southerly flow ahead of the system strengthens near the warm front and veers to west-southwesterly in the warm sector. Across the cold front there is a marked change in direction of the flow. The position of the surface front on this occasion is quite clear even before the aliases are removed, being delineated by a change from the two-vector solution to the four-vector type. SASS data often allow the position of surface discontinuities to be determined more clearly than using conventional reports and infrared satellite imagery. This example illustrates the way in which the scatterometer enhances the spatial sampling of ocean winds. Further use of the data from this and the subsequent pass is made in §4*b*.

A more spectacular example of improved mapping of surface wind occurred on 9/10 September 1978, when coverage was obtained of an intense extra-tropical depression that underwent rapid deepening to reach a minimum pressure of 960 mbar with 30 m s$^{-1}$ winds leading to observed wave heights of 12 m (Guyakum 1980). This storm resulted in the loss of the trawler *Captain Cosmo* on 9 September and caused damage and injuries on the liner *Queen Elizabeth II (QEII)*. Routine analysis seriously underestimated the intensity of the system, over-estimating central pressures by about 20 mbar (though this involved neglecting one ship report which, in fact, turned out to be reliable) and underestimating surface winds by a factor of 2. A reanalysis of the occasion by Cane & Cardone (1981), who included sass data, produced more consistent wind speeds and also placed the depression centre more accurately. Estoque & Fernandez-Partagas (1981), who analysed winds over the Indian Ocean, also found significant improvements in their analysis of the implied streamlines when the more uniform coverage due to sass data was available.

### (b) *Ekman pumping*

An important quantity when considering the response of the ocean to wind forcing or the effect of the atmospheric boundary layer on larger scales is the vertical motion arising from frictional convergence, $W_f$ (often referred to as Ekman pumping), which is related to the wind stress curl by

$$W_f = (1/\rho f)\ \mathrm{curl}_z\ \tau, \tag{2}$$

where

$$\mathrm{curl}_z\ \tau = \frac{\partial \tau_{x0}}{\partial x} - \frac{\partial \tau_{x0}}{\partial y},$$

$$|\tau| = \rho_{\mathrm{air}}\ U_*^2,$$

$\rho$ is the density of air or water as appropriate, $f$ is the Coriolis parameter, $\tau = (\tau_{x0}, \tau_{y0})$ is the surface stress, which is related to the friction velocity, $U_*$, as shown. $U_*$ values are included in the sass data records, calculated from a constant flux layer model (Kondo 1975). (There is some hope (Liu & Large 1981) that it may be possible to relate the backscatter directly to the stress but insufficient high-quality flux data exist to produce a satisfactory algorithm.) The level to which the value of $W_f$ refers is the boundary between the friction layer and the free atmosphere or ocean. Because $W_f$ can be more easily compared with other quantities in the atmosphere (e.g. divergence measurements, cloud and weather types), the examples presented are for the air rather than the mixed layer of the ocean. Ekman pumping in the ocean is smaller by the ratio of the densities (*ca.* 1250).

Guymer *et al.* (1983) mapped the Ekman pumping deduced from sass data in the vicinity of an occluding frontal system in the NE Atlantic. Figure 6 shows their results superimposed on a NOAA-5 infrared image obtained a few hours earlier. Maximum subsidence of 0.8 cm s$^{-1}$ was found some 200–300 km ahead of the occlusion point and to the west of a ridge axis. Upward motion occurs along and just ahead of the surface front, especially near the occlusion point. Much of the warm-sector air is weakly descending in a region of relatively broken cloud as seen on the NOAA-5 infrared image. In general the distribution is consistent with the synoptic situation and the cloud cover. Taylor *et al.* (1982) investigated the horizontal variation of integrated water vapour, liquid water and rain rate derived from smmr measurements on the same pass. Taken together, the results imply that warm, moist air ahead of the cold front is being lifted to produce rain just ahead of the occlusion point. The region with rain rates in excess of 0.4 mm h$^{-1}$ is marked on figure 6. Guymer *et al.* (1983) found similar structure on the subsequent pass. Thus Ekman pumping appears to be related to larger-scale features of the free

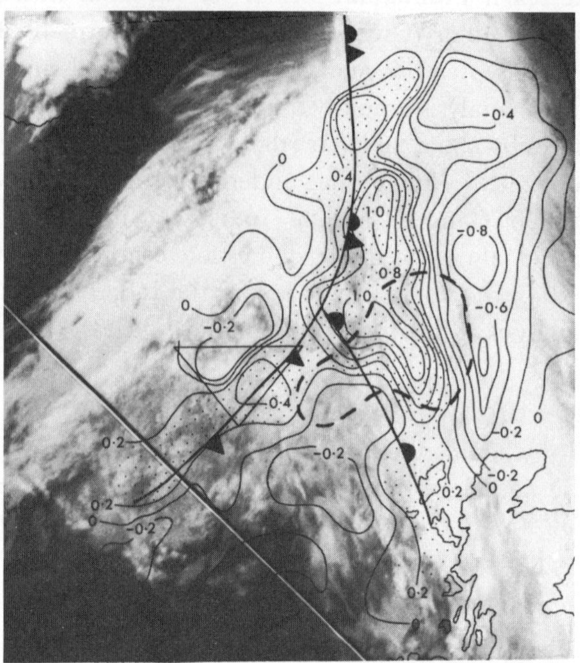

Figure 6. The distribution of vertical velocity (centimetres per second) at the top of the atmospheric friction layer calculated from the curl of the sass wind stress, rev. 929, 23h11 g.m.t., 30 August 1978 (after Guymer *et al.* 1983). Data have been superimposed on a NOAA-5 infrared image (courtesy of University of Dundee) obtained at 18h35 g.m.t.

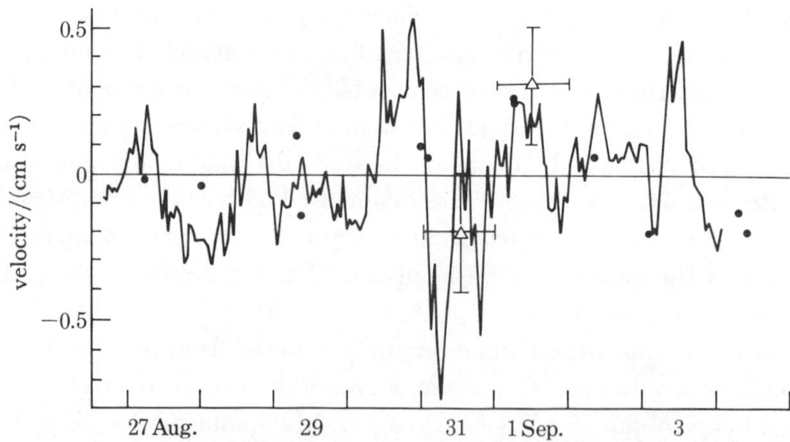

Figure 7. Estimates of the frictionally induced vertical velocity during JASIN Phase 2: solid line, from stresses derived from hourly ship winds; full circles, scatterometer-derived values. For comparisons the vertical motion resulting from the divergence method applied to radiosonde profiles is shown (triangles). (From Guymer & Taylor (1982).)

atmosphere. On another occasion (rev. 1006), however, Guymer (1983 *b*) found little association between the distribution of $W_f$ and the synoptic-scale features.

Several more passes during Phase 2 of JASIN were analysed in a similar way (Guymer & Taylor 1982), and mean values for the JASIN triangle have been plotted as a time series in figure 7. For comparison, hourly values have also been calculated by applying (2) to ship measurements of the surface wind at the corners of the JASIN triangle, after removal of interplatform biases based on careful comparison of results. The two sets of values agree within the uncertainty to be expected from errors in the wind measurements; they are also comparable with vertical velocities obtained from the divergence method applied to JASIN radiosonde winds at the triangle corners. This implies that most of the divergence in the atmospheric boundary layer was frictionally induced.

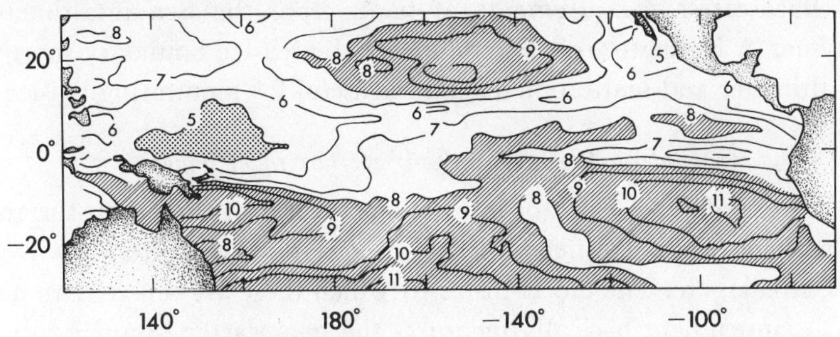


FIGURE 8. Mean wind-speed distribution (metres per second) in tropical Pacific derived from Seasat scatterometer measurements. (From Chelton & O'Brien (1982).)

Vertical motion at the base of the oceanic mixed layer has been inferred for the same period from heat budget considerations (T. Joyce, personal communication) and is an order of magnitude larger than can be accounted for by Ekman pumping. However, these values apply to a limited region near the southern corner of the triangle, where oceanic fronts (Pollard 1982) may have influenced the vertical velocities.

### (c) Time-averaged wind distribution

The 3-month mean wind speed in the tropical Pacific derived by Chelton & O'Brien (1982) from sass data is shown in figure 8. Off-nadir data from the right-hand side of the spacecraft have used so that an atmospheric attenuation correction could be applied, and no attempt was made to retrieve the correct direction. Highest winds were located at about 20° N and 15° S, corresponding to the Trades; those in the southern (winter) hemisphere were stronger and more extensive in the zonal direction. The inter-tropical convergence zone is the light wind region just to the north of the equator and is also marked by a maximum in the vertically integrated water-vapour content (Chelton *et al.* 1981). Low-level convergence of moisture evaporated into the Trades at the Equator is an important mechanism in tropical disturbances and in the general circulation of the atmosphere. Scatterometer wind stresses, when combined with geostrophic currents from a satellite altimeter (Stewart 1982), will enable sea-surface currents to be monitored. Changes in the wind pattern also have important consequences for the variability of tropical ocean currents, e.g. the El Niño phenomenon and the Somali Current, and there is some evidence that there is a relation between current changes in the different tropical ocean basins and the atmosphere. Future scatterometer data have been identified as an important component of a proposed programme (Tropical Ocean and Global Atmosphere (C.C.C.O. 1982)) to study such features.

Chelton & O'Brien (1982) noted a discrepancy between sass and altimeter winds (latter biased low) for this 3 month period and hypothesized that it was due to effects of high atmospheric water content and sea surface temperature compared with their values at the (mid-latitude) calibration sites. It seems to us more likely, in view of §3*b*, that the difference is due to errors in the altimeter wind algorithm above $10 \text{ m s}^{-1}$ combined with sampling errors resulting from its narrow swath.

An example of the wind speeds obtained for a large portion of the world ocean in a 2 day period has been produced by Schroeder *et al.* (1982*b*). In addition to the features described above, several mid-latitude storms can be identified. High wind speeds are also indicated south of 60° S but these are not real, being caused by high backscatter from sea ice, as also found by Boggs (1982*b*). Backscatter measurements at nadir from the Seasat altimeter have been analysed by Guymer & Kennett (1983) to delineate the sea-ice boundary more accurately. A combination of altimeter and scatterometer data is useful for monitoring sea-ice extent.

### (*d*) *Interpretation of synthetic aperture radar imagery*

SAR imagery has revealed a wealth of interesting features at the ocean surface (Vesecky & Stewart 1982) that have been identified with surface gravity waves, internal waves, natural and man-made slicks, streaks, etc. The mechanisms by which these are imaged are not fully understood but since the instrument basically measures the backscatter from short-wavelength (*ca.* 30 cm) surface roughness elements, which are then modulated in some way by larger scales, wind speed must be important. The overlap of sass and sar swaths allows an investigation of the image characteristics in relation to the wind field on scales of 50 km and greater.

Beal (1981) verified from aircraft and sass winds that large dark regions on sar imagery of Chesapeake Bay, corresponding to low backscatter, were associated with light winds. Allan & Guymer (1981, 1983) analysed sass wind-fields for a number of sar images in the NE Atlantic and concluded that the transition between zero and measurable backscatter occurred at $3 \text{ m s}^{-1}$ and was commonly less than 100 m wide. At winds just above the threshold, horizontal variations in image intensity appeared to be more pronounced than in strong winds. Surface gravity waves with wavelengths and directions consistent with surface data were sometimes visible for both cross-track and along-track orientations of the wave-crests, provided that winds were not too strong. For winds greater than $15 \text{ m s}^{-1}$ there was a tendency for streaks to be imaged at the expense of waves (figure 9). These streaks were within $\pm 10°$ of being parallel to the sass winds, had a spacing of 2–3 km and extended over hundreds of kilometres, suggesting that, primarily, atmospheric forcing was modulating the surface backscatter. These observations, together with the fact that neither sar (Jones *et al.* 1981*b*) nor sass backscatter saturates at the low wind speeds (a few metres per second) that theory (Phillips 1977) predicts for short ocean waves, raise important questions about what causes the measured backscatter. One possibility is that the proportion of the sea surface covered by ripples is dependent on wind speed; another is that bubbles at the surface, caused by wave-breaking, are contributing to the backscatter. At high wind speeds, foam patches persist for periods of a minute or longer, cover large areas and are not necessarily organized on the scale of the waves. It is suggested that these mechanisms compete against others (tilt, hydrodynamic and orbital velocity) that have been proposed to explain imaging of surface gravity waves (Tucker, this symposium).

### (*e*) *Other applications*

Numerical weather prediction and storm-surge models require sea-level pressure analyses Endlich *et al.* (1982) have demonstrated the feasibility of deriving pressure fields from sass winds. Wind vectors were computed on a regular grid mesh from three consecutive passes; the pressure distribution was calculated from the balance equation (which in its simplest form

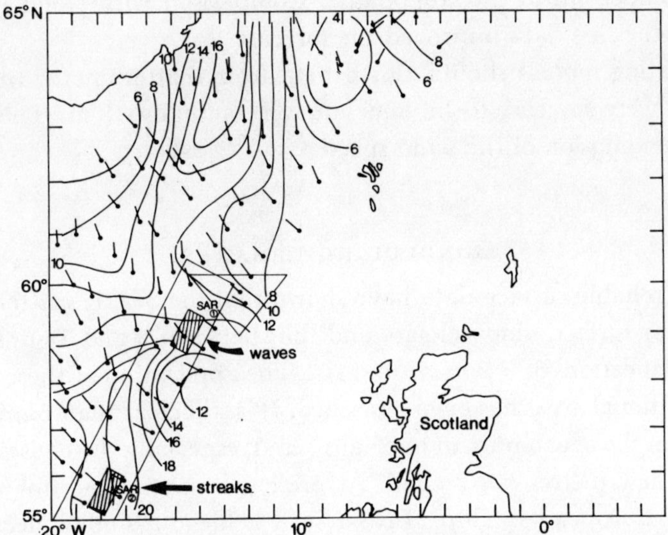


FIGURE 9. Scatterometer wind-velocity field (metres per second) on rev. 1087, 11 September 1978. Locations of two synthetic aperture radar images that have been analysed are also shown. (From Allan & Guymer (1983).)

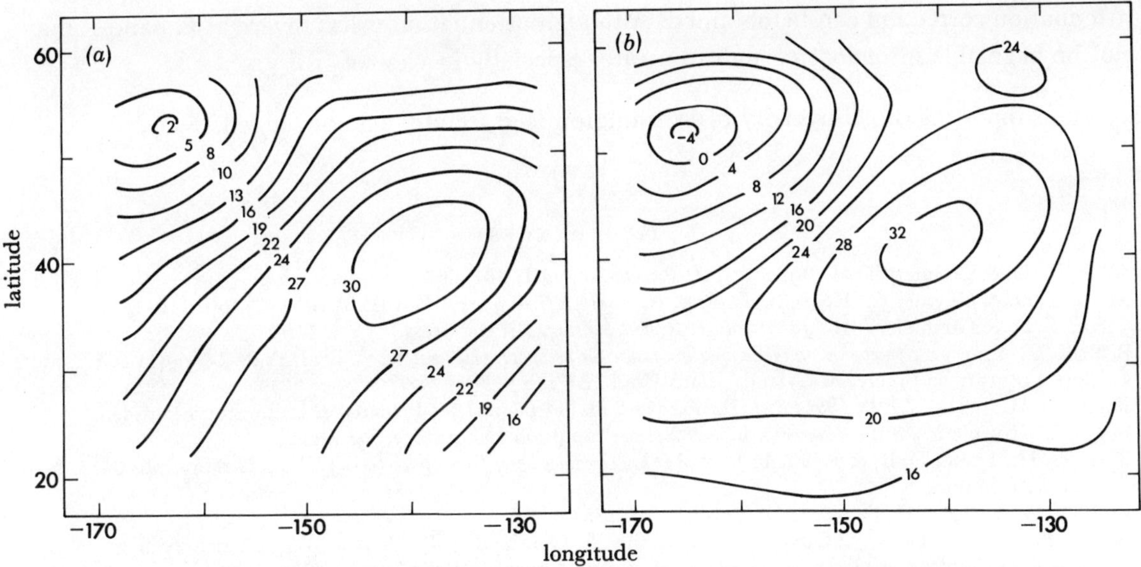


FIGURE 10. Surface pressure field over northern Pacific Ocean as the departure from 1000 mbar on 17 July 1978. (a) Balanced values computed from the non-divergent component of scatterometer winds between 12h00 and 16h00 G.M.T.; (b) corresponding National Meteorological Centre analysis for 12h00 G.M.T. (From Endlich et al. (1981), by permission of the American Meteorological Society.)

reduces to the geostrophic relation) by using the non-divergent component of the wind field. Absolute values were obtained from surface-measured pressure at two locations. The resulting pressure field is shown in figure 10 with the conventional analysis for comparison. The correlation coefficient between the two is 0.91 and the r.m.s. difference is 4.4 mbar; some of the error is due to changes occurring during the 300 min needed to obtain the SASS data.

The incorporation of global SASS data into numerical weather forecasts was investigated for a 2 day period by Yu and McPherson (1981). Largest differences were in the Southern

Hemisphere and extended up to the tropopause. Comparison with cloud imagery suggested that the inclusion of the sass data improved the forecast.

Wind-wave forecasting models should also benefit from scatterometer winds. In the *QEII* storm, wave heights were forecast to be 3 m but were actually 12 m. The discrepancy was mainly due to underestimation of the wind speed.

## 5. Concluding remarks

Comparisons with reliable surface data have shown that the Seasat scatterometer produced useful estimates of near-surface wind velocity and that between 3 and 16 m s$^{-1}$ they were well within the design specifications of $\pm 2$ m s$^{-1}$ and 20°. The application of these data in a number of scientific and operational areas has been described. It is still clear that many questions about radar backscatter from the ocean remain to be answered, especially the causes of its wind-speed dependence above a few metres per second. At present it seems doubtful whether the wind scatterometer on the European satellite, ERS-1, will achieve a similar accuracy to the sass because there are two important differences:

(i) it will operate at 5.3 GHz (C-band), where $\sigma°$ is known to be much less sensitive to the wind;

(ii) ERS-1 will not carry a broad-swath microwave radiometer from which an atmospheric attenuation correction can be obtained. Although attenuation is less severe at C-band it may not be negligible in regions of high rain-rate, especially in view of (i).

The support of the University of Washington is gratefully acknowledged.

## References

Allan, T. D. & Guymer, T. H. 1980 *Int. J. Remote Sensing* **1**, 261–267.
Allan, T. D. & Guymer, T. H. 1981 In *Proc. EARSeL–E.S.A. Symp., Voss* (ESA-SP-167), pp. 119–127.
Allan, T. D. & Guymer, T. H. 1983 *Int. J. Remote Sensing*. (In the press.)
Beal, R. C. 1981 In *Spaceborne synthetic aperture radar for oceanography* (ed. R. C. Beal, P. de Leonibus & I. Katz), pp. 110-127. Baltimore, Maryland: Johns Hopkins Press.
Boggs, D. H. 1982a *J.P.L. N.T.I.S. P.D.* no. 622-230. (44 pages.) Jet Propulsion Laboratory, Pasadena.
Boggs, D. H. 1982b *J.P.L. Report* no. 622-232. Jet Propulsion Laboratory, Pasadena.
Born, G. H., Lame, D. B. & Rygh, P. J. 1981 In *Oceanography from space* (ed. J. F. R. Gower), pp. 3–14. New York: Plenum.
Bradley, G. A. 1971 Ph.D. thesis, University of Kansas.
Brown, R. A., Cardone, V. J., Guymer, T., Hawkins, J., Overland, J. E., Pierson, W. J., Peteherych, S., Wilkerson, J. C., Woiceshyn, P. M. & Wurtele, M. 1982 *J. geophys. Res.* **87** (C5), 3355–3364.
Businger, J. A. & Charnock, H. 1983 *Phil. Trans. R. Soc. Lond.* A **308**, 445–449.
Cane, M. A. & Cardone, V. J. 1981 In *Oceanography from space* (ed. J. F. R. Gower), pp. 587–595. New York: Plenum.
Cardone, V. J. 1970 *New York Univ. Report* no. TR 69-1. University Heights, New York.
C.C.C.O. 1982 *Report of the Joint C.C.C.O./J.S.C. Study Conference on Large-Scale Oceanographic Experiments in the World Climate Research Programme.* (IOC/INF-507.)
Chelton, D. B., Hussey, K. J. & Parke, M. E. 1981 *Nature, Lond.* **294**, 529–532.
Chelton, D. B. & O'Brien, J. J. 1982 *Trop. Ocean–Atmos. Newsl.* **11**, 3–4.
Claasen, J. P., Fung, A. K., Moore, R. K. & Pierson, W. J. 1972 In *IEEE Int. Conf. Engineering in the Ocean Environment, Newport, R.I., U.S.A.*, pp. 180–185.
Cowan, E. W. 1946 *M.I.T. Radiation Lab., Rep.* no. 870. Cambridge, Massachusetts.
Daley, J. C. 1973 *J. geophys. Res.* **78**, 7823–7833.
Dome, G. J., Fung, A. K. & Moore, R. K. 1977 In *Proc. URSI Commission F, La Baule, France, 28 April to 6 May*, pp. 591–596.

Endlich, R. M., Wolf, D. E., Carlson, C. T. & Maresca, J. W. Jr 1981 *Mon. Weath. Rev.* **109**, 2009–2016.
Estoque, M. A. & Fernandez-Partagas, J. 1981 *Tellus* **33**, 463–475.
Fedor, F. S. & Brown, G. S. 1982 *J. geophys. Res.* **87** (C5), 3385–3395.
Grant, C. R. & Yaplee, B. S. 1957 *Proc. Inst. Radio Engrs* **45**, 976.
Grantham, W. L., Bracalente, E. M., Jones, W. L. & Johnson, J. W. 1977 *IEEE J. ocean. Engng* **OE-2**, 200–206.
Guignard, N. W., Ransone, J. T. Jr & Daley, J. C. 1971 *J. geophys. Res.* **76**, 1525.
Guymer, T. H. 1983a I.O.S. internal report (in preparation).
Guymer, T. H. 1983b In *Satellite microwave remote sensing* (ed. T. D. Allan). Ellis Horwood.
Guymer, T. H., Businger, J. A., Jones, W. L. & Stewart, R. H. 1981 *Nature, Lond.* **294**, 735–737.
Guymer, T. H., Businger, J. A., Katsaros, K. B., Shaw, W. J., Taylor, P. K., Large, W. G. & Payne, R. E. 1983 *Phil. Trans. R. Soc. Lond.* A **308**, 253–273.
Guymer, T. H. & Kennett, M. J. 1983 In preparation.
Guymer, T. H. & Taylor, P. K. 1983 In *Proceedings of the Joint C.C.C.O./J.S.C. Study Conference on Large-Scale Oceanographic Experiments in the World Climate Research Programme, Tokyo, 10–21 May 1972.* (In the press.)
Gyakum, J. R. 1980 In Preprints of Eighth Weather Forecasting and Analysis Conference, A.M.S., Boston, pp. 23–28.
Halberstam, I. 1981 *J. geophys. Res.* **86**, 6599–6606.
Hoffman, R. N. 1982 *Mon. Weath. Rev.* **110**, 434–445.
Johnson, J. W., Williams, L. A. Jr, Bracalente, E. M., Beck, F. B. & Grantham, W. L. 1980 *IEEE J. ocean. Engng* **OE-5**, 138–144.
Jones, W. L., Boggs, D. H., Bracalente, E. M., Brown, R. A., Guymer, T. H., Chelton, D. & Schroeder, L. C. 1981a *Nature, Lond.* **294**, 704–707.
Jones, W. L., Delnore, V. E. & Bracalente, E. M. 1981b In *Spaceborne synthetic aperture radar for oceanography* (ed. R. C. Beal, P. de Leonibus & I. Katz), pp. 87–94. Baltimore, Maryland: Johns Hopkins Press.
Jones, W. L. & Schroeder, L. C. 1978 *Bound. Layer Met.* **13**, 133–149.
Jones, W. L., Schroeder, L. C., Boggs, D. H., Bracalente, E. M., Brown, R. A., Dome, G. J., Pierson, W. J. & Wentz, F. J. 1982 *J. geophys. Res.* **87** (C5), 3297–3317.
Jones, W. L., Schroeder, L. C. & Mitchell, J. L. 1977 *IEEE J. ocean. Engng* **OE-2**, 52–60.
Jones, W. L., Wentz, F. J. & Schroeder, L. C. 1978 *AIAA J. Spacecraft Rockets* **15**, 368–374.
J.P.L. 1979a Rep. no. 622-101. Jet Propulsion Laboratory, Pasadena.
J.P.L. 1979b Rep. no. 622-107. Jet Propulsion Laboratory, Pasadena.
J.P.L. 1979c Rep. no. 622-210. Jet Propulsion Laboratory, Pasadena.
J.P.L. 1980 *J.P.L. Publication* no. 80–62. Jet Propulsion Laboratory, Pasadena.
Kerr, D. E. 1951 In *Propagation of short radio waves.* New York: McGraw-Hill.
Kondo, J. 1975 *Bound.-Layer Met.* **9**, 91–112.
Lipes, R. G. 1982 *J. geophys. Res.* **87** (C5), 3385–3395.
Liu, W. T. & Large, W. G. 1981 *J. phys. Oceanogr.* **11**, 1603–1611.
Mitsuyasu, H. 1977 *J. phys. Oceanogr.* **7**, 882–891.
Moore, R. K., Birrer, I. J., Bracalente, E. M., Dome, G. J. & Wentz, F. J. 1982 *J. geophys. Res.* **87** (C5), 3337–3354.
Moore, R. K. & Fung, A. K. 1979 *IEEE Proc.* **67**, 1504–1521.
Moore, R. K. & Young, J. D. 1977 *IEEE J. ocean. Engng* **OE-2**, 309–317.
O'Brien, J. J. (ed.) 1982 *Scientific opportunities using satellite wind stress measurements over the ocean.* Fort Lauderdale, Florida: Nova University/N.Y.I.T. Press.
Offiler, D. 1981 Met. O.19 Branch Memorandum no. 64 (unpublished). Meteorological Office, Bracknell.
Phillips, O. M. 1977 *The dynamics of the upper ocean.* New York: Cambridge University Press.
Pierson, W. J., Cardone, V. J. & Greenwood, J. A. 1974 *The applications of Seasat-A to meteorology* (Technical Report, City University of New York.)
Pierson, W. J. & Stacy, R. A. 1973 *N.A.S.A. Contractor Rep.* no. CR-2247. (128 pages.)
Pollard, R. T. 1982 JASIN News no. 25, unpublished manuscript.
Pollard, R. T., Guymer, T. H. & Taylor, P. K. 1983 *Phil. Trans. R. Soc. Lond.* A **308**, 221–230.
Robinson, A. R. (ed.) 1963 *Wind-driven ocean circulation.* New York: Blaisdell.
Schroeder, L. C., Boggs, D. H., Dome, G. J., Halberstam, I. M., Jones, W. L., Pierson, W. J. & Wentz, F. J. 1982a *J. geophys. Res.* **87** (C5), 3318–3336.
Schroeder, L. C., Grantham, W. L., Mitchell, J. L. & Sweet, J. L. 1982b *IEEE J. ocean. Engng* **OE-7**, 3–14.
Stewart, R. H. 1983 In *Proceedings of the Joint C.C.C.O./J.S.C. Study Conference on Large-Scale Oceanographic Experiments in the World Climate Research Programme, Tokyo, 10–21 May 1972.* (In the press.)
Taylor, P. K., Guymer, T. H., Katsaros, K. B. & Lipes, R. G. 1983 In *Proc. Symp. Variations in the Global Water Budget, Oxford, 10–15 August* 1981. (In the press.)
Vesecky, J. F. & Stewart, R. H. 1982 *J. geophys. Res.* **87** (C5), 3397–3430.
Weller, R. A., Large, W. G., Payne, R. E. & Zenk, W. 1983 *J. appl. Met.* (Submitted.)
Wentz, F. J., Cardone, V. J. & Fedor, L. S. 1982 *J. geophys. Res.* **87** (C5), 3378–3384.

Wurtele, M. G., Woiceshyn, P. M., Peteherych, S., Borowski, M. & Appleby, W. S. 1982 *J. geophys. Res.* **87** (C5), 3365–3377.
Yu, T. W. & McPherson, R. D. 1981 Presented at Fifth Conference on Numerical Weather Prediction, 2–6 November 1981, Monterey, California.

*Phil. Trans. R. Soc. Lond.* A **309**, 415–432 (1983)
*Printed in Great Britain*



# Satellite observations of ocean colour

By I. S. Robinson

*Department of Oceanography, The University, Southampton SO9 5NH, U.K.*

Ocean colour is unique among the properties of the sea that can be measured from satellites because visible wavelength radiation penetrates below the surface skin and contains direct information about the bulk water quality in the upper few metres of the sea. With reference to Nimbus czcs and Landsat mss data, this review examines three aspects of ocean-colour monitoring: the removal of atmospheric effects, which make up a large proportion of the visible signal reaching the satellite; the calibration of the ocean colour signal in terms of more useful ocean parameters such as sediment load, chlorophyll concentration, water depth or the depth of the euphotic zone; and the potential applications of colour-monitoring satellites to oceanography. Algorithms for atmospheric correction are now well developed, but calibration for chlorophyll or sediment is less certain, particularly where inorganic sediments or land-derived yellow substance, as well as the local phytoplankton population, are present in the sea. Oceanographic applications include the estimate of total productivity, the identification of blooms, and the location of productive areas to assist ship surveys. The greatest potential for satellite observations of ocean colour may lie in the synoptic spatial information contained in the images, but this awaits serious exploitation by dynamical oceanographers.

## 1. Introduction

With the launch of the first manned orbiting satellites in the early 1960s it became clear that the colour of the sea not only could be viewed clearly from space, in the absence of cloud cover, but also contained information of potentially great significance to oceanography. What could be viewed with the human eye could also be photographed relatively easily, so that the earliest colour images of the oceans available for scientific analysis were those from visible wavelength cameras on board manned satellites such as Gemini, Apollo and Skylab. Such images gave at best a qualitative impression of patterns of ocean colour variability, suffering from geometric distortions. The development of scanning radiometers made possible measurements capable of geometric registration and radiometric calibration. By using several sensors sensitive to different portions of the visible spectrum, an approximate measure could be made of the spectral signature, or 'colour', of the upwelling radiatian reaching the satellite.

This paper is concerned with the oceanographic use that can be made of both single and multiple waveband radiance measurements in the visible region. Most of the quantitative analyses for marine applications have been performed on visible-wavelength data from the multispectral scanners (mss) carried on the Landsat series of Earth-monitoring satellites, and the Coastal Zone Colour Scanner (czcs) flown on the Nimbus 7 satellite. The sensor and orbital characteristics are listed in table 1. The Landsat mss, designed for land applications, has coarser spectral and radiometric resolution than the czcs, which was designed particularly for ocean-colour observations (Hovis *et al.* 1980). Landsat, with a smaller field of view, has much higher spatial resolution, which permits its use in coastal and estuarine situations, but this is won at the expense of a very long sampling interval. The new thematic mapper (tm) recently launched

TABLE 1. SATELLITE AND SENSOR PARAMETERS

| | CZCS | | MSS | TM |
|---|---|---|---|---|
| platform | Nimbus 7 | | Landsat series | Landsat 4 |
| approx. altitude/km | 955 | | 920 (Landsats 1–3) | 705 |
| orbit | near-polar, Sun-synchronous, northbound pass at local noon | | near-polar, Sun-synchronous, 14 southbound passes per day at approx. 10h00 local solar time | near-polar, Sun-synchronous, $14\frac{1}{2}$ southbound passes per day equator crossing at 09h45 local solar time |
| swath width/km | 2300 at 20° forward tilt; 1600 at no tilt | | 185 | 185 |
| daytime overpass frequency | every day | | every 18 days (2 adjacent days out of 18 in overlap zone) | every 16 days (7 and 9 days in overlap zone) |
| pixel size/m | 825 × 825 | | 56 × 79 (nominally 80 m resolution) | 30 × 30 |
| visible and near visible wavebands/nm | 443 ± 10<br>520 ± 10<br>550 ± 10<br>670 ± 10<br>700–800 | | 500–600<br>600–700<br>700–800<br>800–1100 | 450–520<br>520–600<br>630–690<br>760–900<br>1550–1750 |
| saturation radiance in the above channels/ ($mW cm^{-2} \mu m^{-1} sr^{-1}$) | min. gain<br>11.46<br>7.64<br>6.21<br>2.88<br>23.90 | max. gain<br>5.41<br>3.50<br>2.86<br>1.34<br>23.90 | 24.8<br>24.4<br>20.0<br>17.6 | — |
| radiometric digitizing resolution (no. of digital levels available) | 255 | | 63 | 255 |

on Landsat 4 promises even more spatial detail, at improved radiometric and spectral resolutions.

Three broad research areas can be identified as needing attention before the satellite colour scanner can become a credible and calibrated oceanographic instrument. The first is the problem of removing from the signal received at the sensor all radiation that does not emerge from below the sea surface, i.e. light backscattered by the atmosphere and reflected at the sea surface, and also of correcting for the absorption by the atmosphere of part of the radiation leaving the water. Early visible images of the ocean raised considerable interest in the potential for studying the sea surface by interpreting the sun glint patterns (see, for example, Cox 1974), but more powerful radar techniques have now been developed to study surface-roughness signatures, and the viewing directions of the present scanning visible radiometers are chosen to minimize specular reflexion. I shall therefore treat it here as noise rather than signal.

The second broad study area is the interpretation of the optical signal received from the sea in terms of an oceanographically more meaningful parameter, such as the suspended sediment load, or the concentration of chlorophyll pigmentation. This is strictly the field of optical oceanography, hitherto rather neglected in the U.K., but the coming of the satellite sensor has given it new importance and value, and raised new scientific problems requiring solution.

Finally we need to consider the contributions that satellite observation of ocean colour can make to oceanographic science as a whole. Remote-sensing scientists have been active in the first two areas, but the third has received less attention because the oceanographer has tended to be sceptical of the value of visible remote sensing owing to limitations by cloud cover. It is, however, the only satellite surveillance technique that can penetrate a few metres below the water surface. This paper aims to present to ocean scientists the present capabilities and limitations of the method, and to point towards those areas of oceanography where it promises to make exciting new advances possible.

## 2. ATMOSPHERIC CORRECTIONS

### (a) *Physical principles*

The photons reaching the sensor of a satellite outside the Earth's atmosphere arrive from the incident solar radiation by a variety of routes, including scattering in the atmosphere, scattering within the sea, reflexion at the sea surface and multiple combinations of these possibilities (Sturm 1981 a; Sorensen 1981). The satellite sensor measures the radiant energy in a given spectral window, incident on the aperture area, arriving from a very narrow cone of directions whose projection on the Earth is the instantaneous field of view, defining the pixel area. The observed variable is therefore the radiance as a function of wavelength and viewing direction, $L_t\,(\lambda,\,\theta,\,\phi)$.

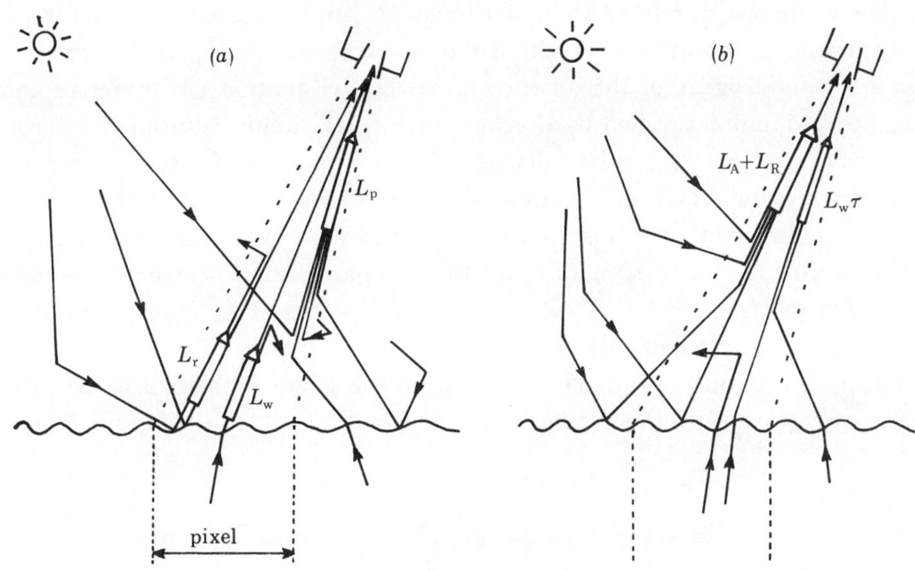


FIGURE 1. Optical pathways.

The optical quantity related to $L_t$ that contains oceanographic information is $L_w\,(\lambda,\,\theta,\,\phi)$, the water-leaving radiance in the direction of the sensor. Some of $L_w$ is scattered or absorbed by the atmosphere before reaching the sensor (figure 1 a). An additional radiance $L_r$ leaves the sea surface owing to specular reflexion of sunlight (sun glint) or direct reflexion of atmospherically scattered light (sky glint). There is also a path radiance $L_p$ contributed by atmospheric scattering into the sensor field of view of photons, some of which may previously have reflected beneath or at the surface outside the pixel area. Thus

$$L_t = L_w\tau_1 + L_r\tau_1 + L_p,\tag{1}$$

where $\tau_1$ is the direct transmittance through the atmosphere. The task of atmospheric correction is to extract $L_w$, given $L_t$.

Given the variety of available paths, the likelihood of multiple scattering, the three-dimensional

[ 175 ]

nature of scattering and the number of unknown quantities such as the optical properties of the atmospheric scatterers and the sea-surface roughness in a given image, it is impossible to determine $L_r$ and $L_p$ from first principles. Progress has been made by simulating the detailed photon-by-photon optical response of typical atmospheres by using Monte-Carlo statistical techniques. These have been used to test the validity of various simplifying assumptions and have led to an approximate model of atmospheric scattering, developed by Gordon (1978, 1981). First, the radiance received at the sensor is divided into those photons that have not penetrated below the sea surface and those that have (figure 1$b$). Given that satellite colour sensors are now designed to avoid direct sun glint, the former are all dependent on random atmospheric scattering processes. The latter will not all have upwelled from the pixel in the instantaneous field of view because of scattering, but provided that the sea properties do not vary strongly with position it is acceptable to relate this part of $L_t$ to $L_w$ from the given pixel. Thus

$$L_t = L_A + L_R + \tau L_w. \tag{2}$$

Here $\tau$ is now the diffuse transmittance of the atmosphere, which can be calculated by using the expression given by Sturm (1983), which includes the effect of ozone absorption. $L_A$ is the radiance that would be due to photons scattered by suspended particles in the atmosphere (aerosol scattering) if the atmosphere itself did not scatter, and $L_R$ is due to the Rayleigh scattering that would occur in the absence of aerosols. Equation (2) therefore contains the simplifying approximation justified by Gordon (1978) that, despite multiple scattering, $L_A$ and $L_R$ are independent. Now $L_R$ can be calculated (see Viollier *et al.* (1980) and Sturm (1981$a$)) for a given viewing angle and solar illumination conditions, whereas $L_A$ cannot be estimated *a priori* for a satellite image because the aerosol content varies with time and space. Since $L_A + L_R$ may contribute up to 80 % of $L_t$ in the blue part of the spectrum, it is essential that $L_A$ be estimated as accurately as possible if small variations of $L_w$ from scene to scene, and within a scene, are to be reliably detected.

The multispectral nature of the information from the sensor is therefore used to achieve an atmospheric correction, with the assumption that

$$L_{A,\lambda_2} = S(\lambda_2, \lambda_1)\, L_{A,\lambda_1}. \tag{3}$$

This appears to be a valid approximation provided that the aerosol phase function (i.e. the function controlling the scattering direction) and the ratio of the aerosol optical thicknesses, $t_A$, at the two wavelengths, does not vary within the scene (i.e. on a length scale of several hundred pixels), even though the aerosol scattering may vary significantly. Furthermore, if the aerosol phase function is independent of wavelength too, then $S$ may be represented by

$$S(\lambda_2, \lambda_1) = \epsilon(\lambda_2, \lambda_1)\, F_{0,\lambda_2}/F_{0,\lambda_1}, \tag{4}$$

where $\epsilon(\lambda_2, \lambda_1) = \omega_{A,\lambda_2} t_{A,\lambda_2}/\omega_{A,\lambda_1} t_{A,\lambda_1}$, $\omega$ is the probability that a photon will backscatter on interaction with the aerosol and $F_0$ is the solar irradiance at the top of the atmosphere.

If the sensor records a waveband centred at $\lambda'$ for which the sea can be assumed to absorb all incident radiation (i.e. $L_{w,\lambda'} = 0$) then $L_{A,\lambda'}$ can be obtained from (2) after calculating $L_{R,\lambda'}$. The significance of (3) is that $L_{A,\lambda}$ can then be obtained from $L_{A,\lambda'}$. Since $L_{R,\lambda}$ can be calculated, the atmospheric correction is determined at wavelength $\lambda$ provided that $S(\lambda, \lambda')$ is known.

[ 176 ]

### (b) Algorithms for czcs

For the czcs, channel 4 (wavelength $670 \pm 10$ nm) is taken as $\lambda'$, for which $L_\mathrm{w}$ is assumed to be zero. Therefore it is required to know $S(\lambda_i, 670)$, with $i = 1, 2, 3$ corresponding to the wavelengths of channels 1 to 3 of the sensor. Several approaches can be used, leading to different algorithms. In each case $S$ is assumed not to vary with position within a scene.

If $t_{\mathrm{A},\lambda_i}/t_{\mathrm{A},670}$ can be measured in one single location simultaneously with the satellite overpass, $S(\lambda_i, 670)$ can be determined from (4) if it is assumed that $\omega_{\mathrm{A},\lambda_i}/\omega_{\mathrm{A},670} = 1$. Alternatively if $L_{\mathrm{w},\lambda_i}$ were measured at one place synoptic with the overpass, $S(\lambda_i, 670)$ could be determined for the whole scene by using (2) and (3), provided that $\tau$ could be accurately estimated. Because $\tau L_\mathrm{w}$ is typically only 20 % of $L_\mathrm{t}$, small errors in $\tau$ would be relatively unimportant, as shown by Gordon (1981). Because in fact most czcs scenes appear to indicate that the aerosol correction varies significantly over the extent of a scene, some of the assumptions concerning the form of $S$ may lead to error and it is preferable to make several independent estimates of $S$ from synoptic observations in different parts of the scene. In this way the validity of assuming $S$ to be constant can be tested, and a mean value of $S$ obtained for the whole scene. Results of such tests remain to be published, although Gordon (1981) has raised questions of the validity of using ground measurements of $t_{\mathrm{A},\lambda}$ to estimate $S(\lambda, 670)$.

A further problem in the adoption of Gordon's approach occurs in that $L_{\mathrm{w},670}$ is not always zero in turbid coastal water (Bukata et al. 1980). Sturm (1981 b) has avoided this by calculating $L_{\mathrm{A},670}$ for the darker sea pixels only, for which $L_{\mathrm{w},670}$ is expected to be zero. This, however, prohibits the pixel-by-pixel calculation of $L_{\mathrm{A},670}$, which has to be estimated for the rest of the scene by taking only the viewing and solar geometry into account, and not the aerosol variability. Smith & Wilson (1981) alternatively adopt $L_{\mathrm{w},670} = 0$ in a preliminary atmospheric correction, after which the chlorophyll concentration is calculated. This can then be used to estimate a likely value of $L_{\mathrm{w},670}$ for each pixel enabling an iterative correction procedure to be followed. This is limited by the accuracy of the chlorophyll algorithms (see §3).

Synoptic measurements of optical quantities at ground level are difficult and costly to obtain at sea, and the goal of a universal atmospheric correction algorithm is that it should operate by using only satellite data. If the form of $S$ could be universally established, the problem would be solved. Gordon & Clark (1980) have used $\epsilon(\lambda, 670) = (670/\lambda)^n$, where $n$ is known as the Ångström exponent, but there appears to be no clear consensus in the literature as to whether the exponent should be positive (as implied by Ångström 1964), negative (Quenzel 1970) or zero (De Luisi et al. 1972).

The most comprehensive self-contained algorithm yet to appear is presented by Sorensen (1981) and Sturm (1983). In this the scene is initially corrected by assuming that $L_{\mathrm{w},670} = 0$ and $\epsilon = 1$. The darkest pixels, remote from land and cloud, and preferably near the centre of the scene, are identified in the resulting $L_{\mathrm{w},\lambda}$ images. Provided that it is oceanographically reasonable, it is assumed that these pixels correspond to clear water with a chlorophyll concentration less than 0.3 µg l$^{-1}$. In this case $L_{\mathrm{w},520}$ and $L_{\mathrm{w},550}$ can be calculated by the method of Gordon & Clark (1981), enabling an accurate estimate of $S(520, 670)$ or $S(550, 670)$ to be made for those dark-water pixels. $S(443, 670)$ is extrapolated from these. These $S$ are then applied to the whole scene for a pixel-by-pixel aerosol correction assuming $L_{\mathrm{w},670} = 0$, and if necessary the Smith & Wilson iteration can be included for the cases where $L_{\mathrm{w},670} \neq 0$. Although it has not yet been exhaustively tested, such an approach promises to work well for open ocean waters where patches of low chlorophyll occur and the water turbidity is dominated by chlorophyll with a reasonably well understood optical signature. Such an approach is not so promising in turbid coastal waters, and research waits to be done on testing the accuracy of these algorithms in a variety of different shallow sea and ocean locations.

### (c) *Application to Landsat* MSS

The same approach to an atmospheric correction should be valid whatever sensor is used. However, attempts to apply the above method to Landsat MSS data over the sea areas reveal that the sensor response, designed for land use, is incompatible with accurate sea observations. Indirectly this serves to illustrate the important design requirements of an ocean colour sensor.

The particular problems are the poor spectral and radiometric resolutions of the MSS. The sensor bands are nominally 100 nm wide, and in practice the radiation entering the sensor is filtered through a spectral window that is not uniformly flat. The equations presented in §2$a$ are appropriate for narrow bandwidths, but because $S$ is probably nonlinearly frequency dependent it is much less precise to apply (3) over a broad band, selecting the central wavelength at which to define $S$. A more accurate approach would require the inclusion of an integral convolving $S$ with the frequency window of the sensor. Landsat has the advantage of a longer-wavelength sensor centred at 950 nm (channel 7), for which $L_{\mathrm{w}}$ must almost always be zero. However, the problem of identifying an appropriate $L_{\mathrm{A},950}$ for use as $L_{\mathrm{A},\lambda}$ is compounded with a 300 nm bandwidth and irregular spectral window in channel 7. Moreover it is not clear that it is appropriate to extend the physical assumptions underlying (2), (3) and (4) so far into the infrared region. MacFarlane (1981) has made an attempt along these lines to develop a haze-removal technique for Landsat sea scenes, but has been restricted by the poor radiometric sensitivity of the MSS, particularly channel 7. Not only is $L_{\mathrm{w},c7}$ zero, but $L_{\mathrm{A},c7}$ is also small and is not resolved by the sensor with sufficient accuracy to determine $L_{\mathrm{A}}$ confidently for the other channels in each pixel.

Attempts to provide atmospheric corrections for Landsat MSS have therefore tended to be much cruder, at best subtracting from a whole scene the radiance level of the darkest pixel (Johnson 1975; Robinson & Srisaengthong 1981) on the assumption that the water-leaving radiance is zero and the signal is entirely due to path radiance. No attempt is usually made to apply a differential atmospheric correction across a scene, and most attempts at quantitative interpretation of Landsat data over sea have incorporated the atmospheric effects in empirically derived regression constants (see, for example, Klemas *et al.* 1974).

It is clear that to facilitate atmospheric corrections to ocean colour data, future sensors must be designed with narrow wavebands well spaced throughout the visible and into the near infrared. The sensitivity must be sufficient at the higher wavelengths to resolve small aerosol-path radiance differences above the instrument noise or digitizing level. Such criteria have influenced the design of the successor to the CZCS (Hovis 1981) and the European OCS (see, for example, Giannini 1981).

## 3. WATER-QUALITY PARAMETERS FROM OCEAN COLOUR

### (a) *Optical principles*

The removal of atmospheric effects is only the first step in making use of satellite radiance measurements, because it still remains to interpret the observations of ocean colour in terms of oceanographically interesting parameters. Techniques in the science of optical oceanography have tended during most of this century to concentrate on measuring absorption and scattering under water (Jerlov 1976) in terms of their effect on productivity or as a measure of suspended material. Only in the last decade has the prospect of space observation provoked much more study of what we can learn by viewing ocean colour from above, as reviewed recently by Sathyendranath & Morel (1983).

The water-leaving radiance $L_w$ is influenced as much by the incident radiation as by the water's optical properties. In fact the parameter of most use is the subsurface reflectance ratio, $R = E_u/E_d$, where $E_u$ is the upwelling irradiance and $E_d$ the downwelling irradiance just below the surface. Now for a flat surface,

$$L_w(\theta, \phi) = RE_d\{1 - \rho(\theta', \phi)\}/Qn^2, \tag{5}$$

where $n$ is the refractive index of water and $\rho$ is the Fresnel reflectivity of the water–air interface ($n \sin \theta' = \sin \theta$ relates the subsurface ray inclination $\theta'$ to the viewing direction $\theta$ of the satellite). Equation (5) is valid for a rough surface provided that $\theta$ is not close to the critical angle of 48° (Austin 1974). Assuming that the ocean is a perfect Lambertian reflector, $Q = \pi$ and is direction-independent. However, observations at sea (Austin 1980) suggest that $Q$ should be approximately 5.0 for near-nadir observations. Given $L_w$ from the satellite sensor, $R$ can then be estimated from (5), by calculating $E_d$ given the known Sun illumination and atmospheric properties based on the atmospheric correction already discussed. To improve accuracy the dependence of $n$ on salinity and temperature should be incorporated (Singh et al. 1983). Since $Q$ and $E_d$ are only weakly wavelength-dependent, it is often more satisfactory to use

$$\frac{R(\lambda_1)}{R(\lambda_2)} = \frac{L_w(\lambda_1)}{L_w(\lambda_2)} \frac{n(\lambda_1)^2}{n(\lambda_2)^2} \frac{[1-\rho(\lambda_2)]}{[1-\rho(\lambda_1)]},$$

i.e. the reflectance spectral ratio is obtained from the remotely observed radiance ratio (i.e. colour) of the water.

$R$ itself depends on the balance of absorption and backscattering of light by the water and its suspended contents. Since particle sizes are somewhat larger than the wavelengths of light, Mie scattering theory applies. Preisendorfer (1961) has shown that the total absorption coefficient, $a$, of the water and the total backscattering coefficient, $b_b$, are the linear sums of the individual coefficients due to the different absorbers and scatterers. It is therefore useful to express $R$ in terms of $a$ and $b_b$. Morel & Prieur (1977) concluded from theoretical modelling studies that

$$R = 0.33\, b_b/a \tag{6}$$

is a satisfactory approximation for most marine situations where $b_b/a$ is small (i.e. less than 0.3). Gordon et al. (1975) had proposed $R = f\{b_b/(a + b_b)\}$, but (6) lends itself better to the study of spectral ratios of reflectance, absorption and backscattering coefficients. Given a knowledge of the spectral absorption and backscattering characteristics of the different water contents, as discussed below, it is therefore possible to predict the water colour that would be viewed by a satellite. The inversion of this process must underlie any physically based algorithm for water quality analysis using satellite colour observations.

### (b) Spectral characteristics of different water types

*Pure sea water* provides the baseline against which to compare other types. Morel & Prieur (1977) and Smith & Baker (1981) present typical observations of the spectral dependence of *a* and $b_b$, summarized in figure 2. Absorption decreases and backscatter increases with decreasing wavelength, leading to the characteristic blue colour of pure seawater.

*Water containing phytoplankton* has much more complex spectral characteristics. This is because along with the living cells, which contain chlorophyll, are the associated detrital products, biogenic particulate material and dissolved organic compounds containing phaeophytin *a*. Each of these has its own absorption and scattering properties, and each is found in different proportions according to the species or even the age of the population. Thus observations of phytoplankton-rich sea water *in situ* (see, for example, Prieur & Sathyendranath 1981; Smith &

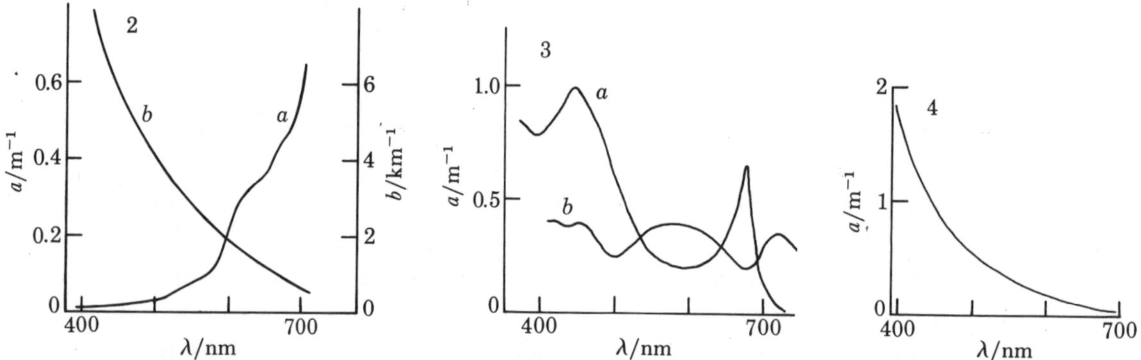

FIGURE 2. Absorption (*a*) and scattering (*b*) of pure sea water.

FIGURE 3. Typical spectral variation of: (*a*) absorption due to chlorophyll, normalized at 440 nm (typical values of specific absorption for chlorophyll at 440 nm vary between 0.01 and 0.1 m⁻¹ (mg m⁻³)⁻¹, depending on age and species) (after Prieur & Sathyendranath 1981); (*b*) specific backscattering coefficient (units are typically $10^{-3}$ m⁻¹ (mg m⁻³)⁻¹).

FIGURE 4. Absorption spectrum of yellow substance.

Baker 1981; Kiefer *et al.* 1979) show a variety of spectra, in agreement with the theoretical predictions of Morel & Bricaud (1981). Certain general characteristics emerge, however, as sketched in figure 3. There is a strong absorption peak at 440 nm and a lesser one at around 675 nm; over the rest of the spectrum in general the absorption decreases with wavelength and the backscatter is fairly uniform apart from decreases corresponding to the absorption peaks.

In some locations such as the Baltic Sea, strong concentrations of *yellow substance* (dissolved organic matter) are found in the absence of phytoplankton. This is believed to be strongly related to land run-off (Højerslev 1980*a*). The absorption spectrum is exponential in form (figure 4) but is the reverse of sea water.

The other, and most diverse, contribution to ocean colour is the presence of *suspended particulate material* not related to phytoplankton. This may be resuspended bottom sediment, river-borne sediment, or eroded coastal and beach material, or it may be due to dumping of waste or dredge spoil. Its spectral characteristics are as diverse as its composition in terms of natural material colour and size distribution, and there appears to be no consensus in the literature about a standard spectrum shape.

Given so many different contributions to the upwelling radiance spectrum, it is apparent that there is little possibility of being able to invert the problem and obtain an analytical relation between chlorophyll content, or suspended solids concentration and the colour radiance or reflectance signal. None the less, the knowledge that has been gained about the optical properties of different substances in sea water does enable a general framework to be constructed from which empirical calibration algorithms may be formulated. A fundamental step is to divide all waters into two categories for optical purposes. Case 1 waters are seas whose optical properties are dominated by phytoplankton and their degradation products only, and case 2 waters have non-chlorophyll-related sediments or yellow substance instead of or in addition to phytoplankton. Figure 5 shows typical reflectance spectra for (a) case 1 water, and for (b) sediment-dominated and (c) yellow-substance-dominated case 2 waters. The combination of absorption and backscatter in case 1 water tends to decrease the reflectance below the clear water spectrum at wavelengths below around 540 nm and slightly increases it at higher $\lambda$. Increasing chlorophyll content enhances this effect with a definite minimum appearing at 440 nm due to the

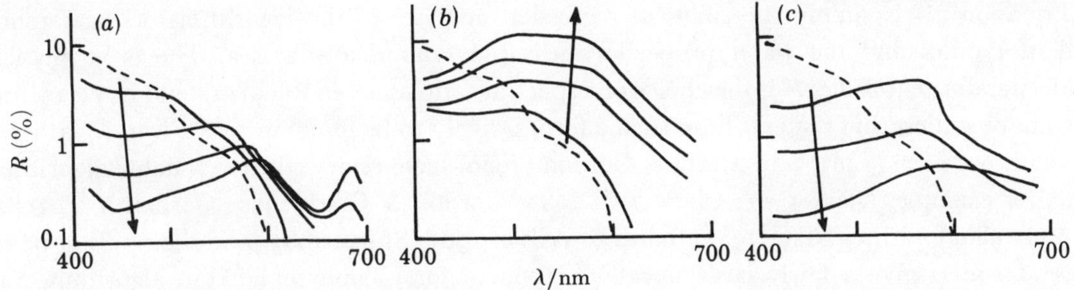

FIGURE 5. Typical reflectance spectra: (a) case 1 water; (b) case 2 water dominated by suspended sediment; (c) case 2 water dominated by yellow substance. – – –, Spectrum for clear blue water; ——→, increasing concentration. (After Sathyendranath & Morel (1983).)

chlorophyll absorption, and at 660 nm, although the red absorption maximum is masked by a reflectance peak at 685 nm due to fluorescence. Although the absolute reflectance values differ with different chlorophyll species, the spectral shape and its variation with chlorophyll concentration remain similar, giving encouragement to the formulation of universal chlorophyll algorithms based on reflectance spectral ratios.

In figure 5b, the backscattering by the sediment produces higher reflectance ratios at the larger wavelengths, although there is some reduction at small wavelengths due to small amounts of chlorophyll absorption. Because the spectral shape does not change much with sediment load, the water colour does not change and a sediment algorithm based on spectral ratios appears less hopeful, except in strongly coloured sediments where figure 5b would not apply. Instead, an algorithm relating suspended load to the absolute reflectance in a single waveband might be more appropriate. The yellow-substance-dominated waters (figure 5c) show a distinct colour change with increasing concentration as the blue reflectance becomes less and less.

From these generalized spectra, it appears that there is sufficient change in optical properties correlated with the concentration of the dissolved or suspended material to permit calibration of colour images in terms of water quality, but only if the water has a single component. This fundamental limitation provides the proper perspective for the development of interpretative algorithms. Optimistic attempts at producing universal algorithms appear to be ill-conceived, but undue pessimism is also out of place. Theoretical models of reflectance such as those by Bukata *et al.* (1981), which confirm that there can be no unique relation between optical and water quality parameters, must be weighed against the fact that we very often have information from sources other than the satellite, which may resolve ambiguities in the remotely sensed data.

### (c) Calibration algorithms for czcs data

Most of the proposed algorithms have been of the form

$$C \text{ or } S \text{ or } K(\lambda) = A(L_{443}/L_{550})^B \quad \text{or} \quad A(L_{520}/L_{550})^B, \tag{7}$$

where $C$ is the total pigment concentration (chlorophyll $a$ plus phaeophytin $a$) (micrograms per litre), $S$ is the dry mass concentration of total suspended solids (milligrams per litre), and $K(\lambda)$ is the diffuse attenuation coefficient of the near-surface water for a given $\lambda$. Although $K(\lambda)$ is itself an optical property rather than an oceanographic parameter, its value lies in that it has been extensively measured at sea and used for the bio-optical classification of sea water (Smith & Baker 1978 $a$, $b$; Baker & Smith 1982).

Equation (7) is simply the result of regression analyses of the logarithms of the satellite radiance ratios and the water properties measured synoptically at sea. The only physical understanding entering (7) is the choice of a spectral ratio as a variable. It should be clear from the above section that the best fit to such a form is likely to be found in case 1 water. Sathyendranath & Morel (1983) and Morel & Gordon (1980) have reviewed the available algorithms (see, for example, Clark 1981; Clark *et al.* 1980; Gordon & Clark 1980; Gordon *et al.* 1982; Morel 1980; Smith & Baker 1982; Smith & Wilson 1981; Sturm 1983) and show this to be so. Case 1 waters give a fairly large negative exponent for the phytoplankton algorithm, and achieve a logarithmic accuracy of $\pm\frac{1}{4}$. By definition, the variables $C$, $S$ and $K(\lambda)$ are not independent in case 1 water, all being controlled by the phytoplankton, and therefore correspondingly good algorithms are achieved for $S$ and $K$, a logarithmic accuracy of one-sixth being claimed for the seston algorithm. In general the use of the $L_{443}/L_{550}$ ratio is preferred, particularly at low chlorophyll concentrations. At high concentrations the absorption reduces $L_w$ (443) to such low levels that the atmospheric correction becomes inaccurate, and the $L_{520}/L_{550}$ ratio has to be used instead, with a further loss of accuracy because the wavelengths are so close to one another.

The success achieved with case 1 waters can be misleading if taken out of context. The only reason that coherence exists between the spectral reflectance ratios and sediment is that the particulate load is controlled by the biomass, which has a chlorophyll colour signature. In case 2 waters, no such correlation can be found between sediment load and reflectance ratios (see, for example, Hojerslev 1974). Indeed, in case 2 waters, where there may be at least three independent components influencing the subsurface optics, it is theoretically impossible to resolve any of them from two spectral ratio measurements, which are not completely independent because they are so close to each other within the spectrum. The only useful algorithms in these circumstances are likely to be limited to a particular location and seasonal condition, and heavily dependent on synoptic sea data. In those case 2 waters where yellow substance is not dominant, calibration for pigment still appears to be possible (Gordon & Morel 1981) and recent observations in the Irish Sea suggest that sediment may be related to the wideband reflectance magnitude whereas the spectral ratio continues to indicate the chlorophyll. None the less it may well be more promising to calibrate the imagery in terms of the attenuation coefficient $K(\lambda)$ (Austin & Petzold 1981), which can then be compared with $C$ and $S$ by ground measurements without the need for data collection at the same time as a clear-weather satellite overpass.

Other types of calibration algorithm, involving sums and differences of different bands, or the ratio of differences, have been tried. Most show some limited success in accounting for a synoptic data set, but their value in predicting calibrations of other scenes than the one they were constructed from is small, because they appear to lack any sound physical basis for their form.

A different and potentially more useful approach, applicable to case 1 and 2 waters, is that promoted by Hojerslev (1980$b$, 1981$a$, $b$). Concerned not only that the colour index (i.e. the spectral reflectance ratio) has no unique relations with $C$ or $S$, but also that it gives no indication of the vertical variability of $C$ and $S$ that often occurs, he draws attention instead to a strong correlation between colour index and the depth of the euphotic zone (i.e. the depth at which the downwelling energy flux is 1 % of the surface value). This observed result is supported by theoretical models, and although it may not be as useful a measure as $C$ or $S$, it is capable of useful exploitation by physical and biological oceanographers.

### (d) Calibration of Landsat data

Given the broader bandwidths and the poorer atmospheric corrections, the attempts to calibrate Landsat data in terms of water quality have been crude compared with the czcs work. It is much harder to pick out the spectral peaks and troughs with a 100 nm band and chlorophyll has therefore been very hard to detect, except in case 1 waters where it is related to particulate scatterers (Le Fèvre $et$ $al.$ 1982). Because of their high spatial resolution, Landsat data have been applied mostly in coastal and estuarine regions where type 2 water predominates. Calibration algorithms are therefore only locally applicable. Because of the infrequent repeat cycle, synoptic sea data have also been very hard to obtain, and it is much harder to test the accuracy of calibrations.

Most attempts have been made to relate the satellite radiances to the load of suspended solids, and algorithms fall into two broad categories, those that use colour and those that use intensity. Since in general non-chlorophyll-related suspended solids do not have a strong colour signature, it is more useful to seek simple relations between $S$ and the reflectance in a single band. Often band 5 has been the most useful because it is less contaminated by atmospheric effects. For low concentrations (less than 60 mg l$^{-1}$) a linear relation is satisfactory (see, for example, Munday & Alfoldi 1979; Robinson & Srisaengthong 1981) whereas over a wider range of loads the reflectance is correlated with $\log_{10} S$ (Klemas $et$ $al.$ 1974).

The most successful use of colour correlation with $S$ has come from the use of chromaticity analysis (Alfoldi & Munday 1978). Radiance values are mapped onto a two-colour chromaticity transform plane and $S$ is correlated with the position along the resulting locus. Atmospheric contamination is treated by shifting the whole locus to a normalized position. The success of this technique in studying the Bay of Fundy is probably due to the highly coloured red sediments which are brought down by local rivers, resulting in a strong colour signature to the concentration.

Because of their high resolution, Landsat mss data have also been used in hydrographic applications to measure shallow water depths. This requires a spectral calibration because in clear water it is the lower wavelengths that penetrate further and hence give a higher reflectance if the bottom is less than about 20 m deep. Cracknell $et$ $al.$ (1982) attempted a quantitative calibration, but the influence of suspended sediment and bottom albedo tend to confine this to a qualitative use.

## 4. Applications to oceanography

The value to science of a new measurement technique is only confirmed when advances in scientific understanding accrue from the new data. When the new technique simply improves the accuracy or precision of a previous method, the new data can usually be applied to modify an existing theoretical framework. In the case of ocean colour, however, the satellite sensor is producing data different in kind from oceanographic information obtained from traditional sources, and a full appreciation of the value of remote-sensing measurements calls for the construction of new theoretical models in which to exploit the data. The unique feature of satellite scanners is their capacity to provide a synoptic view of a large area of sea, with fine spatial resolution, and their ability to repeat that view over months or years. It is in this areal coverage that we should seek to find the principal scientific applications in oceanography, rather than looking to the satellite measurements to duplicate the time resolution and depth penetration achieved by observations based on ships and buoys.

Because the colour scanner is not an all-weather instrument and cannot penetrate cloud it is difficult to use it in the operational role envisaged for some radar sensors, and unreasonable to look to it for observing temporal variability on timescales less than a few days. Of course, where cloud cover is infrequent the daily overpasses of Nimbus 7 can be used semi-operationally. This is true off the California coast where charts showing the position and boundaries of water masses of different ocean colour as detected by czcs are issued to fishing vessels every few days by N.A.S.A./J.P.L. as part of a wider experimental programme to use satellites in support of commercial fisheries.

The obvious application, for which the czcs was designed, is the study of the spatial and temporal variability of primary production in the world's ocean. Already studies such as those by Smith & Wilson (1981), Yentsch & Garfield (1981), Anderson *et al.* (1981) and Gordon *et al.* (1980) have demonstrated the feasibility of mapping synoptic chlorophyll concentrations, in case 1 waters at least, to a good degree of accuracy over areas the size of the Southern California Bight or the Gulf of Maine – Georges Bank region. This would have been impossible to achieve with research vessels because covering such a wide area with even a coarse sampling interval would have taken several days, during which time the plankton populations could be expected to have grown or decayed, and been advected through the region. The accuracy of the satellite calibration of individual scenes also appears to be comparable with that achieved with continuous sampling fluorometers. To the designers of the colour scanner such calibrated images appear to be the successful end-product of the project, whereas to the ocean scientist they represent the beginning of a new type of synoptic oceanography. We find ourselves at the early descriptive phase, discovering new aspects of the spatial distribution of chlorophyll and sediment with every new image, but only slowly beginning to formulate the questions that will lead to a new theoretical understanding of the factors controlling the observed distributions. Although little has yet been published, it is now possible to estimate the total primary productivity of a sea area, and its seasonal variations (Smith & Baker 1982; Smith *et al.* 1982). Because the colour data are most strongly influenced by the near-surface chlorophyll, a model must be used to assess the contribution of the rest of the productive depth (Platt & Herman 1982). With a knowledge of the typical spatial variability it will be possible to evaluate how representative are the ship-based observations that have been used in the past for global productivity estimates. Moreover, the observed spatial distribution of chlorophyll, and its evolution in time, should enable much more to be learned about the way in which physical processes – advection by residual flows and upwelling, mixing across fronts, stirring by tides or storms, etc. – control the biological processes, as for example in a Gulf Stream ring (Gordon *et al.* 1982).

The particular study area that has already benefited significantly from the czcs imagery has been the identification of productive regions in the vicinity of oceanic or shallow sea fronts (see, for example, Yentsch & Garfield 1981; Yentsch 1983; Pingree & Mardell 1981; Pingree *et al.* 1982). Comparison with infrared imagery reveals the relative positions of thermal and colour fronts, or where productive patches occur in the absence of a surface thermal signature. It is worth noting that in the work so far reported, satellite observations have been used to extend the usefulness of, but not to replace, shipborne measurements. Indeed, in process-oriented studies designed to supply data for ecosystem modellers (Esaias 1981) the satellite image can identify the location of the most active areas of productivity where physical and biological parameters require detailed ship-borne study for a period of days to weeks (e.g. during the time of an algal bloom) whereas during the exercise the wider area in which the particular process is set can be regularly surveyed by the satellite. Holligan *et al.* (1983) demonstrate the effectiveness of combined ship and satellite surveying of a dinoflagellate bloom in the English Channel.

Besides the biological science that is developing, satellite ocean-colour data also promise advances in dynamical oceanography. Treating the plankton or suspended sediment as a passive tracer, the satellite images reveal patterns related to residual flows, eddies, frontal instabilities or mixing processes. Many such features may have a colour signature even when they do not appear on thermal infrared imagery. Attempts to obtain dynamical information from colour images have mostly been qualitative, e.g. using czcs images to display sediment movement in headland eddies (Mardell & Pingree 1981) or describing sediment movement from Landsat images of the Louisiana Bight (Rouse & Coleman 1976), the Bay of Fundy (Amos & Alfoldi 1979), the Solent (Srisaengthong 1982) or the Baltic (Horstmann & Hardtke 1981). Others have attempted to map small-scale features in estuaries by using Landsat data, with a view to pollution-control applications (Klemas 1980). However, Gower *et al.* (1980) sought quantitative information from a Landsat image of the chlorophyll–sediment signature of mesoscale eddies in the North Atlantic. By spectral analysis of the image they obtained an estimate of the wave-number of spatial variability.

This last example represents the innovative approach required if the potential of the spatial coverage of satellite data is to be fully realized. In general, dynamical theories have been constructed with time as the principal independent variable. We tend to study temporal evolution and variability because oceanographers have become good at measuring time series cf dynamical variables by using conventional ship-and-buoy sampling. Now that we can sample equally well in the spatial domain, using satellites, there is every reason to develop more theoretical approaches to understanding spatial variations. Hitherto the testing of models of spatial variability has largely been limited to advective systems in which the length scales could be inferred from a point-measured time series, but the satellite scanner has removed that limitation. The further attraction of obtaining the length scales of dynamical processes from images is that the radiometric–ocean-parameter calibration is no longer so important. Provided that spatial variability of atmospheric effects is removed, accurate length scales can be obtained even when there is only a crude calibration algorithm available, in just the same way as frequency peaks can be readily obtained from a time series whose amplitude calibration is suspect. This field of study awaits development, but modern image-processing computers and software are already available for this type of data analysis.

One other application of satellite colour, or visible wavelength, data to dynamical oceanography is the estimation of the radiant shortwave heat flux into the upper ocean. Gautier (1981) has calculated this by using the visible channel on the geostationary GOES meteorological satellite. A development of this may be to use spectral information from the czcs to estimate the depth of energy penetration into the sea, and thence to model more accurately the thermal structure of the upper ocean and its horizontal variability (Simpson & Dickey 1981; Dickey & Simpson 1983).

## 5. CONCLUSIONS

There is no doubt that we now have sufficient understanding of atmospheric scattering and absorption of visible light to be able to determine an accurate ocean-colour signature from space, given a sufficiently sensitive radiometer measuring in narrow bandwidths covering the visible spectrum, and preferably into the near infrared. The czcs has demonstrated the validity of the technique, and its successors should show further improvements in accuracy. Experience with the czcs also suggests that chlorophyll concentration can be measured by satellite to within a factor of two in case 1 waters, although there is a need to test this further with data from a greater variety of locations, whereas waters containing sediment and yellow substance not related to the local plankton population require much more investigation before a universal calibration algorithm can be considered. Looking to the future, it may be possible to adapt aircraft-borne solid-state narrow-band spectrometers for satellite use (Gower & Borstad 1981). By measuring the fluorescence lines these detect the chlorophyll directly, irrespective of the presence of other material.

The application of ocean-colour data is much less developed than the sensor and calibration technologies, which appear to be pushing forward independently of the scientific demand. There is a need for oceanographers to develop a better theoretical framework in which to make use of the spatial data from satellites. The most immediate requirement is for the colour imagery of the oceans to be surveyed and the features revealed on them to be catalogued to stimulate the creation of new hypotheses concerning such spatial problems as the patchiness of productivity and its relation to physical conditions, or the use of chlorophyll and sediment as tracers to reveal the length scales of a variety of dynamical processes. Ocean-colour measurements from space are not to be regarded as a substitute for ship observations, but should be seen as a very different type of data, requiring a different framework for analysis, and with the potential to promote fresh scientific insights in oceanography.

Some of the material for this paper was researched during a study tour of North America funded by the Nato Marine Sciences Panel, and during a visit to the Department of Oceanography, University of British Columbia, supported by the National Sciences and Engineering Research Council of Canada.

## REFERENCES

Alfoldi, T. T. & Munday, J. C. 1978 *Can. J. remote Sensing* **4**, 108–126.
Amos, C. L. & Alfoldi, T. T. 1979 *J. Sedim. Petr.* **49**, 159–174.
Anderson, F. P., Shannon, L. V., Mostert, S. A., Walters, N. M. & Malan, O. G. 1981 In *Oceanography from space* (ed. J. F. R. Gower), pp. 381–386. New York and London: Plenum Press.
Ångström, A. 1964 *Tellus* **16**, 64–75.

Austin, R. W.  1974  In *Optical aspects of oceanography* (ed. N. G. Jerlov & E. Steeman Nielson), pp. 317–334. London: Academic Press.

Austin, R. W.  1980  *Bound.-Layer Met.* **18**, 269–285.

Austin, R. W. & Petzold, T. J.  1981  In *Oceanography from space* (ed. J. F. R. Gower), pp. 239–256. New York and London: Plenum Press.

Caker, K. S. & Smith, R. C.  1982  *Limnol. Oceanogr.* **27**, 500–509.

Bukata, R. P., Brinton, J. E., Jerome, J. H., Jain, S. C. & Zurick, H. H.  1981  *Appl. Opt.* **20**, 1704–1714.

Bukata, R. P., Jerome, J. H., Brinton, J. E. & Jain, S. C.  1980  *Appl. Opt.* **19**, 2487–2488.

Blark, D. K.  1981  In *Oceanography from space* (ed. J. F. R. Gower), pp. 227–237. New York and London: Plenum Press.

Clark, D. K., Baker, E. T. & Strong, A. E.  1980  *Bound.-Layer Met.* **18**, 287–298.

Cox, C. S.  1974  In *Optical aspects of oceanography* (ed. N. G. Jerlov & E. Steeman Nielson), pp. 51–75. London: Academic Press.

Cracknell, A. P., MacFarlane, N., McMillan, K., Charlton, J. A., McManus, J. & Ulbricht, K. A.  1982  *Int. J. remote Sensing* **3**, 113–137.

De Luisi, J. J., Blifford, I. H. & Takamine, J. A.  1972  *J. geophys. Res.* **77**, 4529–4538.

Dickey, T. D. & Simpson, J. J.  1983  *Tellus.* (In the press.)

Esaias, W. E.  1981  *Oceanus* **24**, (3), 32–38.

Gautier, C.  1981  In *Oceanography from space* (ed. J. F. R. Gower), pp. 201–206. New York and London: Plenum Press.

Giannini, L.  1981  In *Oceanography from space* (ed. J. F. R. Gower), pp. 395–402. New York and London: Plenum Press.

Gordon, H. R.  1978  *Appl. Opt.* **17**, 1631–1636.

Gordon, H. R.  1981  In *Oceanography from space* (ed. J. F. R. Gower), pp. 257–265. New York and London: Plenum Press.

Gordon, H. R., Brown, O. B. & Jacobs, M. M.  1975  *Appl. Opt.* **14**, 417–427.

Gordon, H. R. & Clark, D. K.  1980  *Bound.-Layer Met.* **18**, 299–313.

Gordon, H. R. & Clark, D. K.  1981  *Appl. Opt.* **20**, 4175–4180.

Gordon, H. R., Clark, D. K., Brown, J. W., Brown, O. B. & Evans, R. H.  1982  *J. mar. Res.* **40**, 491–502.

Gordon, H. R., Clark, D. K., Mueller, J. L. & Hovis, W. A.  1980  *Science, Wash.* **210**, 63–66.

Gordon, H. R. & Morel, A. Y.  1981  In *Oceanography from space* (ed. J. F. R. Gower), pp. 207–212. New York and London: Plenum Press.

Gower, J. F. R. & Borstad, G.  1981  In *Oceanography from space* (ed. J. F. R. Gower), pp. 329–338. New York and London: Plenum Press.

Gower, J. F. R., Denman, K. L. & Holyer, R. J.  1980  *Nature, Lond.* **288**, 157–159.

Hojerslev, N. K.  1974  *Inherent and apparent optical properties of the Baltic* (*Rep. Inst. Phys. Oceanogr. Univ. Copenhagen*, no. 23). (70 pages.)

Hojerslev, N. K.  1980*a*  *On the origin of yellow substance in the marine environment* (*Rep. Inst. Phys. Oceanogr. Univ. Copenhagen*, no. 42). (16 pages.)

Hojerslev, N. K.  1980*b*  *Bound.-Layer Met.* **18**, 203–220.

Hojerslev, N. K.  1981*a*  In *Application of remote sensing data on the continental shelf* (ed. N. Longdon & G. Levy), pp. 73–76. Paris: European Space Agency, SP-167.

Hojerslev, N. K.  1981*b*  In *Oceanography from space* (ed. J. F. R. Gower), pp. 347–353. New York and London: Plenum Press.

Holligan, P. M., Viollier, M., Dupouy, C. & Aiken, J.  1983  *Continental Shelf Res.* (In the press.)

Horstmann, U. & Hardtke, P. G.  1981  In *Oceanography from space* (ed. J. F. R. Gower), pp. 429–438. New York and London: Plenum Press.

Hovis, W. A.  1981  In *Oceanography from space* (ed. J. F. R. Gower), pp. 213–225. New York and London: Plenum Press.

Hovis, W. A. *et al.*  1980  *Science, Wash.* **210**, 60–63.

Jerlov, N. G.  1976  *Marine optics.* Amsterdam: Elsevier.

Johnson, R. W.  1975  In *Remote sensing of Earth resources*, vol. 4 (ed. F. Shahrokhi), pp. 565–576. Tullahoma, Tennessee: University of Tennessee.

Kiefer, D. A., Olson, R. J. & Wilson, W. H.  1979  *Limnol. Oceanogr.* **24**, 664–672.

Klemas, V.  1980  *Int. J. remote Sensing* **1**, 11–28.

Klemas, V., Bartlett, D., Philpot, W. & Rogers, R.  1974  *Remote Sensing Envir.* **3**, 153–174.

Le Fèvre, J., Viollier, M., Le Corre, P., Dupouy, C. & Grall, J. R.  1982  *Estuar. coast. Shelf Sci.* **16**, 37–50.

MacFarlane, N.  1981  In *Matching remote sensing technologies and their applications*, pp. 499–508. Reading: Remote Sensing Society.

Mardell, G. T. & Pingree, R. D.  1981  *Oceanologica Acta* **4**, 63–68.

Morel, A. Y.  1980  *Bound.-Layer Met.* **18**, 177–201.

Morel, A. Y. & Bricaud, A.  1981  In *Oceanography from space* (ed. J. F. R. Gower), pp. 313–327. New York and London: Plenum Press.

Morel, A. Y. & Gordon, H. R.  1980  *Bound.-Layer Met.* **18**, 343–355.
Morel, A. Y. & Prieur, L.  1977  *Limnol. Oceanogr.* **22**, 709–722.
Munday, J. C. & Alfoldi, T. T.  1979  *Remote Sensing Envir.* **8**, 169–183.
Pingree, R. D. & Mardell, G.  1981  *Phil. Trans. R. Soc. Lond.* A **302**, 663–678.
Pingree, R. D., Mardell, G. T., Holligan, P. M., Griffiths, D. K. & Smithers, J.  1982  *Continental Shelf Res.* **1**, 99–116.
Platt, T. & Herman, A. W.  1982  *Int. J. remote Sensing* **4** (2). (In the press.)
Preisendorfer, R. W.  1961  In *Radiant energy in the sea* (*Int. Union Geophys. Geod. Monogr.*, no. 10), pp. 11–29.
Prieur, L. & Sathyendranath, S.  1981  *Limnol. Oceanogr.* **26**, 671–689.
Quenzel, H.  1970  *J. geophys. Res.* **75**, 2915–2921.
Robinson, I. S. & Srisaengthong, D.  1981  In *Application of remote sensing data on the continental shelf* (ed. N. Longdon & G. Levy), pp. 221–232. Paris: European Space Agency, SP-167.
Rouse, L. J. & Coleman, J. M.  1976  *Remote Sensing Envir.* **5**, 55–66.
Sathyendranath, S. & Morel, A. Y.  1983  In *Remote sensing applications in marine science and technology* (NATO ASI series) (ed. A. P. Cracknell), pp. 323–357. Dordrecht: D. Reidel.
Simpson, J. J. & Dickey, T. D.  1981  *J. phys. Oceanogr.* **11**, 309–323.
Singh, S. M., Cracknell, A. P. & Charlton, J. A.  1983  *Int. J. remote Sensing* **4** (In the press.)
Smith, R. C. & Baker, K. S.  1978*a*  *Limnol. Oceanogr.* **23**, 247–259.
Smith, R. C. & Baker, K. S.  1978*b*  *Limnol. Oceanogr.* **23**, 260–267.
Smith, R. C. & Baker, K. S.  1981  *Appl. Opt.* **20**, 177–184.
Smith, R. C. & Baker, K. S.  1982  *Mar. Biol.* **66**, 269–279.
Smith, R. C., Eppley, R. W. & Baker, K. S.  1982  *Mar. Biol.* **66**, 281–288.
Smith, R. C. & Wilson, W. H.  1981  In *Oceanography from space* (ed. J. F. R. Gower), pp. 281–294. New York and London: Plenum Press.
Sorensen, B. M. (ed.)  1981  *Recommendations of the 2nd International Workshop on atmospheric correction of satellite observation of sea water colour.* (49 pages.) Ispra, Italy: Commission of the European Communities Joint Research Centre.
Srisaengthong, D.  1982  Ph.D. thesis, University of Southampton.
Sturm, B.  1981*a*  In *Remote sensing in meteorology, oceanography and hydrology* (ed. A. P. Cracknell), pp. 163–197. Chichester: Ellis Horwood.
Sturm, B.  1981*b*  In *Oceanography from space* (ed. J. F. R. Gower), pp. 267–279. New York and London: Plenum Press.
Sturm, B.  1983  In *Remote sensing applications in marine science and technology* (NATO ASI series) (ed. A. P. Cracknell), pp. 137–167. Dordrecht: D. Reidel.
Viollier, M., Tanre, D. & Deschamps, P. Y.  1980  *Bound.-Layer Met.* **18**, 247–267.
Yentsch, C. S.  1983  In *Remote sensing applications in marine science and technology* (NATO ASI series) (ed. A. P. Cracknell), pp. 263–297. Dordrecht: D. Reidel.
Yentsch, C. S. & Garfield, N.  1981  In *Oceanography from space* (ed. J. F. R. Gower), pp. 303–312. New York and London: Plenum Press.

## *Discussion*

E. G. Mitchelson (*Department of Physical Oceanography, Marine Science Laboratories, U.C.N.W., Menai Bridge, U.K.*). Dr Robinson has emphasized the severe difficulties of making a satisfactory atmospheric correction in the presence of high levels of suspended solids, which can invalidate the assumption that the water-leaving radiance at 670 nm is zero. In the circumstances, it seems prudent to try to establish whether or not there is a useful colour signature leaving the sea surface when seston concentrations are high.

In collaboration with N. J. Jacob and J. H. Simpson, I have attempted to do this by making a series of observations in the case 2 waters of the northern Irish Sea, with a multichannel irradiance meter. This instrument has the same channels as the czcs, with a bandwidth of $\pm 10$ nm and sufficient sensitivity to permit the measurement of upwelling irradiance even under conditions of low surface illumination. Parallel measurements of total pigments, total seston (organic and inorganic fractions), and yellow substance were made by using standard methods (Strickland & Parsons 1968; Kalle 1963).

Figure D1 shows a comparison of the blue:green ratio of upwelling light just below the sea surface with the values of total pigment concentration, which extend over one decade. These observations were made in the presence of inorganic seston concentrations of up to 8 mg $l^{-1}$ (see figure D2). While the regression analysis shows that the blue:green ratio is strongly dependent on total pigments $C$ ($r = 0.92$, 37 d.f.), including total seston as an additional variable has no significant effect. The negative slope of the regression line, $-3.9$, is somewhat greater than those determined by other workers in case 1 waters (Smith & Baker 1982; More 1980; Gordon & Clark 1980).

Although the colour of the upwelling light is not markedly dependent on the seston concentration, the intensity is. Figure D2 shows a plot of the reflectance at 550 nm against the inorganic fraction of the total seston. There is considerable scatter, but a significant relation between the variables can be detected ($r = 0.77$, 37 d.f.).

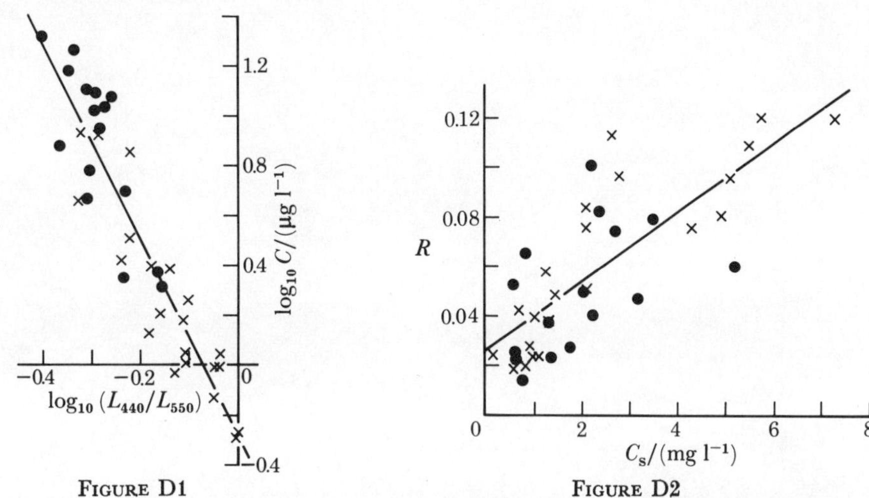


FIGURE D1                FIGURE D2

FIGURE D1. Total pigments, $C$, plotted against the blue:green ratio, $L_{440}/L_{550}$. The line represents the result of a regression analysis, which yields $\log_{10} C = 3.9 \log_{10}(L_{440}/L_{550}) - 0.29$. The data points are from two cruises in the Irish Sea. ×, 7–9 July 1982; ●, 12–14 July 1982.

FIGURE D2. Relative reflectance, $R$, at 550 nm plotted against concentration of inorganic seston, $C_{\mathrm{s}}$. $R$ is the ratio of upwelling radiation reaching the surface normalized by the above-surface downwelling irradiance measured by an uncalibrated broadband sensor. The regression line is $R = 0.014\,C_{\mathrm{s}} + 0.026$. Data points from two different cruises are identified as in figure 1.

These results, admittedly based on a small sample, show that there is useful information leaving the sea surface in the upwelling irradiance signal, even in case 2 waters dominated by suspended sediment. As well as confirming the above relations we would hope that further radiometric measurements at the sea surface will provide a basis for the testing of the atmospheric algorithm, which now appears as a critical problem in the development of this aspect of remote sensing.

### References

Gordon, H. R. & Clark, D. K. 1980 *Bound.-Layer Met.* **18**, 299–315.
Kalle, K. 1963 *Dt. hydrogr. Z.* **16**, 153–166. [In German.]
Morel, A. 1980 *Bound.-Layer Met.* **18**, 177–201.
Smith, R. C. & Baker, K. S. 1982 *Mar. Biol.* **66**, 269–279.
Strickland, J. D. H. & Parson, T. R. 1968 *Bull. Fish. Res. Bd Can.*, no. 167. (31 pages.)

I. S. ROBINSON. Because most of the reflectance measurements reported in the literature are for case 1 waters, these results reported by Miss Mitchelson for case 2 waters are of particular interest. They lend support to the hypothesis that in case 2 waters a chlorophyll algorithm can be constructed on the blue:green spectral ratio of reflectance while inorganic suspended solids can be derived from the broad band magnitude of reflectance.

However, these observations must be treated as location-specific and not necessarily applicable to other sea areas. It will be interesting to see if these results can be repeated in other locations, particularly coastal zones where the inorganic suspended sediment concentration is much greater than the 8 mg l$^{-1}$ encountered here, or in areas where more highly coloured sediment is found. It is observations at sea such as those reported here that will open up the archived czcs data for the British continental shelf seas to quantitative interpretation.

*Phil. Trans. R. Soc. Lond.* A **309**, 433–445 (1983)

*Printed in Great Britain*



# Application of microwave remote sensing to studies of sea ice

By P. E. Gudmandsen

*Electromagnetics Institute, Technical University of Denmark 348,*
*DK-2800 Lyngby, Denmark*

[Plate 1]

Monitoring of snow and ice is of importance for meteorological and climate research and applications, for hydrological purposes and for navigation and offshore activity in polar regions. For some of these applications long-term monitoring on a mesoscale and a synoptic scale is sufficient, whereas other applications require short-term observation on a mesoscale. This applies especially to forecasting of sea ice conditions, for instance. In the latter cases microwave remote sensing is the only technique that may deliver reliable and timely data irrespective of light, weather and cloud conditions. In the polar regions, this feature is of utmost importance.

All known microwave remote-sensing techniques have demonstrated their applicability in polar regions, in particular in connection with observations of sea ice. It has also been shown that a combination of simultaneously acquired data from different sensors may be of advantage in parameter retrieval.

This paper reviews the monitoring requirements and the microwave techniques available for this purpose with a view to snow and sea ice research and applications.

## Introduction

Remote sensing from satellite is particularly useful in the polar regions which to a great extent are unexplored and inaccessible owing to severe environmental conditions. One important object for remote sensing is sea ice. In fact, vast areas of the world's oceans – approximately 10 % in the Northern Hemisphere and 13 % in the Southern Hemisphere – are covered by floating ice that is subject to large seasonal and annual variations in extent and composition. With the increased interest in the polar and sub-polar regions, largely driven by the exploration and exploitation of Earth resources on land and offshore, remote sensing becomes an important tool for surveillance and monitoring with a view to navigation and the safety of the operations.

Also it is understood that the polar regions are important parts of the Earth's heat engine and that the polar climate may have a strong influence on the climate of the Earth as a whole. Any climatic variation, which may occur for one reason or another, may have a large impact on the polar conditions and therefore be more easily observable than in other regions of the Earth. The annual variation of the extent of the sea ice may be an indicator of a climatic trend, for instance.

With such applications in mind, a series of investigations are being undertaken or planned to contribute to the understanding of the environment and to study the phenomena and mechanisms controlling the environmental conditions as expressed by the term air–sea–ice interaction. Remote sensing from aircraft and satellite plays an important role in these studies, which at the same time contribute to the understanding of the interaction between electromagnetic waves and the Earth's surface, i.e. the fundamentals of remote sensing. Many of the

[ 191 ]

investigations are directed towards the development of methods of forecasting of the sea ice conditions, for instance to be based on satellite remote-sensing data as input to numerical models.

AIR–SEA–ICE INTERACTION

In the air–sea–ice system the sea ice plays an important role in a number of ways. Thus a sea ice cover insulates the ocean from the overlying generally much colder air. It reduces the heat flux from the ocean by a factor of about 100, which is of major importance for the air masses travelling over the ice. Also, owing to its high value of albedo, the ice cover insulates the ocean from solar heating, and being radiatively cool with respect to the atmospheric boundary layer it acts as a heat sink that cools the air and reduces the turbulence of the lower atmosphere. In addition, the ice is normally advected towards the open ocean where it melts and thereby cools the upper ocean. This is a major process for heat transport.

Conversely, the atmosphere has a strong influence on a sea ice canopy through the wind's creating two effects, motion of the ice and melting or forming of ice, depending on the relative temperature of the air masses involved. Also, the motion may create a convergence or divergence of the ice floes, thereby changing the ice concentration with a resulting change in the heat flux. In short, in the coupled system of air-sea interaction the sea ice cover will modify the weather system in an important way, one example being the suggested influence of the migration of storms along the edge of a sea ice cover.

In addition to the influence of sea ice on the cold water through melting, the ice floe motion creates a surface stress gradient that modifies the ocean mixed layer. It has been suggested that this modification is the cause of upwelling observed at the ice margin. Also, when the ice is being formed, the resulting brine will increase the salinity of the upper ocean and eventually create a transport of saline (and heavier) water to greater depths. Conversely, when melting, the sea ice is a supply of brackish water forming a surface layer that increases the stability of the ocean. Finally, an ocean current has an impact on an ice field through the resulting motion caused by subsurface drag. An important example is the East Greenland Current, which carries large quantities of sea ice from the Arctic Ocean to the southern tip of Greenland (60° N) influenced by the Coriolis force and to a greater extent by the weather system along its path.

Finally, it is observed that large ocean waves created by storms in the open sea adjacent to an ice field will break the ice canopy into floes, with an effect that diminishes as the waves are attenuated on their way into the ice field. Large waves may be felt at distances of 50 km from the ice margin and may create a divergence of floes into the open sea.

The above is an attempt of describing briefly the processes involved in the air–sea–ice interaction system. Obviously, there is a strong interaction between these processes, and the understanding of the effect may only be obtained through extensive measurements and observations in the field, associated with numerical modelling. An important example is the AIDJEX project (Arctic Ice Dynamics Joint Experiment), which was carried out through several years in the central Arctic Ocean, i.e. in the interior of a large ice field. This experiment demonstrated the capability of modelling the behaviour of the ice under these conditions (Untersteiner 1979). Another example is the NORSEX project (Norwegian Remote Sensing Experiment) carried out in the marginal ice zone north of Svalbard (NORSEX Group 1982). Both experiments are examples of undertakings involving remote sensing and in-situ measurements of atmospheric, oceanographic and ice processes. This will also be so with MIZEX (Marginal Ice Zone Experiment), a drifting experiment which will take place in the Fram Strait from about 80° N and southwards in the East Greenland Current (Johannessen *et al.* 1982). Numerical modelling will be applied to this situation of an open ice edge to arrive at a better understanding of the processes. During a conference at the Nato Advanced Study Institute, Maratea, Italy, in September 1981 the subject of air–sea–ice interaction was discussed in detail by a number of specialists (Untersteiner 1983).

Apart from the scientific interest in understanding the air–sea–ice interaction system there are practical applications to be derived from these experiments (and MIZEX in particular), i.e. the development of methods of forecasting the ice conditions with a view to navigation and offshore activity. In this context, the numerical model(s) will be modified to a format that they may be used solely with data that are relatively easily obtainable, i.e. synoptic weather and remote-sensing data. The subsequent sections in this paper will review the sea ice and ocean parameters that may be derived from satellite remote sensing and the remote-sensing techniques involved.

SEA ICE AND OCEAN PARAMETERS

Remote sensing may in most cases be characterized as observation by means of electromagnetic sensors of surface features, and the interpretation of remote sensing data relies to a great extent upon the differences in form and dielectric properties of the surfaces observed. In the context of sea ice and ocean observations, it is only with multi-year ice that some penetration into the material takes place and thereby reveals special characteristics. These facts are of importance for studies of the air–sea–ice system and for numerical modelling, where for instance some ocean parameters or phenomena will have to be derived from surface features associated with the phenomena.

Lists of pertinent parameters derivable from remote sensing are given in tables 1 and 2 for sea ice and ocean, respectively. These tables were generated during a workshop held in Copenhagen with the purpose of defining future satellite missions with microwave radiometers (Gudmandsen *et al.* 1983). For each parameter the tables state the observation requirements and the satellite instrument(s) likely to supply the parameter in question. The instruments are given in the order of likely importance by letter codes as follows: RA, radar altimeter; SAR, synthetic aperture radar; SCAT, scatterometer; PMR, passive microwave radiometer; VIR, visual and infrared radiometer. The observation requirements for accuracy and for resolution in space and time are based on considerations of their usefulness in basic research, climate research and operational or commercial applications. The frequency of data applications is given in terms of a code (from I to III) as stated. The figures quoted are arrived at on the basis of present knowledge and demonstrated capabilities, and on considerations of the scale of the geophysical variations in space and time. However, very little experience has been obtained with operational application of data and with numerical modelling for forecasting. Future studies in this direction may introduce modifications. In this context, it is worth mentioning that the requirements for time resolution can in most cases only be met with two or more properly phased satellites in orbit. With a near-polar orbiting satellite a frequent coverage is obtained at high latitudes (where sea ice observation is of importance) but with intervals of many days. Also, the frequency of observation is closely dependent upon the swath width of the instrument in question. One extreme is the radar altimeter, which by nature is a profiling instrument, and another is the passive microwave radiometer, which may have a swath width

[ 193 ]

TABLE 1. SEA ICE OBSERVATION REQUIREMENTS

| parameter | category† basic | climate | opn/ comm. | type of observation | accuracy desired | min. | resolution space desired | min. | time desired | min. | code‡ | proposed instrument |
|---|---|---|---|---|---|---|---|---|---|---|---|---|
| boundary | I | I | — | line position | 5 km | 20 km | 5 km | 20 km | 1 d | 3 d | A | PMR |
|  | — | — | II | line position | 500 m | 5 km | 500 m | 2 km | 1 d | 1 d | B | SAR, RA |
| concentration | I | I | — | percentage area | 2% | 5% | 25 km | 25 km | 1 d | 3 d | A | PMR, SCAT |
|  | — | — | I | percentage area | 10% | 20% | 5 km | 25 km | 1 d | 3 d | A | PMR |
|  | — | — | ᾽II | percentage area | 2% | 5% | 1 km | 10 km | 1 d | 3 d | B | PMR, SAR |
| albedo | I | I | — | area average | 0.02 | 0.04 | 25 km | 100 km | 3 d | 2 weeks | A | VIR |
| motion | II | — | — | point displ. | 100 m d⁻¹ | 1 km d⁻¹ | 5 km | 100 km | 1 d | 7 d | A | SAR |
|  | — | — | II | point displ. | 50 m d⁻¹ | 1 km d⁻¹ | 1 km | 10 km | 6 h | 7 d | B | SAR |
| ridging |  |  |  |  |  |  |  |  |  |  |  |  |
| density | — | — | II | number/area | 10% | 50% | 50 m | 100 m | 7 d | 1 month | B | SAR, SCAT |
| orientation | — | — | II | orientation | 10° | 30° | n.a. | n.a. | 7 d | 1 month | B | SAR |
| height | — | — | II | height | 1 m | 1 m | n.a. | n.a. | 1 d | 1 month | C | SAR, SCAT |
| ice type | I | I | II | frac./area, by type | 5% | 10% | 1 km | 25 km | 7 d | 1 month | A | PMR, SCAT |
| leads |  |  |  |  |  |  |  |  |  |  |  |  |
| fractional area | III | — | II | frac./area | 10% | 50% | 50 m | 100 m | 1 d | 3 d | B | SAR |
| orientation | III | — | II | orientation | 10° | 30° | n.a. | n.a. | 1 d | 3 d | B | SAR |
| floe position | — | — | II | point location | 20 m | 100 m | 100 m | 100 m | 6 h | 2 d | B | SAR |
| surface melting | — | II | — | frac./area | wet/dry | wet/dry | 25 km | 25 km | 1 d | 3 d | A | PMR |
| surface temperature | I | I | — | area average | 1 K | 3 K | 25 km | 100 km | 1 d | 3 d | A | PMR, VIR |
| ice thickness | III | III | — | area average | 20 cm | 1 m | 25 km | 100 km | 7 d | 1 month | D | inferred from ice type information |
|  | — | — | III | area average | 20 cm | 1 m | 50 m | 1 km | 1 d | 3 d | D | ice type information |

† Sampling key: I, continuous; II, frequent; III, occasional.
‡ Code: A, desired requirement can be met; B, substantial part of requirement can be met; C, measurement concept, capability not well determined; D, useful measurements, but limited.

TABLE 2. OCEAN OBSERVATION REQUIREMENTS AND SENSOR RECOMMENDATION

| parameter | type of observation | accuracy | | resolution | | | | code† | comments | proposed instrument |
| | | | | space | | time | | | | |
| | | desired | min. | desired | min. | desired | min. | | | |
| **wind speed** | | | | | | | | | | |
| global | area av. | 10% or $\pm 2$ m s⁻¹ | 10% or $\pm 2$ m s⁻¹ | 50 km | 100 km | 6 h | 72 h | A | | SCAT (PMR),‡ PMR, RA, SAR |
| regional (coast, lake, storm, front) | area av. | 10% or $\pm 2$ m s⁻¹ | 10% or $\pm 2$ m s⁻¹ | 10 km | 50 km | 6 h | 72 h | C | | |
| **wind direction** | | | | | | | | | | |
| global | area av. | — | $\pm 20°$ | 50 km | 100 km | 6 h | 72 h | A | — | SCAT (PMR) |
| regional | area av. | — | $\pm 20°$ | 10 km | 50 km | 6 h | 12 h | C | — | — |
| **sea surface temp.** | | | | | | | | | | |
| large-scale | area av. | 0.2 | 0.5–1.0 | 200 km | 300 km | 20–40 d | — | B | 2–6 GHz | PMR, VIR |
| mesoscale | area av. | 0.5 | 0.5–1.0 | 5 km | 10 km | 3–4 d | — | B | wind corr. | VIR |
| small-scale | area av. | 1.0 | 0.5–1.0 | $\leqslant 1$ km | 1–4 km | hours | — | C | — | VIR |
| **waves** | | | | | | | | | | |
| wind waves | area av. | $\leqslant 50$ cm | 1 m | 5–10 km | 100 km | 6 h | — | A | — | RA (PMR) |
| swell | area av. | $\leqslant 50$ cm | 1 m | 5–10 km | — | 6 h | — | A | — | RA (PMR) |
| internal | line | — | detection | 25 m | 50–100 m | <12 h | — | A | — | SAR |
| **spectrum** | | | | | | | | | | |
| direction | line | 20° | 20° | 25 m | 50–100 m | <12 h | — | A | — | SAR |
| dominant $\lambda$ | line | *ca.* 50 m | 50 m | 25 m | 50–100 m | 12 h | — | A | — | SAR |

† For codes, see table 1.
‡ PMR in parentheses implies that the radiometer data are used for atmospheric correction.

of 600 km and therefore may give a hemispherical coverage within 3 days, for instance. Thus operational applications for which processed and interpreted data for a certain area is available frequently and with short delay may not be feasible in the foreseeable future. However, by combining data from different sensors with different swath and perhaps frequency of operation a solution may be found. But this will require a detailed study of the feasibility and of the accuracy obtainable, particularly in forecasting.

One parameter, ice thickness, is important in ice dynamics and for navigation and offshore activity. However, there seems to be no possibility of obtaining this information from space except by a coarse estimate from the ice type classification (thin first-year ice, first-year ice, and multi-year ice) obtainable for instance by means of passive microwave radiometer data. There are in-situ methods of remote sensing available, and an airborne laser profilometer may determine the free board as a measure of the flow thickness.

Other parameters of importance for sea ice studies and forecasting are the surface air pressure and wind velocity. In fact the wind stress field over an ice field is a very important parameter – and for all numerical models of sea ice – which may only be derived from the surface atmospheric pressure field as deduced from data from meteorological stations and buoys.

Another parameter controlling the heat flux through sea ice is the snow cover, but so far there has been no method of estimation of this parameter from space. There may be some possibility of a rough estimate based on the passive microwave radiometer at high frequency (90 GHz and above) coupled to measurement of the air temperature, which approaches that of the snow.

### REMOTE SENSING INSTRUMENTATION

Tables 1 and 2 concentrate mainly on microwave intrumentation, and visual and infrared radiometers are only included when the relevant parameter may be obtained with the required accuracy solely by this instrumentation. Thus sea surface temperature for small-scale and mesoscale applications may only be obtained by the infrared radiometer with its relatively fine spatial resolution, which cannot be obtained by the passive microwave radiometer, the other candidate instrument; likewise the albedo of ice fields can only be determined by means of a visual sensor.

However, the frequent cloud cover over large areas of the polar regions, especially during summer seasons, prevents the acquisition of timely data by visual and infrared techniques, whereas microwave sensors, which are largely insensitive to clouds, will be able to meet the requirements of frequent measurements of the Earth's surface. On the other hand, the very wide swath width of the vir sensors gives a frequent coverage (at least once 1 day) so that this type of data at any rate may be considered complementary to microwave data.

Most of the sensors listed in tables 1 and 2 have been dealt with in other presentations of this Discussion Meeting. Therefore, in what follows I shall only address those problems that are specific for sea ice studies. One exception is the passive microwave radiometers, which will be dealt with in some more detail.

Most of the experience with the active microwave sensors, radar altimeter, scatterometer and sar has been obtained by means of Seasat data. Owing to the low inclination (orbit to 72°) and the period of life (June–October 1978) of this satellite sea ice observation in the Northern Hemisphere is limited to the southern Beaufort Sea and to the East Greenland Current. Also, owing to the limited life time, a series of planned group operations in support of the satellite mission were not carried out. Consequently a great deal of the experience gained is based on previous knowledge of the areas in question.

### Radar altimeter

The use of the radar altimeter in sea sea ice studies is based on the facts that backscatter from sea ice is much stronger than that from water and that the forms of the reflected signals are different. Owing to the roughness of a field of sea ice the return pulses rise slower but to a greater value than those from water and persist generally for a longer period. The altimeter data may therefore be used for detection of the ice–ocean boundary with an accuracy of 1–2 km at the satellite track because of its relatively narrow swath (M. Lefebvre, private communication). Thus the altimeter data may support those of the passive microwave radiometer, which give an accuracy of 10–15 km but over a very much wider area.

Also, because different ice types have different surface characteristics the altimeter data may be used for classification of sea ice based on the return signal characteristics. Again, it will be a profile information in support of the synoptic view obtained by the passive microwave radiometer – and the scatterometer.

The altimeter may prove useful for the observation of ocean waves near to the ice–ocean boundary (Tucker, this symposium) for use in the prediction of the break-up of an ice field, but this has still to be demonstrated. Unfortunately, the data are limited to a narrow area and the direction of the waves will have to be inferred from surface wind data.

### Scatterometer

The scatterometer acquires data which are a measure of the roughness of the surface within the resolution cell of the instrument expressed by the normalized backscatter cross section $\sigma^\circ$. In Seasat the resolution cells have a side of approximately 50 km and the angle of incidence varies between 25° and 65° over the swath of 750 km. The resolution is therefore comparable with that of the passive microwave radiometer, and a comparison of data from the two sensors has been made by Fedor (1982) on a Seasat track in the Beaufort Sea extending from Banks Island to Point Barrow with multi-year and first-year ice and water present. A good correlation between the brightness temperature and $\sigma^\circ$ was observed, but detailed variations require continued research.

Owing to the very different backscatter coefficient of calm water and ice the return signal from ice may be 10–15 dB higher, so that the scatterometer may be useful for ice boundary delineation (Peteherych 1982). Also, there seems to be a seasonal effect, with $\sigma^\circ$ increasing when melt water ponds freeze, so that information may be obtained of such features. Finally, one would expect that directional scatterometer data could be useful for ridging data in areas with rather regular large-scale ice dynamics.

This application for sea ice studies requires further detailed studies and so far no results are available about the ice fields around the Antarctic, which went through their maximum extent during the Seasat period.

Application of scatterometer data for the open ocean is described by Guymer (this symposium) but it should be pointed out that, because of the large resolution cell, wind data may not be useful near to the ice–ocean boundary but will certainly serve as a complement to the wind calculated from the air pressure field.

*Synthetic aperture radar*

A synthetic aperture radar gives a synoptic image of an ice field with very fine spatial resolution. In Seasat synthetic aperture radar the data may be processed to a resolution of 25 m; the swath width is 100 km with an angle of incidence of about 20°.

A great number of Seasat orbits over the Beaufort Sea has been recorded, but only a few of them are processed digitally. From the East Greenland Current only two orbits were recorded and processed optically (spatial resolution about 100 m).

From the processed images it may be concluded that the main morphological features of sea ice can be clearly discerned, such as leads, polynyas, pressure ridges, shore-fast ice, multi-year floes, mixtures the multi-year and first-year floes and shore leads (Campbell 1983). There is an indication that this identification of sea ice type is due to a difference in the speckles (R. W. Larson, private communication). A statistical analysis involving third and fourth moments seems to reveal this feature, which is of importance for automatic analysis of synthetic aperture radar images.

The images from the Beaufort Sea so far processed and analysed were apparently acquired during calm conditions. However, the two orbits covering the East Greenland current were recorded under different weather conditions (Søndergaard 1979): one under relative calm conditions and the other under strong wind and large waves in the open sea (Denmark Strait). From the optically processed images it is not possible to distinguish sea ice from waves, while the ice boundary may be easily delineated during the preceding (3 days) passage. (An airborne sea ice reconnaissance 6 days after the storm showed clearly the presence of sea ice.)

From seven sequential overlapping images it was ascertained that ice motion could be determined with an accuracy of $\pm 150$ m when using ground control points located in each end of an image 1000 km long. With control points in only one end, the accuracy was only $\pm 500$ m (Leberl *et al.* 1981), but by combining ground control points and satellite orbit data, a motion accuracy of the order of 50–100 m may be obtainable. For Landsat image processing a ground control point correlation technique has been developed so that two images may be registrated with a relative accuracy of 0.2–0.3 pixels (Hansen 1982). If this were also possible with radar images the 50 m accuracy would cetainly be feasible. However, with an angle of incidence of 20° or more, ground control points viewed from different aspect angles may be difficult to correlate, in particular in mountainous areas. At any rate, the accuracy attainable is much better than is possible today with drifting buoys and stations using present satellite navigation systems.

Satellite synthetic aperture radar images may also detect swell in an ice field as it has been observed by means of an airborne system (Overgaard 1980) and for satellites the interpretation is simpler owing to the high relative velocity, which results in an almost 'one-shot image'.

There is no doubt that spaceborne synthetic aperture radar will give a wealth of information useful for sea ice studies. It will provide unique data for detailed studies of ice models at different scales – site-specific, mesoscale and macroscale – under all weather conditions and during both day and night. However, considering the enormous amount of data – even after the initial compression of the radar data – some method of automatic (possible interactive) analysis will be needed if the data are going to be used even semi-operationally.

*Passive microwave radiometers*

A great deal of experience with passive microwave radiometer data for sea ice applications has been obtained from the electronically scanning microwave radiometer (ESMR) on Nimbus-5 (Wilheit 1972) and the scanning multichannel microwave radiometer (SMMR) system on Nimbus-7 (Gloersen & Barath 1977). ESMR operated on a single frequency, 19.35 GHz (1.55 cm wavelength), in a scan at right angles to the orbit, whereas SMMR operates on five frequencies, 6.6, 10.7, 18, 21 and 37 GHz, in a conical scan at vertical and horizontal polarization with an Earth incidence angle of about 50°.

Radiometers measure the ground's natural thermal emission in the microwave region, as expressed by Rayleigh–Jeans law. The intensity incident on a radiometer antenna is a measure of the brightness temperature of the area observed modified by the atmosphere, which causes an attenuation and an additional intensity contribution according to the atmosphere opacity and its temperature. Also, the radiated microwave intensity is weighted by the antenna pattern, the main lobe and the sidelobes. Furthermore, because an antenna creates a certain amount of cross-polarization and the emitted intensity in general is different at the two polarizations, a cross-polarized component will be added to the desired microwave signal. Thus although simple in principle – and also in equipment – microwave radiometer measurements become relatively complicated when corrections for the antenna pattern, the cross-polarization and the atmospheric effects have to be carried out to derive the brightness temperature from the raw data. This last effect is dependent on the content of atmospheric water vapour and liquid water, for which an estimate may be obtained for the SMMR, which has a channel tuned to near the water vapour absorption line (22.235 GHz).

Being a passive system, the field of view or 'footprint' is limited by diffraction, i.e. determined by the size of the antenna measured in wavelengths. For the ESMR the field of view is of the order of 35 km (actually varying over the swath), whereas the SMMR has a field of view ranging from about 30 to 150 km. Microwave radiometry may therefore only be useful for studying large-scale phenomena such as sea ice, for instance. This has been demonstrated by Gloersen *et al.* (1978), by Carsey (1982) and lately by Zwally *et al.* (1983).

These papers are based on data from the single-frequency radiometer, ESMR, on Nimbus-5 and rely upon the observations that the emissivity of first-year ice has a temperature near 240 K (only varying by a small percentage), multi-year ice near 220 K and open (calm) water 140 K, at the frequency of 19.35 GHz. Also, it is found (assumed) that the atmospheric influence in the Arctic region may be accounted for by a constant-brightness temperature contribution. Based on the work leading to the paper by Gloersen *et al.* (1978), animated cine films have been produced that clearly show the large annual variability of the sea ice in the Northern Hemisphere.

Carsey (1982) analysed the summer–autumn microwave variation of the Arctic pack ice in the period 1973–6 and, assuming that at the time of minimum ice extent (end of summer) there was no first-year ice, multi-year ice emissivity and ice concentration could be determined. It was found that the emissivity showed regional variations in accordance with knowledge of the ice composition in the Arctic Ocean. Zwally *et al.* (1983) analysed data for the same period, 1973–6, in the southern ocean and determined the ice extent through the period, assuming that only first-year ice was present; they found that the maximum ice area around the Antarctic decreased during the period by the order of 10%. Also, Cavalieri & Parkinson (1981) demonstrated the large-scale air–sea–ice interaction by relating the advance and retreat of the ice–ocean boundary to the mean surface air pressure.

Early results from the SMMR on Nimbus-7 and the parameter-retrieval algorithms developed were presented by Gloersen *et al.* (1981 *a, b*), but it is only after extensive further development of algorithms and calibration procedures that the results from the first SMMR observations could be presented (Gloersen *et al.* 1983). In addition to other global data (sea surface temperatures, near-surface winds, atmospheric water vapour, cloud liquid water and rainfall rates), results from the polar regions comprise sea ice concentration, multi-year ice fraction and radiating temperatures. The results are presented as polar maps and the distribution of first-year and multi-year sea ice seems to be in accord with historical experience and localized observations from underflights. It has been demonstrated (NORSEX Group 1982) that the ice–ocean boundary can be located to within 10 km, that temporal and spatial changes in ice concentration can be observed at a 25 km scale (Cavalieri *et al.* 1982 *b*), and that the spatial distribution of the multi-year ice fraction can be monitored (Cavalieri *et al.* 1982 *a*). A distinct patch of multi-year ice about 100 km in extent was tracked over a 24 day period as it drifted into the Fram Strait and the East Greenland Current from north of Svalbard (NORSEX Group 1982).

These results look very promising for large-scale sea ice studies as a contribution to the understanding of the air–sea–ice interaction process. However, further developments are necessary to improve the accuracy of parameter retrieval, in particular in the marginal ice zone. Here the temperature is near the melting point of the ice, so it is important to include temperature variations in the sea ice emissivities. If this is done (from ice buoy temperature data, for instance) the accuracy of the sea ice concentration may be better than 5 % (NORSEX Group 1982). Another improvement may result from including the snow cover.

The development of general parameter retrieval algorithms stems from the desire of being able to process satellite data routinely. It is to be expected, however, that by proper regional

---

### Description of Plate 1

Figure 1. Example of brightness temperatures determined by means of the passive microwave radiometer system (SMMR) on NIMBUS-7 at the frequencies 6.6, 10.7, 18, 21 and 37 GHz, vertical polarization. The figures represent a mosaic of data from seven orbits recorded on 13 March 1979, based on the so-called cell-tapes, i.e. digital brightness temperature data in standardized Earth-located cells. These images of Greenland consist of 128 × 128 picture elements covering an area of approximately 3000 km × 3000 km. Black indicates that data were not available on that day but gaps will be filled if data for the following recording day – 2 days later – were included.

Based on these data and those from the horizontal polarization an unsupervised classification has been made into five categories of clouds and sea ice, as shown in the lower right image. The ice classification is rather coarse distinguishing only between 'west ice', i.e. sea ice in Baffin Bay and Davis Strait, and 'east ice', i.e. ice in the East Greenland Current and the Greenland Sea. The west ice is predominantly first-year ice whereas the east ice is a mixture of first-year ice and multi-year ice, the latter transported by the East Greenland Current southwards through the Fram Strait (between Greenland and Svalbard, which is seen in the upper right corner). A great deal of the ice east of Greenland is first-year ice and the large advective fan extending into the Greenland Sea is a mixture of the two types of ice. It is also seen that it is apparently difficult to distinguish between light clouds and low-concentration sea ice, labelled drift ice, whereas the heavy clouds of a frontal system southwest of Greenland stand out clearly.

At all five frequencies a pattern of brightness temperatures is noted on the Greenland ice cap. This has been attributed to the snow pack on the ice with variations in snow crystal size, which is dependent upon accumulation and temperature. In fact a principal component analysis based on this multi-frequency data has shown patterns approximating existing contour maps of accumulation rate and annual mean temperature.

This figure is part of a study carried out at the Electromagnetics Institute, Technical University of Denmark, to explore the capabilities of using satellite microwave radiometer data for Arctic research and surveillance.

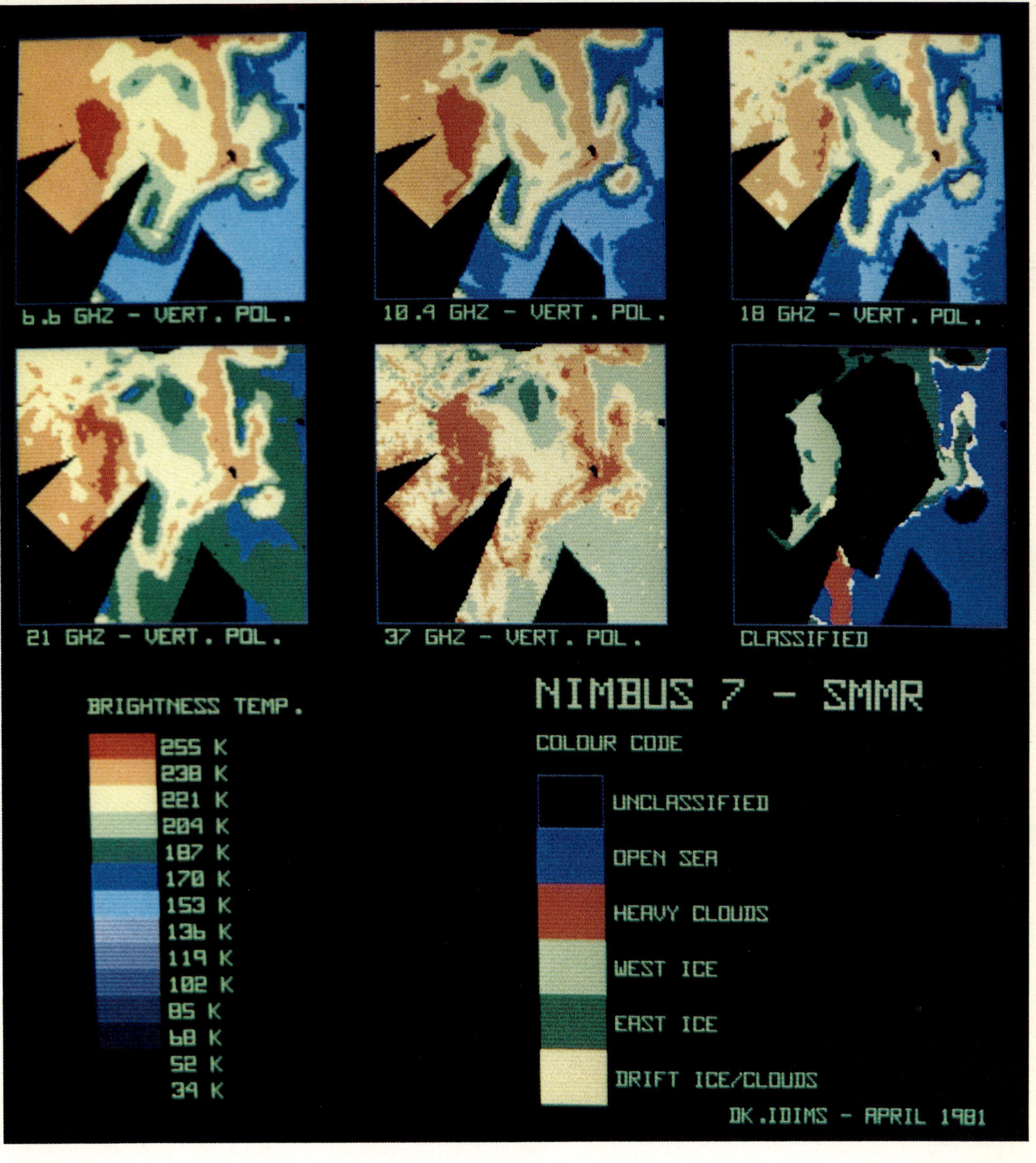


FIGURE 1. For description see opposite.

tuning of the algorithm better retrievals may be obtained. In fact the 5 % accuracy quoted above was obtained in this way. Also, statistical analysis procedures, as employed successfully with the multispectral Landsat data, may prove very useful, but an appreciable increase in computer time may rule this out except for regional studies. For ice studies, data from other satellites (active microwaves and visual–infrared) may prove useful for tuning of retrieval algorithms in an interactive process; also it may be necessary to account for seasonal and regional variations.

As outlined previously, data from the adjacent open ocean are also of interest in these studies. An analysis of the retrieval of sea surface temperature and near-surface winds from SMMR data shows that by temporal and spatial smoothing of the data ($\frac{1}{2}$ month and 200 km) a standard deviation of about 1.5 K (compared with climatological data, FGGE and WMO) and 2 m s$^{-1}$ results. However, data are only reliable at distances of 600 km from coastlines owing to antenna sidelobe interference, and the usefulness of this information for studies of the marginal ice zone is subject to further research.

## FUTURE SATELLITE MISSIONS WITH MICROWAVES

The only microwave sensor instrument in operation today is the scanning multichannel Microwave radiometer on Nimbus-7, which was launched in November 1978. It has by far exceeded its projected lifetime and there seems to be no important sign of fatigue. Global data have been acquired since the launch and a huge data base is awaiting detailed geophysical interpretation, which certainly will take place when well calibrated and corrected data become available. The data are incorporated in the plans for MIZEX, where there will be great possibility for conferring with ground control and underflight measurements (Johannessen *et al.* 1982).

In 1984 a satellite of the U.S. Air Force Meteorological Satellite Program will be launched with a microwave radiometer system under the name of special sensor microwave/imager (SSM/I). It is a seven-channel, four-frequency conically scanning radiometer with two channels (dual polarization) at 19, 37 and 85 GHz and a single channel at 22 GHz. The fields of view of this instrument will be comparable with those of the SMMR, and it is hoped that the data may be available to non-military agencies so that the success of the SMMR may be continued. In particular, the high frequency of 85 GHz may prove very useful because of the better spatial resolution, but also because there is hope that it may supply data on snow cover.

A follow-on to Seasat will be seen first with the launch of the European Remote Sensing satellite, ERS-1, in 1987. With the promising results with Seasat data there are good reasons to expect that the active microwave instrument (synthetic aperture radar and wind scatterometer) and the radar altimeter will supply data of great value to ocean and ice studies as concluded by the EARSeL/E.S.A. Workshop on microwave ocean and ice. However, in view of the great amount of data to be acquired from this satellite, great emphasis will have to be placed on data processing.

## CONCLUSIONS

In this brief review of microwave remote sensing in ice studies, emphasis has been placed on satellite observations. However, the work reported was based on a number of extensive investigations with equivalent aircraft instruments supported by ground control measurements. From the experience gained in this way in many areas of the polar regions and under various conditions it has proved possible to work out processing algorithms to make a useful parameter retrieval for sea ice research, for instance. A great deal of knowledge about interpretation of microwave information and images was likewise obtained.

To obtain more experience in this direction and better retrieval accuracy, more field investigations are needed, however, and a large-scale programme like MIZEX may prove very important in this respect. It is a programme involving many research disciplines, which will require exchange of data between scientists to achieve their goals; in this connection it is interesting to note the different attitudes to remote sensing by the people concerned. The geophysicist tends to look to remote sensing as a 'research tool' whereas the remote-sensing scientist considers his discipline as still being in a research and development phase.

Nevertheless there is already one operational application of satellite microwave remote-sensing data, previously from ESMR and now from SMMR. The data are used in almost real time for sea ice mapping and forecasting in both polar regions, taking advantage of the fact that they are insensitive to clouds and light conditions.

The European remote sensing satellite, ERS-1, has been announced as a semi-operational satellite, but judging from the experience gained by several groups in reaching the present level of understanding and utility of microwave remote-sensing data it can be stated that a great amount of interest and work is needed to reach this goal. It is not just a question of establishing a system for processing and quick dissemination of data: it is an equally large task to develop the parameter retrieval algorithms and the necessary processing facilities with sufficient throughput. In this context it is worth noting that whereas most investigations have been concerned with only one type of instrument, there seems to be a tendency to explore the symbiosis of several instruments.

## REFERENCES

Campbell, W. J. 1983 Presented at the EARSeL/E.S.A. Workshop 'Microwave ocean and ice' Igls, April 1982 (ed. W. J. Campbell & P. Gudmandsen). (In preparation.)

Carsey, F. D. 1982 *J. geophys. Res.* **87**, 5809–5835.

Cavalieri, D. J. & Parkinson, C. L. 1981 *Mon. Weath. Rev.* **109**, 2323–2336.

Cavalieri, D. J., Gloersen, P. & Campbell, W. J. 1982a (Submitted for publication.)

Cavalieri, D. J., Martin, S. & Gloersen, P. 1982b (Submitted for publication.)

Fedor, L. 1982 Presented at the EARSeL/E.S.A. Workshop 'Microwave ocean and ice', Igls, April 1982 (ed. W. J. Campbell & P. Gudmandsen). (In preparation.)

Gloersen, P. & Barath, F. 1977 *IEEE J. ocean. Engng* **OE-2**, 172–178.

Gloersen, P., Campbell, W. J. and Cavalieri, D. 1981a In *Oceanography from space* (ed. J. R. F. Gower), pp. 777–783. New York: Plenum Press.

Gloersen, P., Cavalieri, D. & Campbell, W. J. 1981b In *Oceanography from space* (ed. J. R. F. Gower), pp. 823 829. New York: Plenum Press.

Gloersen, P., Cavalier, D. J., Chang, A. T. C., Wilheit, T. T., Campbell, W. J., Johannessen, O. M., Künzi, K. F., Ross, D. B., Staelin, D., Windsor, E. P. L., Barath, F. T., Gudmandsen, P., Langham, E. & Ramseier, R. O. 1983 (Submitted for publication.)

Gloersen, P., Zwally, H. J., Chang, A. T. C., Hall, D. K., Campbell, W. J. & Ramseier, R. O. 1978 *Bound–layer Met.* **13**, 339–359.

Gudmandsen, P., Skou, N. & Wolff, B. (eds) 1983 *Microwave radiometer requirements, a workshop report.* Electromagnetics Institute, Technical University of Denmark. (In preparation.)

Hansen, P. 1982 *Image-to-image registration on a subpixel level using cross correlation.* (49 pages.) Report no. R264, Electromagnetics Institute, Technical University of Denmark.

Johannessen, O. M., Campbell, W. J., Dunbar, M., Dyer, I., Hasselmann, K., Hibler, W. D. III & Wadhams, P. (eds). 1982 MIZEX. A plan for a summer Marginal Ice Zone Experiment in the Fram Strait/Greenland Sea. Unpublished manuscript.

Leberl, F., Roggam, J., Elachi, C. & Campbell, W. J. 1981 (Submitted for publication.)

NORSEX Group, NORSEX 1 1982 *The Norwegian Remote Sensing Experiment in the marginal ice zone.* Geophysical Institute, University of Bergen. (Submitted for publication.)

Overgaard, S. 1980 *Multiyear sea ice investigation in the East Greenland Current.* (69 pages.) Report no. R228, Electromagnetics Institute, Technical University of Denmark.

Peteherych, S. 1982 Presented at the EARSeL/E.S.A. Workshop 'Microwave ocean and ice', Igls, April 1982 (ed. W. J. Campbell & P. Gudmandsen). (In preparation.)

Søndergaard, F. 1979 In *Proceedings of the workshop on SEASAT/SAR Processor* (ESA SP-154), pp. 69–75.

Untersteiner, N. 1979 *Polar Rec.* **19**, 363–367.

Untersteiner, N. (ed.) 1983 *Air–sea–ice interaction.* NATO Advanced Study Institute. (In preparation.)

Wilheit, T. T. 1972 In *NIMBUS-5 user's guide* (ed. R. R. Sabatini), pp. 59–105. Greenbelt, Maryland: N.A.S.A. Goddard Space Flight Center.

Zwally, H. J., Comiso, J. E., Parkinson, C., Carsey, C. L., Campbell, W. J. & Gloersen, P. 1983 *Antarctic sea ice cover 1973–1976 from satellite passive microwave observations.* Washington, D.C.: N.A.S.A. (Submitted for publication.)

*Phil. Trans. R. Soc. Lond.* A **309**, 447–461 (1983)
*Printed in Great Britain*



# Satellite observations of polar ice fields

By G. de Q. Robin, D. J. Drewry and V. A. Squire
*Scott Polar Research Institute, University of Cambridge, Lensfield Road,
Cambridge CB2 1ER, U.K.*

This paper outlines the research needed to advance the knowledge of polar ice masses that could be or has been provided by satellite techniques. Inland ice sheets, ice shelves and pack ice on polar oceans are discussed in separate sections. Because a satellite radar altimeter can cover wide areas of polar ice economically and rapidly, the interpretation of altimeter results is discussed in detail. Theoretical considerations are used to explain various characteristics of pulse shapes recorded over Antarctica by Seasat. Over inland ice, surface undulations cause the pulse shape to change rapidly with location, and the leading edge to migrate rapidly. Altimetry over ice shelves should be accurate to better than $\pm 1$ m and provide new methods for studying their mass balance. Nearly all sea ice returns that have been studied show a characteristic glistening quite distinct from returns from the open ocean, inland ice or ice shelves. Reasons for glistening are presented and the need for ground truth stressed because pulse shapes should provide an additional source of data to define pack ice characteristics from satellite observations.

## 1. Introduction

Ice sheets in Greenland and Antarctica constitute 10 % of the surface land area of the Earth. Their accumulated $36 \times 10^6$ km$^3$ of ice comprises 90 % of global freshwater reserves. Sea ice covers a maximum global surface area of $24 \times 10^6$ km$^2$ during the northern summer, *ca.* 7 % of the world ocean surface. These ice masses are coupled closely with other aspects of the natural environment through complex exchanges of energy and mass (Radok 1978; Sergin 1979). An understanding of their dynamics and thermodynamics is fundamental to realistic global modelling of climatic, oceanographic and lithospheric processes. The investigation of ice masses, principally in polar areas, is hampered by extreme cold, long periods of darkness and persistent cloud cover in some areas. Remote sensing has and will play a critical role in the gathering of data and their routine surveillance.

This paper briefly reviews the most important research needs for the study of polar ice and points out where research can benefit from satellite techniques. Table 1 lists these needs for ice sheets and ice shelves and the contributions that have been made by satellite missions or those being planned. Table 1 of Gudmandsen (this symposium) lists similar requirements for the study of sea ice. These cover two types of data sets; one acquired within a time domain to study time-dependent changes such as sea-ice distribution and glacier dimensions, the other within a spatial domain where very detailed geographic coverage is required. Because of their importance to both types of study, techniques of radio altimetry from satellites are discussed in some detail.

## 2. Ice sheets

### (a) Surface form and ice dynamics

It is not the continued existence of major ice sheets in remote polar regions that is of importance to mankind, but rather the possible global environmental effects of changes in these ice sheets on sea level, ocean circulation and climate. A first requirement in studying polar ice masses is

[ 205 ]

determination of significant changes. Landsat and other types of satellite imagery provide a rapid and accurate means of horizontal fixing and continued monitoring (to better than 100 m) of prominent surface features such as crevasse systems and the coastal boundaries of ice sheets. Imagery, whether at visual or infrared wavelengths or by radar methods, is not, however, satisfactory for mapping more critical parameters such as surface elevations or for providing data for glaciologists to assess ice sheet mass budgets.

Satellite altimetric methods, using radar altimetry with an accuracy of $\pm 1.0$ m or better and laser altimetry with an accuracy of $\pm 0.1$ m, have the potential to provide within one or two

TABLE 1. OBSERVATIONS REQUIRED FOR RESEARCH ON POLAR ICE SHEETS AND ICE SHELVES

| glaciological parameter | required resolution | satellite measurement | satellites flown | satellites planned |
|---|---|---|---|---|
| 1. surface morphology | | | | |
|   (i) large-scale | 1.0 m | radar, laser altimetry | Geos-C Seasat | radar altimeter on ERS-1, TOPEX, Poseidon |
| | | doppler retro-reflector ranging | Transit–Tranet | — retro system on ERS-1 |
|   (ii) small-scale | 0.5 m | laser altimetry, SAR MSS, TM, RBV | Seasat Landsat series | SAR on ERS-1, Radarsat, TOPEX, SPOT |
| 2. ice thickness | 5.0 m | — | — | future prospects limited |
| 3. temperature field | 1 K | inferred from microwave emissivity: ESMR, SMMR, LSR, VHRR, NEMS | Nimbus series, Tiros series, NOAA series; | radiometers (IMR) on ERS-1, MOS, Radarsat |
| 4 accumulation or ablation | | | | |
|   (i) surface | 0.1 m | (tentative relationship with microwave emissivity from ESMR) | Nimbus series NOAA series | ATSR/M on ERS-1 Radarsat |
|   (ii) bottom | 0.1 m | surface elevation changes of ice shelves | — | radar altimeter on ERS-1 |
| 5. velocity field | | | | |
|   (i) surface | 0.5 m a⁻¹ | doppler ranging retro-reflector ranging | Transit/Tranet ARGUS | retro system on ERS-1 radar altimeter in ERS-1, TOPEX, SPOT, Poseidon |
| | | altimeter backscatter matching: MSS, RBV | Landsat series | |
|   (ii) internal or bottom | 0.10 m a⁻¹ | | | future prospects limited |
| 6. dimensional fluctuations† | $h = 0.1\text{–}0.5$ m a⁻¹ | $h$ by radar + laser altimetry | Geos-C, Seasat | radar altimeter on ERS-1, SPOT, TOPEX, Poseidon |
| | $L = 10$ m a⁻¹ | $L$ by imaging MSS, TM, RBV, SAR | Landsat series, Seasat | SAR on Radarsat, TOPEX, SPOT, ERS-1 |
| 7. snowline position | 10 m | microwave emissivities: ESMR, SMMR, VHRR | Nimbus, Tiros, NOAA, Landsat | ATSR/M on ERS-1, MOS, Radarsat |

† $h$, surface elevation; $L$, margin position.

decades much more accurate and extensive information than present methods on changes in ice sheet volume. Once available, such information is simple to interpret and provides a straight-forward answer to the present effect of ice sheets on the global hydrological cycle.

Simple monitoring of ice sheets by satellites will not, however, advance knowledge of ice dynamics to the point where it is possible to forecast future changes in ice sheets that result from climatic changes. For this we need a better understanding of ice sheet dynamics.

Full description of ice sheet dynamics requires knowledge of the seven parameters listed in table 1, and a starting point is the relation between driving stress ($\tau$), ice thickness ($h$) and surface slope ($\alpha$):

$$\tau = \rho_i g h \sin\alpha, \tag{1}$$

where $\rho_i$ is ice density and $g$ is gravitational acceleration. Equation (1) is only effective when mean values of parameters over distances at least an order of magnitude greater than the ice thickness are used. Over short distances stress variations caused by ice flow over bedrock irregularities affects surface relief considerably. Satellite altimetry can determine values of $\alpha$ with greater accuracy, at much greater speed and at lower cost per unit area than by any other method (Robin 1966; Brooks *et al.* 1978). The other major variable, ice thickness, has been routinely measured for two decades by airborne radio echo-sounding techniques (Robin *et al.* 1969; Drewry 1981). The possibility of measuring ice thickness from satellites is not promising.

An early study by Nye (1951) treated ice as a plastic material with a constant yield stress ($\tau_0$). Then for an ice sheet on a roughly horizontal bedrock with low surface slope $\alpha = \mathrm{d}h/\mathrm{d}x$ ($x$-flow direction), integration of (1) gave

$$h = \{2\tau_0 (L - x)/\rho_i g\}^{\frac{1}{2}}, \tag{2}$$

where $L$ is ice-sheet half width. By 1958 this model did not fit new information on the form of the Greenland and Antarctic ice sheets. On the assumption that outflow was due to internal deformation in the absence of basal sliding, Vialov (1958) derived an improved expression for the surface profile of an ice sheet as

$$(h/H)^{2+2/n} + (x/L)^{1+1/n} = 1, \tag{3}$$

where $H$ is thickness at the ice sheet centre and $n$ is the viscoplastic parameter in ice flow law of Glen (1955), so with $n = 3$ we get

$$(h/H)^{2.7} + (x/L)^{1.3} = 1. \tag{4}$$

Weertman's (1957) equation for basal sliding velocity ($U_b$) in terms of basal shear stress $\tau_b$, and a roughness parameter $R$ is

$$U_b = C\tau_b^{\frac{1}{2}(n+1)}/R^{n+1}, \tag{5}$$

gave

$$(h/H)^{2.5} + x/L^{1.5} = 1. \tag{6}$$

According to Paterson (1981) both equations give a satisfactory fit to an observed profile for 800 km inland of Mirny on the Antarctic ice sheet.

Because $n \approx 3$, Weertman's theory suggests that the velocity due to sliding increases with the square of $\tau_b$, whereas the surface velocity due to internal deformation ($U_d$) used to derive (3) from the vertical shear rate $\dot\epsilon_{xz}$ is given by

$$U_d = \int_0^z h\dot\epsilon_{xz}\,\mathrm{d}z \tag{7}$$

gives

$$U_d/h \propto \tau_b^3.$$

Logarithmic plots of $\tau$ against $U_d/h$ for ice sheets in Budd & Jenssen (1975), Budd & Smith (1981) and Cooper *et al.* (1982) show a quasi straight-line relation with $n$ ranging from 2.5 to 4, whereas corresponding plots of sliding velocity against $\tau_b$ show much greater scatter than would be expected if (5) applied.

It therefore appears that internal deformation explains the motion of interior regions of ice sheets reasonably satisfactorily, but it is clear that in outer regions sliding is the dominant mechanism in fast-flowing glaciers and ice streams that discharge a large proportion of the ice.

[ 207 ]

Surface undulations, over distances of a few times the ice thickness, superimposed on the mean slope (equation (3)) have been explained in terms of varying longitudinal and shear stresses in the ice mass (Robin 1967; Collins 1968; Budd 1970). Theory is now at the point where detailed surface profiling from satellites may make useful estimates of ice velocities possible.

### (b) Velocity field

Surface velocities have been successfully measured by satellite systems by using Transit–Tranet doppler ranging. While the accuracy of this method is limited to 1 or 2 m., and is thus not suitable for short-term monitoring of slow-moving interior regions of Antarctica and Greenland, it has proved adequate for ice shelves where velocities greater than $0.1$ km a$^{-1}$ are usual (Thomas 1979). The technique is restricted in that a ground station has to be deployed on the ice on at least two occasions, and occupied for a minimum of 2–3 days for maximum accuracy. Further needs are information on the variation of velocity with depth, and basal sliding velocities. Both are as yet unobtainable by satellite.

### (c) Temperature field

The surface and internal temperature of ice sheets play important roles in ice dynamics. Ice creep rate is temperature-controlled. Whether or not basal sliding takes place is largely determined by the balance between geothermal and frictional heat and cold carried down from the upper surface by ice motion.

Approximate temperatures in the upper few metres of ice sheets may be estimated, given adequate calibration, by passive microwave radiometry. Measurements may be made at single or multiple frequencies. An initial application of satellite radiometry to ice used the ESMR (150 mm wavelength) on Nimbus-5 (Zwally & Gloersen 1977). A major problem in interpretation is the correlation of physical temperature with emissivities, which are also affected by crystal size or grain size due to scattering and absorption.

### (d) Accumulation rate

This is a basic input parameter to all studies of ice-sheet modelling. No effective airborne or satellite methods of measuring accumulation rates are known, although there does appear to be a link between microwave emissivities mentioned above and accumulation rates through a relation with crystal size and fabric of surface layers of cold ice sheets. Such information is largely qualitative in suggesting areas of very low accumulation rates in central Antarctica (Zwally & Gloersen 1977).

### (e) Dimensional changes

Satellite radar altimetry with an accuracy of $\pm 1.0$ m is adequate to map within several years the major changes in vertical height of ice sheets that may be associated with surging of a basin on an ice sheet, or slower changes of the order of $1.0$ m a$^{-1}$ over a decade, (suggested by Mae & Naruse (1978)). In central Antarctica changes of $0.01$ m a$^{-1}$, which represent a significant imbalance of around $35\%$ total accumulation, require a relative accuracy of elevation measurement of around $0.1$ m a$^{-1}$ over two decades for detection. Surface elevation changes at Camp Century in Greenland and Byrd Station, Antarctica, for example, may be $+0.03$ m a$^{-1}$ and $-0.03$ m a$^{-1}$ respectively (Budd et al. 1971) and should be detected by laser altimetry to $\pm 0.1$ m over one decade.

[ 208 ]

### 3. Ice shelves

Floating ice shelves exist around *ca.* 55 % of the periphery of the inland ice sheet of Antarctica (Drewry *et al.* 1982). Ice shelves are formed principally from inland ice discharged into the sea, and by the accumulation of snow on fast ice. The flow of ice shelves takes place by spreading and forward motion over a virtually frictionless base. The driving stress governing spreading is proportional to the weight of ice above sea level, $\bar{\rho}_1 gE$, where $E$ is surface elevation and $\bar{\rho}_1$ is mean ice density above sea level, or equally by the buoyancy upthrust on ice below sea level, $|(\rho_w - \rho_1) gh_w|$, where $\rho_1$ and $\rho_w$ are the mean densities of ice and sea water and $h_w$ is the ice draught below sea level. Ice thickness $h_1 = E + h_w$, and for hydrostatic balance

$$\int_0^{h_1} \rho_1 \, gh \, dh = \rho_w gh_w. \tag{8}$$

The driving stresses produce horizontal strain rates $(\dot{\epsilon}_x, \dot{\epsilon}_y)$ that may be partly balanced by accumulation rates on top and bottom surfaces – the latter by ice frozen from the sea, and by the flow of ice from inland. Ablation on Antarctic ice shelves is mainly bottom melting and also affects strain rates. Drag at side boundaries and due to local grounding has to be considered. Normally ice shelves thicken inland from the ice front.

Bottom melting and freezing takes place beneath ice shelves, but relevant observations are few and mainly apply near the ice front, where melting is expected to be high (Thomas 1973; Crary 1962; Swithinbank 1962; Zotikov *et al.* 1980; Jacobs *et al.* 1979). Indirect estimates of the presence or absence of bottom melting based on the nature of radio echo reflexions from the ice–water interface (Neal 1979; Robin *et al.* 1983) indicate wide variations in bottom melting–freezing with location.

We expect (8) to apply over most of the ice shelf, exceptions being areas close to ice shelf boundaries, to local grounding, and where extensive shear between ice streams of different velocities and thickness occurs. Even though satellites may not measure ice thickness directly, accurate measurements of surface elevation by satellite altimetry and use of (8) should make it possible to map ice shelf thickness. This should be based on comparison of extensive radio echo values of thickness $(h_1)$ of Antarctic ice shelves with surface elevations from satellite altimetry, because the variation of $\rho_1$, $\rho_w$ and $g$ with location is small. Measurement of ice shelf surface elevations from satellites to an accuracy of $\pm 1.0$ m should then make it possible to determine ice thickness accurate to $\pm 10$ m, a useful accuracy for ice shelf dynamic studies. Increasing the accuracy to $\pm 0.1$ m in elevation, as may be possible over level ice shelves, may give ice thicknesses accurate to $\pm 1.0$ m, an accuracy sufficient to determine significant changes of ice thickness from bottom melting at a specific geographical point in a few years, but this demands more rigorous knowledge of $\bar{\rho}_1$ and $\rho_w$. Satellite altimetry should thus open up a new method of studying the mass balance of ice shelves. A further example may be given. Unlike inland ice, where surface undulations caused by ice flowing over bedrock irregularities appear to be fixed in space, surface rolls and chasms on ice shelves are carried forward by the motion of the ice. Accurate fixing of these features by satellite imagery should enable measurement of ice velocity from satellites to be made.

The study of the calving of icebergs should benefit from repeated collection of MSS-type imagery over a number of years, and provide enough basic data for the development of adequate theory. Already lower-resolution imagery from Nimbus and other satellites has provided information on the calving of some major icebergs and produced evidence that collision of large icebergs with glacier tongues and ice shelves can be a relatively frequent cause of fracture and formation of icebergs (Swithinbank *et al.* 1977). Tracking of large icebergs on these images also shows the major oceanic circulation. More extensive information has been obtained by tracking transponders on smaller Antarctic icebergs (Tchernia 1974).

## 4. Sea ice

Sea ice is spatially and temporally one of the most variable natural materials found on the Earth's surface. Its character and morphology depend critically on its age, and on climatic and oceanographic factors prevalent during its growth and subsequent lifetime. Its growth is well documented in the polar literature. First, minute spherical ice crystals form at the sea surface, which quickly evolve into thin discoids of 1–3 mm diameter and 1–10 μm thickness. These rapidly spread laterally to form a suspension, which gives the sea a soupy appearance, called *frazil* ice. Unless oceanic conditions are remarkably calm, ocean waves and currents break up the frazil to form either *pancake* ice or a thick polycrystalline slurry of frazil platelets in concentrations between 20 and 40 % known as *grease* ice (Martin 1981). Once formed, sea ice growth proceeds rapidly under the control of prevailing atmospheric and oceanic conditions.

Young ice (less than 0.3 m thick) and first-year ice (0.3–2 m) are more saline than older sea ice and are often made up of three layers of different crystallographic structure. Beneath the snow cover but near to the upper surface, the ice is polycrystalline, with typical salinities between 5 and 15‰. Under this layer lies a region of vertical columnar crystal growth with significantly lower salinity (4–5‰). During growth the ice in contact with the underlying sea water forms a thin layer, of high salinity (30‰) and low mechanical strength, known as the skeletal layer.

Second-year and multiyear ice are very different from younger sea ice. Pressure ridges with sails up to 10 m high and keels down to 40 m depth can be formed during its growth, due to compressional and tensile stresses in the ice cover brought about by changes in ocean currents and winds. The summer melt cycle also serves to alter the properties of Arctic sea ice by metamorphosis of the ice surface into an array of refrozen meltponds and hummocks. Beneath this, the columnar structure and skeletal layer are preserved, but salinities are substantially less (due to desalination within the bulk of the ice) and crystal sizes substantially larger. This broad classificatory scheme for sea ice overlooks several important factors that can change its basic properties. Rate of growth, for example, will determine salinity and also crystal fabric. Furthermore, grease ice crystals can sinter to form irregular clumps of ice, which grow very rapidly to produce a sea ice of different character that may be much more widespread in Antarctic sea ice than in the Arctic. The presence of a significant snow cover during growth can also change the composition of sea ice by insulating its upper surface or by depressing the ice sufficiently to cause infiltration by brine. Again this is probably more frequent in Antarctic waters than in the Arctic Ocean. In altering the character of sea ice these effects are significant for the interpretation of satellite remote-sensing imagery.

Gudmandsen's table 1 (this symposium) lists a set of key glaciological needs for the study of sea ice and its relations with climate and the oceans. In this context, the main role of sea ice is to inhibit the exchange of heat, moisture and momentum between atmosphere and ocean.

The location of the ice edge and the morphology and distribution of ice floes within the Marginal Ice Zone (MIZ) are important and little-understood parameters in both climatological and oceanographic modelling studies. Within the MIZ the distribution of floe sizes will be determined by, *and will determine*, the ocean wave energy found at any location. Waves penetrating the pack from the open sea will fracture ice floes if the floes are too large to exist in waves of that height (Robin 1968; Goodman *et al.* 1980). Waves are also attenuated by the pack, with the energy decaying as a function of wave period and ice characteristics encountered en route (Robin 1963; Wadhams 1975, 1978; Squire & Moore 1980). Thus the ultimate floe size at any position within the MIZ is under the control of waves from the open sea and of the local oceanographic and meteorological effects that redistribute the ice cover. Once determined, floe size distribution is crucial to a thorough understanding of MIZ oceanic processes.

The concentration of ice floes in any ice-covered region represents a delicate balance of wind, waves and ocean currents that will be upset by meteorological or oceanic disturbances. We therefore see concentration as a sensitive marker of change and as an indicator of zones of divergence or compressive motions within the pack ice.

The concept of an individual ice floe is difficult to apply to the central Arctic Ocean where hundreds of square kilometres are totally ice-covered with little evidence of open water except for occasional sinuous leads or polynyas. Pressure ridges form the only relief and determine surface roughness characteristics in a particular region. A detailed knowledge of their distribution is essential if measured values of form drag coefficients for different ice surfaces are to be incorporated into the momentum balance equations so that realistic and reliable global weather predictor models may be routinely implemented.

No remote sensing method has yet been able to measure thickness accurately for all types of sea ice. Initial successes with impulse radar-type measurements are limited to ice of low salinity, and new ice of any reasonable thickness has evaded measurement. Satellite imagery at certain wavelengths, passive microwave emission data, scatterometry, altimeter pulse shapes and synthetic aperature radar all combine to provide information that helps to identify the ice type and concentration present in any region. From this a qualitative estimate of ice thickness may be made and estimates made of the heat exchange between the ocean and atmosphere.

## 5. Satellite altimetry

Mapping the Antarctic ice sheet by satellite radar altimeter was proposed by Robin (1966), whereas Zwally (1975) suggested that the technique could be used to determine changes in elevation of ice sheets. Results from Geos-C showed a repeatability better than 3 m over Greenland (Brooks *et al* 1978), and better than $\pm 1$ m from Seasat data over Antarctica (Brooks 1982). The figure for repeatability is relevant to determination of changes of surface elevation, but for mapping, satellite altimetric data should be corrected for deformation of the geoid.

The altimeters on Geos-C and Seasat were designed for the study of ocean surfaces. Systems developed for tracking echoes have not proved entirely satisfactory over ice sheets, where the lock frequently failed. Wide variations in the returned echo pulse shape and strength found with Geos-C and Seasat data also provide useful information about the nature of the reflecting surface, so we discuss the factors controlling pulse shape at some length.

The main characteristics of satellite radio altimeters and of an airborne system used to sound polar ice thickness are shown in table 2. While the airborne system uses a lower frequency and longer pulse length than the satellite altimeters, the geometry of the echo from the base of the ice has many features that are relevant to the interpretation of satellite radar altimetry.

We consider a pulse of square form with a pulse width of 3 ns transmitted from a satellite at 800 km height. This spreads as a spherical wave within limits governed by the antenna beamwidth. When the end of the pulse reaches a flat surface, the pulse will cover a circular area of diameter around 1.6 km. Thereafter the pulse will spread out as an annulus around the nadir point until limited by the antenna beamwidth to a diameter $ca.$ 22 km for Seasat. The pulses shown later, in figure 1, are limited by the time gate to a diameter around 9 km on a flat surface.

Three types of reflecting surface discussed in Robin $et\ al.$ (1969) should be considered.

TABLE 2. RADAR ALTIMETER CHARACTERISTICS

| | Skylab | Geos-C (intensive mode) | Seasat | ERS-1 | N.S.F.– S.P.R.I.– T.U.D. radio-echo sounder† | |
|---|---|---|---|---|---|---|
| mean altitude/km | 435 | 843 | 800 | 777 | 0.3 | operating height above terrain/km |
| antenna beamwidth/deg | 1.5 | 2.6 | 1.6 | 1.2 | $20 \times 120$ | antenna beamwidth/deg (sideways, fore-aft) |
| frequency/GHz | 13.9 | 13.9 | 13.5 | 13.7 | 0.06 and 0.3 | frequency/GHz |
| peak power/kW | 2 | 2 | 2 | 0.5 | 1–10 | peak power/kW |
| pulse width/ns (uncompressed) | 130 | 1000 | 3200 | $ca.$ 5000 | 60–1000 | pulse width/ns (uncompressed) |
| pulse width/ns (compressed) | 10 | 12.5 | 3.1 | 3 | 63 | pulse width/ns (compressed) |
| pulse repetition frequency/Hz | 250 | 100 | 1020 | $ca.$ 1000 | 12.5–25 | pulse repetition frequency/kHz |
| pulse-limited footprint diameter/km | 8 | 3.6 | 1.6 | 1.6 | $0.05 \times 0.5$ | beam-limited footprint/km (at surface) |
| altimeter precision/m (r.m.s.) | < 1 | < 0.05 | < 0.01 | 0.01 | 1.25 | altimeter precision/m (r.m.s.). |
| orbit/deg | 50 | ± 65 | ± 72 | ± 82 | — | — |

† National Science Foundation – Scott Polar Research Institute – Technical University of Denmark.

1. *Plane polished reflector.* In this case any departure from a plane surface is small compared with the radio wavelength ($\lambda$). If $P_t$ is the transmitted power, $g_t$ the antenna power gain along its axis, $S_r$ the effective absorption cross section of the receiving antennae, $r_0$ the range to the reflecting surface and $R_n$ the power reflexion coefficient of the reflecting surface at normal incidence, the received power $P_r$ is given by

$$P_r/P_t = R_n g_t S_r / 16\pi r_0^2. \tag{9}$$

2. *Perfect diffuse reflector.* This reflects a fraction ($R_d$) of incident radiation into a hemisphere according to Lambert's law. The flux per unit angle reflected in a direction $\phi$ to the normal to the surface is proportional to $\cos\phi$ so that, in optical terms, the surface has equal brightness from all viewing directions. For near-normal angles, a plane reflecting surface, and with certain qualifications concerning polarization, we have within the pulse-limited footprint

$$P_r/P_t = R_d g_t S_r vt / 4\pi r_0^3, \tag{10}$$

where $v$ is the radio wave velocity and $t$ is the time after arrival of the leading edge of the pulse. $P_r$ rises linearly until time $t_p$, the pulse length, after which the reflecting annulus is of approximately constant area so that $P_r$ is approximately constant within the beam-limited footprint.

3. *Extended rough surfaces.* We consider a surface in which slopes are uniformly distributed up to some maximum value ($\beta_{max}$), which is not large. Backscattering in the direction of incidence

has been analysed by Beckmann & Spizzichino (1963), who consider that this scarcely varies at normal incidence between models of varying complexity. We consider two cases.

(i) When the surface consists of random irregularities of all sizes down to $\lambda/2\pi$ or less in the horizontal plane, Booker *et al.* (1950) showed that the surface behaves like a Lambert-law reflector.

(ii) When individual facets of the surface are flat to a small fraction of a wavelength over dimensions large compared with the wavelength, the problem becomes one of geometrical reflexion as with the optics of a 'glistening surface' defined by $\phi_p < \phi < \beta_{max}$, where $\phi_p$ is the angle of the pulse-limited footprint. We assume that the power reflected by individual facets is in a random phase relation to its neighbours and the average power received is the sum of the individual powers. For a square pulse, the echo rises to an initial maximum given by

$$P_r/P_t = \frac{R\lambda^2 g_t^2}{64\pi^2 r_0^{2.5}} \frac{(vt_p)^{\frac{1}{2}}}{\beta_{max}}, \tag{11}$$

after which the power falls approximately in proportion to $t^{\frac{1}{2}}$ until a time $t_\beta$ when it will fall abruptly to zero; $t_\beta$ is related to $\beta_{max}$ by

$$\beta_{max} = (2vt_\beta/r_0)^{\frac{1}{2}}. \tag{12}$$

When $\beta_{max} = \phi_p$, $t_\beta = t_p$, and since $Sr = g_t\lambda^2/4\pi$ we have, from (11) and (12),

$$P_r/P_t = \frac{Rg_t S_r}{16\pi r_0^2} \frac{1}{\sqrt{2}}, \tag{13}$$

which differs from (9), for example, by only $\frac{1}{\sqrt{2}}$, or around 1 dB. Thus for $\beta_{max} < \phi_p$ in this case it appears reasonable to treat the maximum returned power as equal to that given by specular reflexion.

*Fading range and fading rate.* Owing to satellite motion, phase changes between reflecting facets that are not resolved in range cause a calculated fading rate on Seasat records up to around one cycle per metre of path length compared with a spacing along track of *ca.* 6.6 m between pulses. This suggests a random distribution of returned energy with a Rayleigh distribution, in which case the fading range will be 13.4 dB, which is of the order of the Seasat observations over the oceans. To reduce this scatter the returned energy in each gate was averaged over 50 pulses and two 50 pulse averages were combined for data transmission. This gave statistical smoothing by a factor of 10. This was effective over the oceans, but over 600 m of an undulating ice sheet systematic drift in energy returns may be of magnitude comparable with statistical fluctuations.

## 6. Seasat pulse data

### (a) Introduction

Some typical Seasat pulse shapes obtained during a single crossing of Antarctica in 1978 are shown in figure 1. The horizontal scale of time is linear and divided into 60 intervals, for each of which the mean power is plotted on a linear scale. However, the plot has been normalized to fill the vertical scale, and figures on the left-hand scale indicate how much this scale has been expanded between successive plots.

Within each group (*a*) to (*e*) the recorded pulses are shown at one second (6.6 km) intervals, although data are available at one tenth of that spacing. The pulse shapes of groups (*a*) open ocean, (*b*) ice shelf and (*c*) inland ice approximately fit our description for diffuse (Lambert's law) reflectors, whereas returns typical of a glistening surface are seen in groups (*d*) and (*e*).

### (*b*) *Open ocean*

In figure 1 *a* the open ocean surface is best described as an extended rough surface with random irregularities down to $\lambda/2\pi$ (3 mm) that produce reflexions that follow Lambert's law. The initial pulse rise time is governed by longer wavelengths as discussed by Cartwright and Tucker (both this symposium).

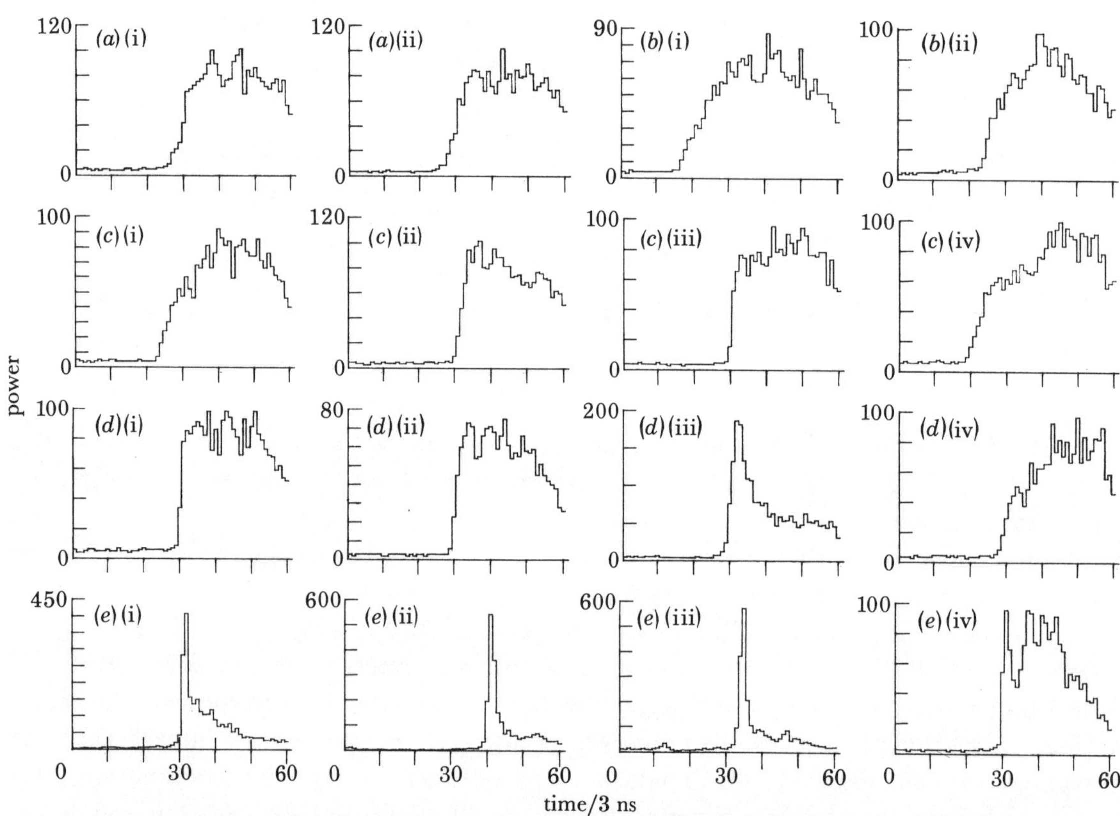

FIGURE 1. Altimeter pulse shapes recorded by Seasat during one crossing of Antarctica, showing power (linear scale) against time (60 intervals of *ca.* 3 ns). (*a*) Open ocean, 55.245° S, 56.794° E (i) to 55.196° S, 56.739° E (ii); (*b*) ice shelf, 67.537° S, 81.671° E (i) to 67.502° S, 81.544° E (ii); (*c*) ice sheet, 69. 363° S, 89. 539° E (i) to 69.277° S, 89.196° E (iv); (*d*) pack-ice–Ocean, 61.977° S, 67.036° E (i) to 61. 845° S, 66.781° E (iv); (*e*) pack ice, 66.165° S, 77.105° E (i) to 60.052° S, 76.768° E (iv).

### (*c*) *Ice shelves*

In figure 1 *b* only two pulse shapes are shown as the orbit just crossed a corner of the West Ice Shelf. The similarity of pulse shape to those of the open sea in figure 1 *a* is even more apparent from longer series available on other orbits. The diffuse-type pulses are attributed to sastrugi on a near-horizontal and flat surface. There is no evidence of the 'glistening' effect found on sea ice. Over vast areas of ice shelves, the surface irregularities are small, so the resultant radio altimetric errors could be as small as those over the ocean, i.e. around 0.1 m. The use of such accurate ice shelf altimetry, discussed in § 3, should open up new approaches to research on ice shelves.

### (d) *Ice sheets*

The pulse shapes obtained over the Antarctic ice sheet shown in figure 1 c again indicate a diffuse reflector. There is, however, a greater variability of pulse shape over the ice sheet than in figure 1 a, b. The rate of rise of the pulse and the position of the strongest return vary considerably and often the pulse ramp migrates rapidly across the time gate with a resultant inability to maintain tracking. These effects all result from large-scale undulations of the surface of the ice sheet.

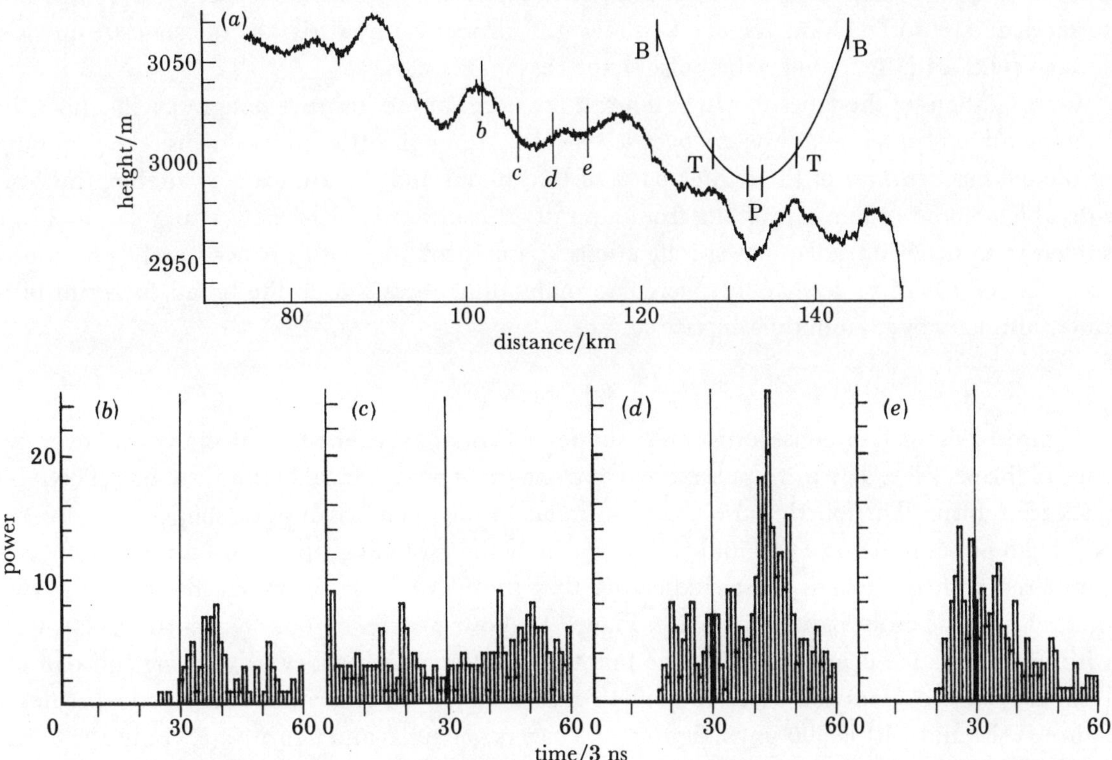

FIGURE 2. (a) Surface profile over the East Antarctic Ice Sheet from 88.96° S, 17.50° W to 88.32° S, 28.52° W from an aircraft radio altimeter record corrected by the aircraft pressure altimeter record by N. McIntyre. The shape of the altimeter pulse front from Seasat is shown in the upper right section, allowing for vertical scale exaggeration. Footprint widths over horizontal surfaces are shown for the beam-limited case (BB), for the time-gate limits used in figure 1 (TT) and for the pulse-limited footprint (P). The pulse length is about the same as the line thickness.

(b–e) Radio altimeter pulse shapes computed by M. R. Gorman for a two-dimensional model by using the profile in (a) and other characteristics as for Seasat and ERS-1. Resultant pulse shapes shown are those for corresponding nadir points (b) to (e) along the profile (a).

In figure 2 a we show a typically undulating surface profile of the East Antarctic Ice Sheet near the South Pole from radar altimeter and pressure altimeter records of the N.S.F.–S.P.R.I.–T.U.D. flights. Also shown is the satellite altimeter pulse shape for a satellite height of 800 km. The pulse length (Seasat and ERS-1) of 1 m is similar to the line width. We see that large-scale curvature of the ice surface is comparable with and at times is greater than that of the altimeter pulse, so that over a valley the first return to the satellite will come from the valley sides rather than the bottom of the valley.

A two-dimensional computer analysis of the pulse shapes that would be returned to a satellite (Seasat or ERS-1) by the surface is shown in figure $2b$–$e$. For a flat surface the two-dimensional model returns a pulse that will fall in proportion to $\sqrt{t}$ instead of remaining approximately constant as in the three-dimensional model. Nevertheless, the analysis shows similar effects of surface undulations to those shown in figure $1c$. The analysis was centred on the nadir point from the satellite, so the pulse migrates as surface slopes change.

In figure $2b$ the nadir point on a convex upward surface produces a weak return that falls rapidly with time. Figure $2c$, over a sloping surface, causes the first return to migrate off the time window. This could cause a tracking failure. Figure $2d$ shows an initial rise somewhat stronger than in figure $2b$ followed by a substantial rise due to the combined effects of concave upward surfaces at 110 and 114 km. Figure $2e$ shows the strongest initial rise, as the concave upward surface relief initially matches the pulse shape reasonably closely.

Computations of the types shown in figure 2 are necessary to improve design specifications for the operation of satellite altimeters over ice sheets. Although little three-dimensional mapping of the ice sheet surface of the higher parts of Greenland and Antarctica is available, the considerable amount of linear profiling from aircraft and surface vehicles now available should be sufficient to draw up adequate specifications. Laser profiling with an accurately positioned footprint of 100 m or less would overcome many difficulties due to the broad footprint of a radar altimeter over an undulating ice surface.

### (e) Sea ice

Figure $1d$ shows four consecutive pulse shapes 6.6 km apart as the satellite moves out over the edge of the pack ice. The third pulse shows a strong early peak typical of the glistening shown by pack ice returns. The fourth pulse appears similar to the open ocean pulse shapes of figure $1a$. Although pulses in figure $1d$ (i) and (ii) do not show the glistening, they show a rapid pulse rise compared with the fourth pulse, indicating that the wave height is much lower for the two pulses inside the outer band of pack ice. Figure $1e$ shows a series of four consecutive pulses well within the pack ice near the coast. The first three pulses show particularly strong 'glistening' returns typical of pack ice while the fourth shows very little glistening. The vertical scale of figure $1e$ (ii) and (iii) is 600 units, figure $1e$ (i) covers 450 units and figure $1e$ (iv) only 100 units. Apart from the enhanced glistening effect on figure $1e$ (i)–(iii) the remainder of the pulses are all of similar form and amplitude to figure $1e$ (iv).

From inspection of all 94 pulse shapes recorded from the coast to the open ocean, only 8 showed no clear glistening characteristic. These included five of the outermost ten records, where one may expect belts of pack ice alternating with stretches of open water. Variations of glistening with location and with season appear likely and require 'surface truth' observations. However, it appears reasonable to attribute the glistening effect to the absence of ripples or wavelets that give ocean surface pulses a diffuse character. Such ripples are damped out completely by any thin film of ice, leaving long wavelengths to govern surface slopes that produce glistening returns. Few direct slope measurements within pack ice are available, but figure 3 shows slopes calculated from wave-energy spectra measured by Squire & Moore (1980) by using vertical accelerometers on ice floes in the Bering Sea for periods above 4 s (25 m wavelength). Even at this wavelength, the outer floes of 10 m diameter damp out most wave energy within 5 km, as shown also in figure $1d$.

Sea surface slopes result from the combined effect of all frequency bands of figure 3 and can be estimated from the period of maximum energy and significant wave heights derived from figure 3. Resultant slopes range from $1.6 \times 10^{-2}$ at the ice edge, $1.2 \times 10^{-2}$ at 5 km to $2.7 \times 10^{-3}$ at 65 km from the edge. The last slope is close to the value of $3 \times 10^{-3}$ derived from the width of the pulse in figure 1$d$ (iii) and equation (12), which should be typical of regions with broken floes in moderate weather.

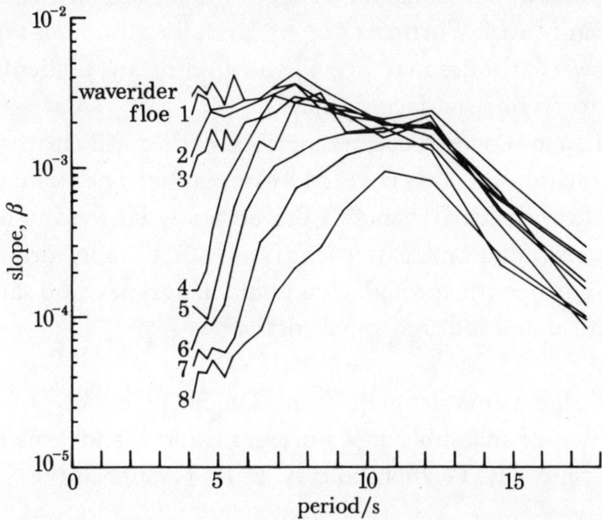


FIGURE 3. Surface slopes of the marginal ice zone of the Bering Sea derived from measured wave-energy spectra in frequency bands given in Squire & Moore (1980). The waverider buoy record was obtained just outside the ice edge, that from floe no. 1 at the ice edge, no. 2 at 0.7 km, no. 3 at 2.2 km, no. 4 at 5.3 km, no. 5 at 10.3 km, no. 6 at 18.9 km, no. 7 at 38.6 km and no. 8 at 65.1 km from the ice edge. Ice floes were typically 10 m in diameter from 0 to 5 km, then increased gradually to 40 m at 30 km, after which the floe size increased abruptly to greater than 100 m.

Not all pack ice consists of thin newly frozen ice; older ice typically has a rough surface, new ice is usually covered after a few days by snow, and open water between floes will be rippled by any wind of appreciable strength. Because these surfaces may produce a diffuse echo the reason for the dominant glistening echo from sea ice needs further explanation. This comes from considering the relative strengths of the two types of return. If equal areas contribute to the power of glistening ($P_{\mathrm{rg}}$) and diffuse ($P_{\mathrm{rd}}$) returns, we can calculate the ratio of energies involved by dividing (11) by (10), and putting $S_{\mathrm{r}} = g_{\mathrm{t}} \lambda^2 / 4\pi$. This gives

$$P_{\mathrm{rg}}/P_{\mathrm{rd}} = \frac{R_{\mathrm{g}}}{R_{\mathrm{d}}} \frac{r_0^{\frac{1}{2}}}{4(vt)^{\frac{1}{2}}} \frac{1}{\beta_{\max}}. \tag{14}$$

For a satellite height of 800 km, $\beta_{\max} = 3 \times 10^{-3}$ and other parameters for Seasat and ERS-1 in table 2, we get $P_{\mathrm{rg}}/P_{\mathrm{rd}} = 8.5 \times 10^4 R_{\mathrm{g}}/R_{\mathrm{d}}$. Because the power reflexion coefficient for thin sea ice is typically 7 dB higher than for a snow/firn surface, for equal areas of glistening and reflexion $P_{\mathrm{rg}}/P_{\mathrm{rd}} \approx 5 \times 10^5$.

The glistening character of the pulse shape will still be quite obvious for a ratio $P_{\mathrm{rg}}/P_{\mathrm{rd}}$ of 50 when less than 0.01 % of the surface is producing the glistening type of echo. Thus, provided that a small area of new ice or calm water is exposed, as is likely at the locations and season of pulses shown in figure 1$d, e$, the dominant glistening characteristic is explicable.

When surface slopes due to wave action are small well inside the pack ice zone, we expect intermittent returns of a specular nature from new sea ice and open water leads (when calm).

Strong specular returns, when averaged over 100 pulses, should still dominate the pulse shape if any significant specular reflexion is present as in the first three pulses of figure 1 *e*. However, with large floes many kilometres across, the glistening effect will be intermittent if the remainder of the surface follows Lambert's law, as shown in the fourth pulse of figure 1 *e*. The presence of large ponds on the surface of Arctic pack ice in summer, when newly frozen or in calm weather, may also produce glistening. Indeed, our calculation suggests that glistening could be much stronger still than that shown in figure 1 *d, e*. Furthermore, a search for glistening-type returns from very calm seas and lakes in lower latitudes may provide a rapid means of identifying regions where wind speeds are too low to produce surface ripples.

The varying surface of an ice-covered ocean described earlier will clearly give wide variations in reflected energy, often within the area covered by one pulse. The evaluation of such surfaces from satellite-derived data will clearly benefit from an analysis of many different parameters, such as passive microwave radiation, scatterometry, synthetic aperture radar and altimeter pulse shapes, all of which can operate through clouds and in darkness and supplement the higher-resolution imagery at optical and infrared wavelengths.

The authors acknowledge valuable help from Dr H. J. Zwally of N.A.S.A.–Goddard Spaceflight Center, who made available tapes of Seasat data, and assistance with aspects of this study by N. McIntyre, M. R. Gorman and A. P. R. Cooper of the S.P.R.I.


## References

Booker, H. G., Ratcliffe, J. A. & Shinn, D. H. 1950 *Phil. Trans. R. Soc. Lond.* A **242**, 579–607.

Brooks, R. L. 1982 *Ann. Glaciol.* **3**, 32–35.

Brooks, R. L., Campbell, W. J., Ramsier, R. O., Stanley, H. R. & Zwally, H. J. 1978 *Nature, Lond.* **274**, 539–543.

Budd, W. F. 1970 *J. Glaciol.* **9**, 19–27.

Budd, W. F. & Jenssen, D. 1975 *Int. Ass. Sci. Hydrol. Publ.* no. **104**, pp. 257–291.

Budd, W. F., Jenssen, D. & Radok, U. 1971 *Nature, phys. Sci.* **232**, 84–85.

Budd, W. F. & Smith, I. N. 1981 *Int. Ass. Sci. Hydrol. Publ.* no. **131**, pp. 369–409.

Beckmann, P. & Spizzichino, A. 1963 *The scattering of electromagnetic waves from a rough surface.* London: Pergamon Press.

Collins, I. F. 1968 *J. Glaciol.* **7**, 199–204.

Cooper, A. P. R., McIntyre, N. F. & Robin, G. de Q. 1982 *Ann. Glaciol.* **3**, 59–64.

Crary, A. P. 1962 *IGY Glaciol. Rep.* no. 6. New York: American Geographical Society.

Drewry, D. J. 1981 In *Remote sensing in meteorology, oceanography and hydrology* (ed. A. P. Cracknell), pp. 270–284. Chichester: Ellis Horwood.

Drewry, D. J., Jordan, S. R. & Jankowski, E. J. 1982 *Ann. Glaciol.* **3**, 83–91.

Glen, J. W. 1955 *Proc. R. Soc. Lond.* A **228**, 519–538.

Goodman, D. J., Wadhams, P. & Squire, V. A. 1980 *Ann. Glaciol.* **1**, 23–27.

Jacobs, S. S., Gordon, A. L. & Ardai, J. L. Jr 1979 *Science, Wash.* **203**, 439–443.

Mae, S. & Naruse, R. 1978 *Nature, Lond.* **273**, 291–292.

Martin, S. 1981 *A. Rev. Fluid Mech.* **13**, 379–397.

Neal, C. S. 1979 *J. Glaciol.* **24**, 295–307.

Nye, J. F. 1951 *Proc. R. Soc. Lond.* A **207**, 554–572.

Paterson, W. S. B. 1981 *Physics of glaciers*, 2nd edn. Oxford: Pergamon.

Radok, U. 1978 *Climatic roles of the ice.* Paris: Unesco.

Robin, G. de Q. 1963 *Phil. Trans. R. Soc. Lond.* A **255**, 313–339.

Robin, G. de Q. 1966 *Can. J. Earth Sci.* **3**, 893–901.

Robin, G. de Q. 1967 *Nature, Lond.* **215**, 1029–1032.

Robin, G. de Q. 1968 In *Symposium on Antarctic oceanography* (ed. R. I. Currie), pp. 191–197. Cambridge: Scott Polar Research Institute for SCAR.

Robin, G. de Q. 1983 *The climatic record in polar ice sheets.* Cambridge University Press.

Robin, G. de Q., Evans, S. & Bailey, J. T. 1969 *Phil. Trans. R. Soc. Lond.* A **265**, 437–505.

Sergin, V. Ya. 1979 *J. geophys. Res.* **84** (C6), 3191–3204.

Squire, V. A. & Moore, S. C. 1980  *Nature, Lond.* **283**, 365–368.
Swithinbank, C. W. M. 1962  *Norsk Polarinst. Årb.* **196**, 28–31.
Swithinbank, C. W. M., McLain, P. & Little, P. 1977  *Polar Rec.* **18**, 495–501.
Tchernia, P. 1974  *C.r. hebd. Séanc. Acad. Sci., Paris* D **278**, 667–670.
Thomas, R. H. 1973  *Br. antarct. Surv. sci. Rep.* no. 79.
Thomas, R. H. 1979  *J. Glaciol.* **24**, 167–177.
Vialov, S. S. 1958  *Int. Ass. Sci. Hydrol. Publ.* no. 47, pp. 266–275.
Wadhams, P. 1975  *J. Geophys. Res.* **80**, 4520–4528.
Wadhams, P. 1978  *Deep Sea Res.* **25**, 23–40.
Weertman, J. 1957  *J. Glaciol.* **3**, 33–38.
Zotikov, I. A., Zagorodnov, V. S. & Raiskovsky, J. V. 1980  *Science, Wash.* **207**, 1463–1465.
Zwally, H. J. 1975  *J. Glaciol.* **15**, 444.
Zwally, H. J. & Gloersen, P. 1977  *Polar Rec.* **18**, 431–450.

## *Discussion*

A. S. Laughton, F.R.S. (*Institute of Oceanographic Sciences, Wormley, Godalming, U.K.*). Dr Robin reported the inability of the altimeter on ERS-1 to cope with the rough topography of the margins of the polar regions, and in particular its inability to record the bottoms of steep-sided valleys. The geometrical problems of altimetry are essentially similar to those of standard echo-sounders in the ocean, which have been studied for decades by marine geophysicists. The rise time of the first return and the subsequent reverberation time are both measures of surface roughness, the surface in echo-sounding being the sea floor. Larger-scale sea-floor topography is represented by the superposition of hyperbolae arising from the multiplicity of point reflectors, which give a record that can be analysed, by migration techniques now highly developed in the oil prospecting industry.

The major difference between altimetry and echo-sounding is that whereas the former uses gating techniques to lock onto first arrival echoes and rejects any subsequent data, the latter records all returns. Could the coarse topography over the marginal polar regions be better measured by widening the receive gates on the altimeter and relaying back to Earth the multiple pulses, which could then be analysed to give the required information?

G. de Q. Robin. In principle the answer is yes. At the Scott Polar Research Institute we have developed these migration techniques for improving our analysis of bedrock topography obtained by radio echo-sounding of the ice sheet. With radar altimetry from satellites the problem is one of data acquisition and handling. If the time gate for data acquisition is lengthened we have to decrease the accuracy of elevation measurement because of constraints on data transmission for ERS-1 as currently planned. The alternative of greatly increasing the amount of data transmitted would add very considerably to the costs. This may only be justified if extra information obtained on the surface form of the ice sheet is of critical importance to scientific knowledge.

*Phil. Trans. R. Soc. Lond.* A **309**, 463–464 (1983)

*Printed in Great Britain*



# Concluding remarks

By Alan H. Cook, F.R.S.

*Cavendish Laboratory, Madingley Road, Cambridge CB3 0HE, U.K.*

In opening this Discussion Meeting, Professor Houghton said that it was the aim of the organizers to put most emphasis on the consideration of problems of the interpretation of data from remote sensing. That aim does indeed seem to have been achieved and we have had some very useful studies of the significance of data and of the principles and methods used in interpretation. Two points have in particular occurred to me. In the first place, as with other types of satellite observation, the techniques of observation and the capacity for obtaining very large amounts of data are outrunning the understanding of the results and the ability to extract significant conclusions; it should, however, be noted that whereas in research it is the new, ill-understood observation that leads to further progress, the many, well-understood, routine observations are the ones that are needed in applications. The second point, and it is related to the first, is the great complexity of some processes that produce detectable signals, so that the interpretation of the response of a remote sensor may only be possible in some limited, well-defined circumstances. A clear instance was given by Dr Curran in his discussion of the determination of leaf area index, while Dr Tucker showed how incompletely the imaging of sea waves is understood.

Today has been devoted to oceanography and glaciology, where the questions being addressed by remote sensing have been with us for a long time. Dr Cartwright showed us some records of tides in the English Channel – the first systematic survey of those tides was made by Edmond Halley in 1701. I suspect that many in the meeting may not fully realize the great importance of the polar ice fields in determining global flows in the oceans and atmospheres, whether in cooling the air or water, or establishing flows around or out of the polar regions, or damping waves and tides and perhaps slowing down the spin of the Earth. The ability of remote sensors to determine the thickness of an ice cap, the extent of sea ice and the proportion of open water will surely lead to important developments in our understanding of the polar ice fields and of this global effect.

Professor Woods reminded us of the dominant influence of the oceans on climatic change, both because the ocean absorbs two thirds of all heat absorbed from the Sun, and because it stores in just the uppermost 5 m or so as much heat as the atmosphere stores. Thus climate cannot be understood without a good knowledge of the general circulation of the oceans, but hitherto that has depended on sampling at isolated points and the results have been grossly distorted by incompleteness and aliasing. For this reason, remote sensing is especially attractive in oceanography, but its use is beset by many problems. The measurement of the height of the sea surface, potentially a most powerful technique, depends upon the very accurate determination of satellite orbits, and thus upon a detailed knowledge of the gravity field, required also in estimating the hydrostatic surface of the sea, and on sufficiently frequent tracking to reduce the uncertainties due to the drag of the air upon a satellite to an acceptable level. Again, waves and wind stress can be estimated from satellite observations, but the mechanism by which wave signals in particular are produced seems still to be obscure. Lastly, as Sir John Mason mentioned, the temperature of the surface of the sea can be measured, but what is it that is measured: the temperature of the bulk liquid, or that of spray or foam? Thus there are major problems to be analysed, and I hope that this meeting will bring them to the attention of physicists, applied mathematicians and engineers who might work upon them. At the same time, as Dr Robinson pointed out, some of the data that are being received are of novel sorts, and instead of trying to interpret them in preconceived ways, they should be looked at on their own terms to see what they tell us about the natural world. One of the uses of remote sensing will be to alert us to changes in the environment and to predict natural catastrophes. It was made clear that, as has been said elsewhere, the effects of an increase in carbon dioxide in the atmosphere and oceans of the amounts foreseen in the next half century or so are less than the uncertainty of, for example, measurements of sea temperature. Satellite surveillance should help to follow the effects of pollution, but its capacity should not be overestimated. As for catastrophes, satellite observations can already detect incipient hurricanes; whether it will ever be possible to recognize areas in which earthquakes might be about to occur must at the moment remain speculative.

There are indeed great prizes to be looked for in studying the world about us by remote sensing but many of them are still far off, and will only be attained by deep, thorough and subtle analysis.

It remains for me to offer to all our contributors the thanks of the organisers for their participation and ready cooperation; our thanks go likewise to those who have contributed extempore. We all of us, organizers, contributors and participants, are indebted to the officers and staff of the Royal Society for enabling this meeting to take place and for providing the facilities of the Society with their usual efficiency.